Minimal Cells: Design, Construction, Biotechnological Applications

Alvaro R. Lara • Guillermo Gosset
Editors

Minimal Cells: Design, Construction, Biotechnological Applications

Springer

Editors
Alvaro R. Lara
Departamento de Procesos y Tecnología
Universidad Autónoma
Metropolitana-Cuajimalpa
Ciudad de Mexico, Mexico

Guillermo Gosset
Departamento de Ingeniería Celular
y Biocatálisis
Instituto de Biotecnología, Universidad
Nacional Autónoma de México
Cuernavaca, Morelos, Mexico

ISBN 978-3-030-31896-3 ISBN 978-3-030-31897-0 (eBook)
https://doi.org/10.1007/978-3-030-31897-0

This Springer imprint is published by the registered company Springer Nature Switzerland AG.
The registered company address is: Gewerbestrasse 11, 6330 Cham, Switzerland

Preface

The goal of industrial or white biotechnology is the generation of products and services by employing living organisms or their components. The aim is the development of commercially viable processes for the transformation of renewable raw materials into useful products, thus replacing technologies based on the use of nonrenewable fossil feedstocks whose processing and utilization generates toxic by-products. While most organisms can be employed in biotechnological processes, microbes are widely used since they can be grown and genetically modified with relative ease. The development and optimization of a biotechnological process entail diverse goals; one of them is to improve the performance of a microbial cell as a factory. This objective is mainly achieved by employing a wide array of techniques, collectively known as genetic engineering, which enable the modification of the information in the genome of the organism. The capacity for genetic modification started as modest changes in specific gene regions and the transfer and expression of genes among species. Over the last few years, genetic engineering techniques have improved considerably, allowing precise and extensive genetic modifications. These methodologies have been recently complemented by the emergence of synthetic biology. This new field applies engineering principles to biology and is based on mathematical modeling to design and create novel biological parts, devices, and systems.

The genome of each organism specifies all the cellular functions required for its growth and survival. A fraction of the genes in the genome is required for essential functions, while many other genes become active only under specific conditions in the continuously changing natural environment. The gene set required for self-replicating life is considered the minimal genome. Approaches such as the generation and analysis of mutants, the comparison of sequenced genomes, the generation and analysis of genome-scale metabolic models, and the study of bacteria with small genomes have enabled the estimation of the minimal number of genes that could sustain life. These methods place the number of essential genes around two to three hundred. In addition to its value for basic science, the minimal cell concept has implications in biotechnology. In contrast to a natural environment, in industrial production processes, physicochemical conditions are highly controlled to remain at specific values. In this context, many of the genes in the genome are not required for production performance. The replication and expression of genes required for survival in the natural environment could represent a waste of energy in an industrial bioreactor. In addition, the presence of repeated and self-replicating genetic elements in the genome can result in production strain instability. Therefore, it is expected that the elimination of genome regions that are not required for cell replication and the synthesis of the desired product could result in minimal cells with improved production performance. A cell with a reduced or minimal genome can be considered as a chassis where natural or synthetic functions could be added to generate a cell specialized for the synthesis of a specific biotechnological product. Moreover, a reduced genome may decrease the rate of interactions between the synthetic genetic program and the chassis genome (e.g., insertions of genes of the genome into an expression vector plasmid).

The objective of this book is to provide reviews on the current knowledge related to the design, characterization, and use of minimal cells. Leading experts wrote the book chapters and included up-to-date information as well as the in-depth analysis of current issues and challenges on this topic. This book aims to become a source of reference for researchers and students working in this field in academia and industry. Chapters in this book describe the specific approaches employed for the generation of minimal cells of the microbes *Escherichia coli*, *Pseudomonas* species, *Bacillus subtilis*, *Corynebacterium glutamicum*, *Lactococcus lactis*, *Streptomyces* species, *Schizosaccharomyces pombe*, and *Saccharomyces cerevisiae*. These organisms are employed as production strains in industry and as models in basic biological research. Genome reduction in these organisms has resulted in strains with improved productivity and genetic stability. Specific chapters also address the various methods employed to define the minimal gene set required for the generation of minimal cells. Genome-scale metabolic models can be employed to predict gene essentiality based on computational simulations and also to determine the growth phenotypes expressed from the minimal genomes. Specific chapters also address the concepts of gene persistence based on metabolic functions and optimal cellular resource allocation as alternatives to define the minimal gene set for the generation of chassis strains.

Support from the Universidad Autónoma Metropolitana and Universidad Nacional Autónoma de México is gratefully acknowledged by the editors.

Ciudad de Mexico, Mexico — Alvaro R. Lara
Cuernavaca, Mexico — Guillermo Gosset
July 1, 2019

Contents

Reduced and Minimal Cell Factories in Bioprocesses: Towards a Streamlined Chassis

Martin Ziegler and Ralf Takors

Abstract The rapid advances in molecular genome engineering, systems biology and synthetic biology over the past decade have laid the ground for extensive engineering of bacterial and fungal genomes and the rational setup of synthetic biological systems. In order to optimize the production host for biotechnical processes, the genomes of many industrial workhorse microorganisms such as *Escherichia coli*, *Bacillus subtilis*, *Corynebacterium glutamicum*, *Streptomyces* species, *Pseudomonas* species, *Saccharomyces cerevisiae*, and *Lactococcus lactis* have been successfully reduced. Here, we evaluate this progress in the context of biotechnical application for the production of industrially attractive products. Based on the view of microbial cells as factories, we discuss the concept of relevant genes. We attempt to estimate the theoretical benefits of genome reduction which form the basis of target selection. Subsequently, we comprehensively discuss existing studies on genome-reduced strains. The current limits of beneficial genome reduction and potential future developments in both prokaryotic and eukaryotic systems are considered.

Keywords Genome reduction · Minimum genome · Microbial cell factory · Chassis · Essential genes · Relevant genes

1 Introduction

In 1982, the complete sequencing of the genome of Bacteriophage Lambda marked the beginning of a new era in Biosciences (Sanger et al. 1982). The extensive availability of sequence information promised to revolutionize our understanding of life. Technical progress such as shotgun sequencing technology enabled sequencing of much larger genomes in the following decade with the *Haemophilus influenzae* genome being the first published genome of a free-living organism

M. Ziegler · R. Takors (✉)
Institute of Biochemical Engineering, University of Stuttgart, Stuttgart, Germany
e-mail: martin.ziegler@ibvt.uni-stuttgart.de; ralf.takors@ibvt.uni-stuttgart.de

A. R. Lara, G. Gosset (eds.), *Minimal Cells: Design, Construction, Biotechnological Applications*, https://doi.org/10.1007/978-3-030-31897-0_1

(Fleischmann et al. 1995). The complete sequence of *Saccharomyces cerevisiae* assembled in 1996 was not only the first eukaryotic sequence but also the first of an economically relevant organism (Goffeau et al. 1996). Only 1 year later, the sequences of model organisms *Escherichia coli K-12* and *Bacillus subtilis* were reported (Blattner et al. 1997; Kunst et al. 1997), and in 2001 sequencing of the human genome was finished (Venter et al. 2001). Since then thousands of genomes have been sequenced and are publicly available through resources such as GenBank (Benson et al. 2013).

At the same time, the scientific community initiated first projects yielding cells with a reduced genome. The aim was the generation of deletions up to the point of a minimal set (Koob et al. 1994). Already then a major motivation was to simplify the model organism *Escherichia coli* for research purposes and to improve its accessibility and predictability as an industrial workhorse organism. Since then, advances in molecular biology have enabled genome modifications and functional genomics in a vast variety of microorganisms and eukaryotic cell lines. With the availability of CRISPR-Cas technology, this process has once again accelerated (Ford et al. 2018; Salsman and Dellaire 2017). Complementary research has targeted our understanding and annotation of genomic information as well as the application of this knowledge for the purpose of metabolic engineering (Stephanopoulos 1999; Tao et al. 1999).

This contribution aims to provide an overview of the current progress in genome reduction both from the viewpoint of the molecular biologist and with the eyes of the metabolic engineer. Our considerations will always be directed towards the utilization of microbes in an industrial production scenario. The target conditions assumed are thus controlled cultivations in bioreactor systems. We will begin with a short presentation of the cell factory concept, followed by an in-depth look at the theoretical gains expected from genome reduction. Modern methods from the field of molecular biology for the generation of genome-reduced organisms will be addressed only superficially as they have been extensively reviewed elsewhere (Adli 2018; Freed et al. 2018; Guha et al. 2017; Nakashima and Miyazaki 2014; Oesterle et al. 2017). Instead we thoroughly summarize and examine published studies on genome-reduced strains. We critically consider their potential benefits for industrial production.

2 The Concept of Microbial Cell Factories

In the early 2000s, the first generation of genome reduction studies focusing on *Escherichia coli* were published (Kolisnychenko et al. 2002; Yu et al. 2002). Remarkably, although a significant portion of the *E. coli* genome was deleted the resulting strains did not show growth defects. It was proposed that the combination of these approaches with complete essentiality libraries like the KEIO collection might even form the basis for accurate modeling of cellular responses to real-time changes in its environment (Baba et al. 2006; Smalley et al. 2003). Following this concept the idea of creating cells with a minimal genome resulted in various projects to reduce the genomes of several industrially relevant microorganisms with the purpose of improving their basic production parameters (Ara et al. 2007; Giga-Hama et al. 2007; Mizoguchi et al. 2007). On the contrary, molecular biology and synthetic biology showed interest in finding the essential set of genes needed to sustain life and to artificially create it (Burgard et al. 2001; Fraser et al. 1995).

Both approaches have in common that the cell is regarded as a complex but defined self-replicating factory. Similar to a macroscopic industrial facility, there are core components that cannot be replaced and auxiliary components for special situations. On a molecular level, the scientific community generally accepts the interpretation of complex multi-protein structures as biological machines (Alberts and Miake-Lye 1992). A cell is consequently an assembly line consisting of such biological machines. However, given that the number of some reaction partners in a cell is very low—in the extreme case of genomic information it can equal one—metabolism is also inherently stochastic (Kiviet et al. 2014; Kurakin 2005). The inherent stochasticity of growth, enzymatic reactions, and gene expression leading to phenotypic variability of single cells in a clonal population can be assessed based on single-cell data and computational simulation (Kiviet et al. 2014; Thomas et al. 2018). Under the conditions of an industrial large-scale reactor another layer of stochasticity is added: variations in the local environment of cells due to imperfect mixing. The extracellular stimuli from fluctuating substrate gradients enhance existing phenotypic variability through their interaction with cellular regulation (Delvigne et al. 2009). The concept of regarding cells as microbial factories holds though—as long as we consider cellular individuality and plasticity.

From the simplest point of view, cells convert Gibbs free energy of substrates into biomass and products, thereby linking Gibbs free negative catabolic reactions with Gibbs free positive anabolism. It should be noted that net Gibbs free energy is negative which usually coincides with an increase in systemic entropy. Thus, the cell works very similar to a chemical refinery—with the bonus of being able to duplicate itself by self-assembly. From a chemical engineer's perspective such systems are called autocatalytic: One of the reaction products (biomass) is a catalyst for the reaction (substrate turnover). For a simple conversion process observed from maximum distance we can describe this behavior as if an entire population of cells was working as a single catalytic unit. This black-box approach leads to an unstructured, unsegregated model (Weuster-Botz and Takors 2018). The empiric description of the resulting process kinetics results in the well-known Monod model (Monod 1949).

Looking closer into what happens inside the cell reveals the complex metabolic network and its regulation behind this behavior. The flow of carbon through the central metabolism alone involves dozens of conversions, forks, and joints. Models integrating this information are called structured. In analogy to a refinery, enzymes can be seen as the workforce, information carrier proteins serve as managers and the genome represents the administrative headquarter. If we now consider that not all cells in a bioreactor are exactly identical, we reach a segregated model including microbial individuality (Weuster-Botz and Takors 2018). Population heterogeneity can arise on various levels: Bioreactor populations are clonal expansions from single cells, but random mutations and plasmid loss occur frequently. If mutants grow faster than their parent strain they can become a significant fraction of total biomass or even overtake the population. Frequently such mutations lead to the generation of non-producer populations. Speaking in macroscopic terms such cells are on strike—they refuse to produce the desired product. In recent years, other sources of heterogeneity have been increasingly investigated. Imperfect mixing leads to the generation of spatial gradients in bioreactors and can be the cause for heterogeneity on the transcriptional level and in production parameters (Delvigne et al. 2009; Simen et al. 2017).

A real-world factory is planned and managed by intelligent beings. Microbes on the other side are naturally only optimized in a trial-and-error fashion: through mutation and selection. The evolutionary pressure experienced by microorganisms over billions of years on earth has unquestionably converted them into complex, adaptable, and versatile living beings. However, they were not selected for serving modern humanity's need for efficient conversion of chemicals on an industrial scale. There are rare exceptions: Various strains of *Saccharomyces* species, which were domesticated hundreds of years ago, have evolved favorable traits for ethanol production such as utilization of maltotriose (Gallone et al. 2018). In general, however, we can expect microbes to behave imperfectly in the artificial production scenarios imposed on them. For instance, *Bacillus subtilis*, traditionally utilized for *natto* fermentation and the major workhorse for industrial enzyme production (Schallmey et al. 2004), has an inherently suboptimal flux distribution which directs carbon and energy towards adaptive responses even under optimal growth conditions (Fischer and Sauer 2005). Moreover it displays rigid regulation at the flux distribution ratio of glycolysis and pentose phosphate pathway which points towards excess production of NADPH independent of process conditions (Fischer and Sauer 2005). *Bacillus* species thus appear predestined for metabolic improvement through genome reduction and flux rerouting.

The optimization of cellular factories through addition of novel production streets, removal of wasteful auxiliary systems and optimization of pathway flux is the *raison d' être* of the metabolic engineer (Stephanopoulos 1999). The goal is to redesign cellular metabolism and regulation in an intelligent way (Bailey 1991; Takors et al. 2007).

The interpretation of cells as living factories is helpful for this purpose, because similar concepts like in real world factories can be applied. Similar to macroscopic production facilities they have an inherently modular structure embedded in a complex framework. Modularity does not mean orthogonality. In the chaotic environment of a crowded cell all molecules interact with each other and all fluxes are interdependent. Bailey (1999) pointed out that the metabolic and regulatory framework of a cell interacts with all its modules. Thorough understanding is consequently only possible at the systemic level (Bailey 1999). For example, knockouts of pyruvate kinase in *E. coli* have surprisingly little impact on global flux ratios as the lack of this enzymatic activity is subsequently compensated on systemic level by the activation of phosphoenolpyruvate carboxylase and malic enzyme (Emmerling et al. 2002). Flux rigidity is often observed in single knockout strains and based on the evolutionary evolved regulation of enzyme activities and cellular concentrations. Minimization of metabolic adjustment (MOMA) provides a method for estimating flux distributions in a rigid flux ratios scenario (Segrè et al. 2002). Compare these findings to the situation in a real-world factory. Imagine what would happen if you removed a single workstation in a manufacturing line due to safety issues. There are two possible outcomes: Either the workstation is critical for the entire process and production will completely shut down or workers will quickly find a bypass by taking an alternative route or using a similar workstation to compensate for loss. Overall production will likely be lower but the ratios of manufactured good fluxes on the level of the entire facility will hardly change. It will take time to find a new production optimum. Management might be involved and order more workers to the alternative workstation. Changes in preceding or following stations might be appropriate. Going back to the cell adaptive and evolutionary mechanisms will finally optimize disturbed fluxes to reach optimal growth in single knockout and genome reduction strains (McCloskey et al. 2018; Nishimura et al. 2017).

Global interdependence of all cellular components is a fact, but many operons and the pathways their encoded enzymes partake in are organized as modules. A prime example of cellular modularity is the synthesis of terpenoid-derived compounds. The primary precursor for terpenoid biosynthesis, isopentenyl diphosphate (IPP), is accessible either through the methylerythritol-phosphate (MEP) pathway native to *E. coli* or through the mevalonate (MVA) pathway native to *S. cerevisiae* (Schempp et al. 2018). A chassis for the production of terpenoids in *E. coli* can thus be designed by two approaches: debottlenecking and deregulation of the native MEP pathway or heterologous expression of MVA pathway enzymes (Du et al. 2014; Willrodt et al. 2014). These two modules are embedded in the complex network of cellular reactions, and they do not interact identically with it. Stoichiometric calculations indicate that with glycerol or glucose as the primary carbon source utilization of the MEP pathway to IPP may be advantageous as it is balanced in reducing power and offers higher carbon conservation (Ajikumar et al. 2010; Dugar and Stephanopoulos 2011). If IPP is converted into more complex terpenoids, the excess NADH generated by the MVA pathway might be advantageous though by providing necessary reducing power or ATP through respiration. The intelligent choice of production strategy lies in the hands of metabolic engineers.

In the end, improvements can be made on both scales: in single modules like aromatic amino acid production pathways and on a global scale using systematic approaches (Bailey et al. 2002; Lee and Kim 2015; Takors et al. 2007). By adopting a holistic view, unnecessary systems that decrease the overall efficiency can be removed, converting the cell into a *lean* system (Leprince et al. 2012; Valgepea et al. 2015). The primary characteristics of such a cell would be a reduced maintenance demand and an increase in productivity while preserving other physiologically relevant parameters (Fig. 1). Vickers et al. (2010) proposed that starting from minimal cells *chassis* organisms can be derived through the addition of reactions or modulation of metabolic fluxes. Such a chassis may serve as a platform organism for the production of sets of chemicals or proteins (Vickers et al. 2010).

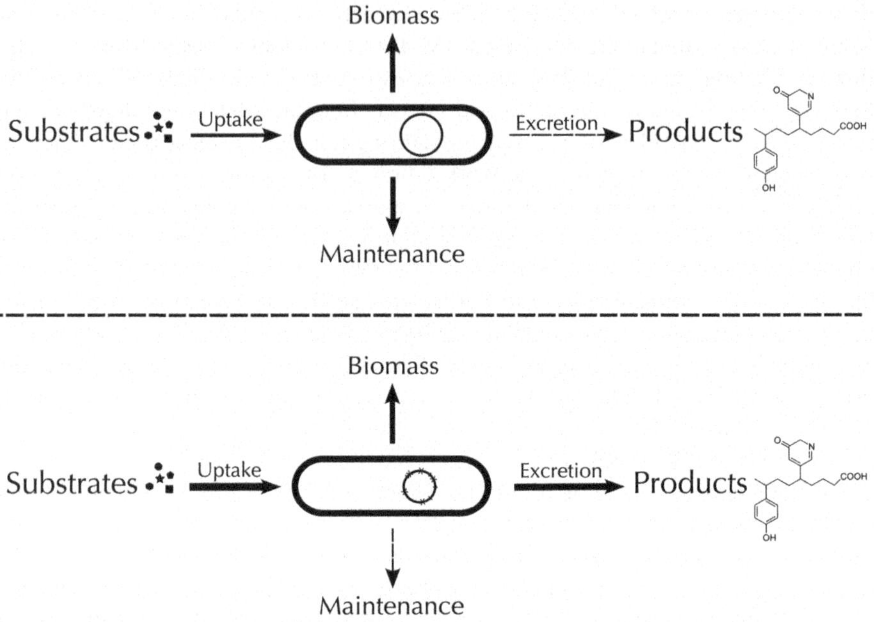

Fig. 1 Desired traits of lean microbial cell factories. In comparison to wild-type cells (upper part), the genome-reduced lean cell (lower part) directs a greater proportion of nutrients to product synthesis. The maintenance demand is reduced while substrate uptake rates may improve. Nutrient versatility is conserved

The concept of microbial cell factories has been adapted by authors from diverse research fields (Gong et al. 2017; Pscheidt and Glieder 2008; Villaverde 2010). While new genetic hosts are on the rise, the majority of industrial bioprocesses, for instance, biopharmaceutical production, utilizes a small set of reoccurring species (Ferrer-Miralles et al. 2009). The construction of chassis organism tailored towards the production of a subset of chemicals appears as a logical way to simplify research and development processes and increase process performance (Esvelt and Wang 2013; Sauer and Mattanovich 2012). An already heavily optimized chassis is *Corynebacterium glutamicum*, one of the leading microbes for amino acid production (Becker and Wittmann 2012; Lee et al. 2016). With the use of systems biology and synthetic biology tools plus the availability of efficient genome-engineering methods, simplification through intelligent genome reduction becomes a realistic target as another layer of improvement (Colin et al. 2011; Gao et al. 2010). Or, to stay with the analogy of a factory: The aim is to cut the operating costs of living cells.

3 Theoretical Considerations on Genome Reduction

What is *genome reduction*? In the context of this contribution, we will run with the following working definition:

> Genome reduction is the repeated deletion of irrelevant genes by methods of genetic engineering with the purpose of constructing a functionalized cell for a selected application.

This somewhat bulky definition attempts to distinguish genome reduction from other deletion attempts. Clearly, simple deletions targeted at the understanding of gene function such as the Keio collection (Baba et al. 2006) cannot be considered genome reduction as they lack the cumulative nature of the approach. On the other hand, reductive evolution, which may very well be consecutive in nature (Wolf and Koonin 2013), must be excluded, hence the requirement for genetic engineering methods.

It is difficult to accurately delimit genome reduction from metabolic engineering. Looking at the definition of metabolic engineering proposed by Gregory Stephanopoulos provides us with a basis: "Here we define metabolic engineering as the *directed* improvement of product formation or cellular properties through the modification of *specific* biochemical reactions or the introduction of new ones with the use of recombinant DNA technology" (Stephanopoulos 1999). Genome reduction as defined by us differs in that specificity is not necessarily assumed and single knockouts to reroute intracellular fluxes are excluded. However, recent approaches towards genome reduction discussed later in this contribution are increasingly knowledge-driven and specific in their target selection. Consequently, there is a gray area where metabolic engineering and genome reduction overlap. If we adopt a broader interpretation of Stephanopoulos's definition of metabolic engineering and integrate the concept of cell factories within it, then genome reduction becomes a subset of metabolic engineering.

The Concept of Relevant Genes Our definition of genome reduction specifically mentions *irrelevant genes* in an attempt to circumvent the fuzzy term *(non-) essential genes*. For years, there has been an ongoing discussion on which genes are *essential* for life (Commichau et al. 2013; Juhas et al. 2014; Koonin 2000; Szathmáry 2005). Unfortunately, definitions of the term vary and are complicated by inconsequent use and addition of descriptive operators like *critical*, *core* or *superessential* (Barve et al. 2012; Gil et al. 2004; Hooven et al. 2016). Simulated minimal metabolic networks—albeit being valuable tools—provide no uniform answer either (Ye et al. 2016). Elementary mode analysis can be used to compute all minimal sets of reactions or enzymes necessary to support a metabolic network at steady state, but there is typically more than one solution for a given network (Schuster and Hilgetag 1994; Trinh et al. 2008). Additionally, inferring a biologically meaningful minimal set of enzymes from such calculations is difficult as enzymes may be promiscuous or their regulation may not allow an elementary mode to be actually realizable in vivo. Some authors restrict the term *essential* to only the most basic cellular functions such as DNA replication and protein biosynthesis implying that many metabolites are readily available for uptake in an artificially defined system (Gil et al. 2004). Others emphasize the fact that essentiality depends on the outer circumstances in which a living system thrives (Burgard et al. 2001).

For the application of microbes in an industrial setting these definitions are insufficient. In the defined niche of a bioreactor it is necessary to individually consider which genes are required for survival and replication. Next, genes which are beneficial towards economically relevant parameters such as product tolerance are identified. Following a recently proposed definition, the union of these genes is termed *relevant genes* and entirely dependent on the specific needs of a process (Noack and Baumgart 2018; Unthan et al. 2015). For example, in an *E. coli* process substrate uptake must be fulfilled by a suitable transporter. Depending on the choice of carbon source the associated transporter is a relevant gene: Regarding a glycerol-based process *glpF* is a major relevant gene (Hénin et al. 2008). In a glucose-based process, it falls into the category of irrelevant genes and is in consequence a potential deletion target (Noack and Baumgart 2018). In summary we limit genome reduction to a point where the cell retains its ability to replicate on its own while displaying advantageous production traits in a specific and predefined environment.

In the previous sections, we have discussed the concept of microbial cell factories. Genome reduction from this perspective attempts to simplify the host reducing its carbon and energy waste. The question arises what improvements in terms of measurable parameters can be expected from a genome-reduced organism and how to identify promising deletion targets.

Reduction of Genomic DNA Content DNA replication is not a free lunch. In minimal medium nucleotides are synthesized de novo from glucose and salts, and polymerization requires further energy input. Theoretical calculations on biomass composition alone estimate the cost of DNA synthesis under these conditions at around 3% of total cellular ATP expenses (Stouthamer 1973). Since even under optimal conditions about half of the available ATP is apparently not channeled into biomass but lost in membrane potential upkeep and other processes (Farmer and Jones 1976), this value decreases to approximately 1.5% of total ATP generated. The potential to save energy merely from reduced DNA synthesis is thus relatively small. On the other hand, even small contributions can be impactful under the conditions of an economically strained large-scale process. One of the largest successful deletion series in *E. coli* comprises roughly 22% of the genome (Mizoguchi et al. 2008) which by itself would be estimated to reduce ATP expenses by about 0.33%. From a carbon balance point of view, this strain would save roughly 0.5% of carbon molecules per division assuming 2% DNA content in dry biomass at growth rates close to 0.3 h^{-1} (Taymaz-Nikerel et al. 2010).

Are there other potential benefits accompanying reduced genomic DNA content? Choe et al. (2015) proposed that a smaller genome might result in faster doubling times since DNA replication time exceeds doubling times under fast-growing conditions in *E. coli* (Choe et al. 2016). However, *E. coli* naturally balances this situation by multifork replication (Fossum et al. 2007). Studies on different *E. coli* deletion series did not yield uniform outcomes, and none conclusively showed an increase in the maximum specific growth rate μ_{max}. In fact several groups independently observed detrimental effects in μ_{max} over cumulative deletions based on different criteria (Hashimoto et al. 2005; Karcagi et al. 2016; Kurokawa et al. 2016). Simply assuming benefits of genome reduction by saving of energy on DNA replication ignores side effects that may be caused by multifunctionality of genes or promiscuity of encoded enzymes. In contrast, if deletion regions were specifically selected based on neutrality towards growth no decrease, but also no increase, in μ_{max} could be observed (Mizoguchi et al. 2008). The probably most comprehensive study on process relevant parameters of a genome-reduced organism showed an increase in μ_{max} from 0.38 to 0.53 in *Pseudomonas putida* with deletions comprising only 4.3% of the genome (Lieder et al. 2015). Obviously, this increase cannot be caused by the reduced DNA content alone, and it is impossible to accurately attribute potential contributions. Perhaps only a reversed approach can solve this issue: artificially burdening a microbial wild-type genome with non-expressible junk DNA.

Reduction of RNA and Protein Synthesis Costs While ATP used for DNA replication appears to be a minor fraction, things change if we look at the synthesis costs for RNA and proteins. Under the same assumptions as mentioned in the prior section, and regarding the exemplary case of *E. coli*, monomer synthesis and polymerization make up about 35% of total cellular ATP expenses (Farmer and Jones 1976; Stouthamer 1973). Comprehensive work on the *lac* operon showed that the process of expressing useless genes burdens cells considerably, resulting in a growth disadvantage in competition experiments (Stoebel et al. 2008). The actual presence of useless protein in turn does not confer a competitive disadvantage—at least under the conditions of aforementioned study. These findings are well in line with theoretical estimations that appraise polymerization of proteins as the most costly cellular process (Noack and Baumgart 2018; Stouthamer 1973). However it remains unclear whether occupation of RNA polymerases and ribosomes contributes significantly to the observed disadvantage (Stoebel et al. 2008).

In consequence, if we free a strain from the expression of certain proteins it will have more resources available for the formation of biomass or target products. Gene expression is heavily regulated though raising the question to what extent expression of unnecessary proteins occurs. The well-known *lac* operon is strongly repressed up to 1300-fold by Lac repressor in the absence of lactose (Oehler et al. 1990). Under such circumstances its products are regularly below the detection limit of proteomic methods. To achieve its strong repression, Lac repressor is sufficient in minimal quantities, typically in the range of ten molecules per cell (Schmidt et al. 2016, supplementary material). Compared to the total number of proteins per cell, which is estimated at roughly 3,000,000 (Milo 2013), the impact of single deletions targeted at unused operons does not appear promising. Even the combined benefit of dozens of deleted operons with similarly tight regulation is unlikely to lead to measurable advantages.

In this respect Valgepea et al. (2015) proposed to delete proteins with high synthesis costs, in other words proteins that are both abundant and relatively large (Valgepea et al. 2015). Proteomic studies in *E. coli* have shown that under the majority of conditions tested, about two thirds of the total proteome mass belonged to proteins grouped into only four COG categories: *translation, ribosomal structure, and biogenesis, energy production and conversion, amino acid transport and metabolism, and carbohydrate transport and metabolism* (Schmidt et al. 2016). In all categories many but not all genes are required for normal growth in minimal medium. Therefore, handpicking of relevant and irrelevant genes is a necessity. Regulation of fundamental features such as growth rates is very complex as recently observed for *C. glutamicum* (Baumgart et al. 2018). Accordingly, deleting members of complex modulons is a tedious work and likely initially compensated by counteracting gene expression to keep the previous cellular state. Interestingly, complementary experimental studies indicate that about 22% of the *E. coli* proteome have no measurable benefit in glucose minimal medium (Price et al. 2016). A substantial fraction of the unused proteins is attributed to preadaption to changing nutrient availability and stress resistance (O'Brien et al. 2016). Given that the major external stressors pH, temperature, and medium composition are controlled in bioprocesses anyway the underlying genes are promising targets for impactful deletion series.

Can we generalize these findings to other organisms than *E. coli*? The vast majority of superfluous proteins are present at extremely low levels in *Saccharomyces cerevisiae* (Ghaemmaghami et al. 2003). Deletion studies have regularly failed to identify beneficial phenotypes from single deletions (Sliwa and Korona 2005). Mutations leading to the loss of mating functions by silencing of 23 genes were shown to confer a small but measurable growth advantage (Lang et al. 2009). In conclusion, the situation appears to be very similar in the case of yeast and potentially also in other organisms.

Reduction of Secondary Cost Secondary costs accompany many cellular processes: ATP is hydrolyzed, reduction equivalents are consumed or the potential energy of gradients is used. One of the most prominent examples of secondary cost is the flagellar apparatus. Its biosynthesis is already quite resource intensive, resulting in an estimated fitness cost of about 2% (Macnab 1992). Once a flagellum is assembled, its operation is energy dependent. Macnab (1992) estimated the value for energy consumed by rotating flagella at approximately 10^{-15} W per cell, thus arriving at a fraction of total energy expenditure of 0.1% (Macnab 1996). From our point of view, this estimation is quite conservative, and we present an alternative calculation here.

The number of flagella ranges from about 1 to 10 in *E. coli* (Mears et al. 2014). Each rotating flagellum uses an influx of roughly 1200 protons per revolution and operates under free-swimming conditions slightly above 100 Hz (Berg 2003; Lowe et al. 1987). Assuming a constant H+/ATP ratio of 4 (Turina et al. 2003), this translates into an ATP hydrolysis of around 1.5×10^5 molecules per second for an exemplary *E. coli* with five flagella rotating at 100 Hz each (Lowe et al. 1987). How does this compare to the total ATP turnover in *E. coli*? Assuming exponential growth in glucose minimal medium oxygen uptake rate is in the range of 27 mmol O_2 per gram biomass dry weight and hour (Jain and Srivastava 2009). Literature data on P/O ratio vary considerably, but for the purpose of this calculation, we will assume a constant P/O ratio of 1.5 (Noguchi et al. 2004; Taymaz-Nikerel et al. 2010). Given that a single *E. coli* roughly weighs 3×10^{-12} g, we arrive at a net ATP production of about 4×10^6 ATP molecules per cell and second. Under the conditions chosen, flagellar motion can thus consume about 3.75% of the total aerobic cellular ATP-generating potential.

This number is a rough estimate and not necessarily representative for specific conditions. The number of flagella varies, we have neglected swimming and tumbling behavior and flagellar rotation is temperature and voltage dependent (Berg 2003; Lowe et al. 1987). Our calculation for net ATP production per cell is based on the assumption of exponential aerobic growth. If, in the absence of external electron acceptors, fermentative metabolism is used, ATP productivity is expected to be substantially lower. Macnab (1996) reported that the proton motive force in such a scenario is subsaturating, which results in lower rotation speed. Given that our calculation assumes five flagella per cell and a rotation speed of 100 Hz, our result is higher than previously reported values (Macnab 1996). Also we calculate the fraction of ATP-generating potential used by flagella, not the fraction of total energy expenditure. To sum up, the true fraction of ATP-generating potential consumed is likely in the range of 0.1–10%.

Note that it is unclear whether the flagellar motor proteins actually transport protons under the conditions of a bioreactor. Flagella are prone to shearing and, at least in small-scale bioreactors with high power input per volume, shearing forces are regularly sufficiently high to displace the outer parts of *E. coli* flagella. Our calculation is supported by experimental evidence from pilot-scale reactors though. Studies on *Pseudomonas putida* deletion strains lacking flagella showed that the deletion of flagellar genes led to an increase in biomass yield of 1–7% depending on growth rate (Lieder et al. 2015). Lieder et al. (2015) collected this data under the controlled conditions of a small-scale bioreactor operating in chemostat mode.

Aside from this example the accurate attribution of secondary costs of metabolic processes is often difficult. Particularly in the case of global regulation it is hardly possible. Löffler et al. (2016) and Simen et al. (2017) attempted to measure metabolic costs caused by oscillatory regulatory patterns in scale-down systems mimicking gradients in large-scale reactors. They estimated that the secondary costs due to aberrant gene expression increase maintenance energy demand by about 15–50% depending on the limitation scenario (Löffler et al. 2016; Simen et al. 2017). The calculated increase is distributed over dozens of genes though. Moreover, *E. coli* rapidly accumulate inactivating mutations in the major stress sigma factor *rpoS* under the conditions of a glucose-limited chemostat indicating potential inherent inefficiencies or trade-offs (Ferenci 2005; Notley-McRobb et al. 2002). While the aforementioned studies indicate potential deletion targets on a comprehensible basis, the observed growth or fitness advantages are entirely cryptic in other cases. In strains of *Escherichia coli*, B silencing deletions of the *rbs* operon responsible for ribose degradation conferred a measurable competitive fitness advantage of about 1–2% per generation in glucose minimal medium (Cooper et al. 2001). The authors concluded that absence of ribose catabolism itself is supposedly advantageous in glucose media, but the molecular basis of this effect is unknown. Also the possibility remains that the energy savings from silencing or deleting the *rbs* operon are responsible for the competitive advantage. Another study found an association of loss of *cspC* expression with an unusually high increase in fitness in complex medium (Rath and Jawali 2006). The absence of a clear molecular basis for *cspC* functions makes it difficult to deduce a robust deletion strategy in such a case.

Side Effects of Genome Reduction So far this section has mainly covered possible benefits of genome reduction. For a complete picture, potential imponderables must be considered as well. Besides the obvious pitfalls—poor choice of target genes or unexpected side effects after deletion of uncharacterized genes—there are two issues we would like to briefly discuss here: transcription factor DNA binding sites and chromosome organization.

Current information in RegulonDB (Gama-Castro et al. 2016) estimates the number of transcription factors in *E. coli* to be slightly more than 200. Only 30 of these transcription factors have more than 20 binding sites on the *E. coli* chromosome. Given that gene expression is inherently stochastic (Raj and van Oudenaarden 2008), even small reductions in the balance of transcription factor availability and number of genomic binding sites may cause off-target effects in the form of stronger repression or activation on the remaining sites. This pitfall can likely be circumvented by deleting coding regions only and leaving regulatory DNA elements intact. Alternatively, if entire operons or pathways are chosen as deletion targets, their regulatory transcription factors could be deleted simultaneously.

The chromosomal structure of microorganisms serves a function, and its disruption must be carefully considered. Evidence that caution needs to be exercised when potentially tampering with nucleoid architecture is provided by various studies. It has been shown that changes in the level of expression are correlated between genes at defined long-range distances in both *E. coli* and *B. subtilis* (Carpentier et al. 2005). Moreover, there is a periodicity in the expression pattern of the *E. coli* genome in accordance with a solenoidal nucleoid architecture (Képès 2004). Consequently, when planning the deletion of large regions, potential side effects on nucleoid architecture must be taken into consideration. To complicate this issue, large discrepancies in the size of the two replichores lead to severe growth defects (Esnault et al. 2007). While deletions of single genes or small operons are unlikely to be problematic on the scale of nucleoid architecture, large deletions should be designed to be neutral towards replichore length and expression periodicity.

4 Achievements in Genome Reduction: Organisms and Process Parameters

This section covers studies on the construction and evaluation of genome-reduced organisms organized by species. While for some projects short remarks on the construction of the deletions are included, we focus on the phenotypic properties of the deletion strains. Table 1 contains an overview of the studies examined.

Escherichia coli The Gram-negative bacterium *Escherichia coli* is the most well-studied prokaryotic model organism and often the first choice for industrial production hosts (Baeshen et al. 2015; Dong et al. 2011; Terpe 2006). The idea to derive a genome-reduced or even minimal cell was born with *E. coli* in mind (Koob et al. 1994). The advantages of working with *E. coli* are numerous: rapid growth on a variety of carbon and nitrogen sources in both minimal and complex media, the availability of many sequenced variants, sophisticated molecular biology tools, well-defined functional annotation, and the knowledge required for scale-up and commercialization of processes.

One of the first reported attempts at constructing a genome-reduced *E. coli* MG1655 was based on a transposon library to identify non-essential regions (Yu et al. 2002). The combination of some of these regions culminated in a strain lacking about 6.8% of its parent's genome. While the molecular biology involved in this study was highly sophisticated, the resulting strains were only rudimentarily characterized—a reoccurring issue in genome reduction studies.

Transposon-based approaches are appealing from a number of viewpoints: their random nature requires little sequence knowledge, they work in a vast array of hosts, and some enable the capturing of deleted genomic regions on conditional plasmids (Goryshin et al. 2003). Ultimately, they did not prevail, as the wide array of genomic tools and the sequence information available in *E. coli* made other systems, particularly recombineering based on phage λ and P1 transduction, more attractive.

Another early attempt at constructing deletion strains culminated in a 29.7% genome-reduced strain (Hashimoto et al. 2005). The strains displayed reduced fitness compared to their parent MG1655. All of them grew substantially slower in LB or similar rich media. Moreover, the deletion strains had aberrant cell size and nucleoid structure. While these phenotypes are certainly not suitable for production, the study nevertheless yielded valuable information. The construction of large genomic deletions per se is laborious but manageable. The difficulty lies in target selection. Target regions must be chosen very carefully and tested for neutrality towards growth or other desired characteristics.

The most famous series of deletion mutants is probably the "MDS" series based on the strain *E. coli* K-12 MG1655 with a native genome size of about 4.6 Mb (Blattner et al. 1997). Initially, comparative genomics between phylogenetically distinctly related *E. coli* strains lead to the identification of strain-specific islands within a core genome of about 3.7 Mb (Kolisnychenko et al. 2002). Deletions of 12 strain-specific regions were performed with a phage λ-derived homologous repair

Table 1 Overview of genome reduction projects

Scope of Genome Reduction	Results	Source
Escherichia coli K-12 MG1655		
Up to 6.8% non-essential regions identified by transposon insertion	Normal growth in LB	Yu et al. (2002)
Up to 29.7%	Reduced growth in rich media; aberrant nucleoid organization and cell size	Hashimoto et al. (2005)
8.1% 12 strain-specific regions	Normal growth in minimal medium	Kolisnychenko et al. (2002)
15% strain-specific regions; IS elements	No differences in growth or protein production; higher electroporation efficiency; no IS-related mutations; increased plasmid stability	Pósfai et al. (2006)
Strains from Pósfai et al. (2006)	No differences in fermentation parameters for MDS40 and MG1655	Sharma et al. (2007b)
Strains from Pósfai et al. (2006)	Expression of recombinant CAT leads to acetate accumulation in MDS41 and MDS42 but not in MG1655; no differences in fermentation parameters for inactive protein	Sharma et al. (2007a)
Strains from Pósfai et al. (2006)	Engineering of threonine production: About twofold increase in titer compared to WT, final titer in fermentation of 40.1 g/ll	Lee et al. (2009)
Strains from Pósfai et al. (2006)	Increased plasmid stability when toxic ORF is present	Umenhoffer et al. (2010)
Strains from Pósfai et al. (2006)	Increased plasmid stability when large direct repeat regions are present	Chakiath and Esposito (2007)
Strains from Pósfai et al. (2006)	Chromosomal periodicity of MDS42 reduced to 6, transcriptional changes partly overlap with heat shock response	Ying et al. (2013)
Up to 20% based on strains from Pósfai et al. (2006)	MDS strains are overgrown by the wild-type in competition assays and have reduced biomass yield as well as elevated *rpoS* levels	Karcagi et al. (2016)
23% deletion of IS elements and non-essential regions	MS56 stably maintains plasmids and heterologous protein production	Park et al. (2014)
Escherichia coli BL21(DE3)		
Up to 9%	Slightly altered growth phenotypes; substantially stabilized genome and plasmids	Umenhoffer et al. (2017)
Escherichia coli K-12 W3110		
22%	MGF-01 reaches 1.5 times higher cell density in M9 minimal medium and produces twice as much threonine as the wild-type (unpublished results by H. Mizoguchi)	Mizoguchi et al. (2007), Mizoguchi et al. (2008)

(continued)

Table 1 (continued)

Scope of Genome Reduction	Results	Source
Up to 36% based on strains from Mizoguchi et al. (2008)	Reduced lag phase and faster growth to higher optical densities in corn steep liquor medium	Hirokawa et al. (2013)
Strains from Mizoguchi et al. (2008)	Decreased maximum specific growth rate and biomass yield with increasing deletion size	Kurokawa et al. (2016)
Bacillus subtilis		
7.7% prophages and AT-rich islands	No major differences in phenotypical parameters; alterations in motility	Westers et al. (2003)
Up to 24% prophages, polyketide synthesis, several non-essential regions	Slightly reduced growth rate in minimal and complex media; comparable protein production	Ara et al. (2007)
20.7% prophages, polyketide synthesis and 11 non-essential regions	Strain MGB874: Reduced maximum growth rate; no sporulation; identical growth under production conditions; increased sugar consumption rate and protein production; altered transcriptional profile	Morimoto et al. (2008)
Strains from Morimoto et al. (2008)	Strain MGB874: Identification of *rocDEF-rocR* regulatory effects and elevated transcript levels of heterologous gene	Manabe et al. (2011)
Strains from Morimoto et al. (2008)	Strain MGB874: Deletion of *rocG* further increases heterologous protein production in ammonia controlled pH-auxostat	Manabe et al. (2013)
Up to 19.7% prophages, antibiotic gene clusters, non-essential regions	Minimal medium: Reduced growth rate and specific glucose uptake rate, but increase in biomass yield; reduced autolysis, sporulation rate and transformation efficiency; reduced maintenance coefficient; higher transcript levels of overexpressed genes and elevated production of thymidine and guanine	Li et al. (2016)
36% non-essential regions, based on strains from Westers et al. (2003)	Reduced growth rate; filamentous phenotype; transcriptional and proteomic changes; reduced flux through glycolysis; altered amino acid metabolism	Reuß et al. (2017)
Corynebacterium glutamicum R		
11 individual SSIs	Slightly improved growth in minimal medium	Suzuki et al. (2005a)
5.7% eight SSIs previously identified	No significant differences detected	Suzuki et al. (2005b)
Corynebacterium glutamicum ATCC13032		
6% prophages CGP1, CGP2, CGP3	Higher specific growth rate under CPG3 inducing conditions otherwise no differences in growth; increased	Baumgart et al. (2013)

(continued)

Table 1 (continued)

Scope of Genome Reduction	Results	Source
	transformability and plasmid copy number	
13.4% several irrelevant genomic regions	Identical growth of *C. glutamicum* C1∗ and its parent in minimal and rich medium	Baumgart et al. (2018)
Deletions of IS elements families	Improved plasmid stability resulting in increased production of fluorescent protein, PHB and GABA	Choi et al. (2015)
Pseudomonas putida KT2440		
1.1% flagellar genes	Reduced lag phase when growing on fructose; higher oxidative stress resistance but failure to use various carbon sources; sensitivity to antibiotics and low pH stress	Martínez-García et al. (2014b)
2.6% all annotated prophage regions	Increased tolerance to UV light and DNA damaging agents	Martínez-García et al. (2015)
4.3% flagella, prophages, transposons, DNA restriction-modification system	Combinatorial phenotype similar to previous studies; reduced maximum specific growth rate in complex medium but increase in minimal glucose medium	Martínez-García et al. (2014a)
4.3% strains from Martínez-García et al. (2014a)	Phenotyping in bioreactors: Increased maximum specific growth rate and biomass yield in glucose minimal medium; reduced maintenance coefficient; higher AEC; higher plasmid stability and specific protein productivity; reduced organic acid formation	Lieder et al. (2015)
Pseudomonas chlororaphis		
Up to 10.3% 22 regions containing non-essential genes (rich medium)	Some strains show impeded growth and production, others display substantially increased phenazine production	Shen et al. (2017)
Lactococcus lactis NZ9000		
2.83% four non-essential regions containing prophage proteins	Improved growth parameters and higher final cell densities, higher mRNA and protein production levels	Zhu et al. (2017)
Streptomyces coelicolor		
2% four native antibiotic gene clusters	Normal growth and sporulation of strain M1146; increased production of chloramphenicol and congocidine; reduced side peaks in HPLC analysis	Gomez-Escribano and Bibb (2011)
Up to 14% ten secondary metabolite clusters and a 900 kb subtelomeric region	Normal growth of all strains; some strains show potentially increased actinorhodin biosynthesis	Zhou et al. (2012)

(continued)

Table 1 (continued)

Scope of Genome Reduction	Results	Source
Streptomyces avermitilis		
Up to 18.5% 1.4 Mb subtelomeric region, several secondary metabolite clusters	3–5-fold higher titers of streptomycin in SUKA5 strain than its parent; absence of secondary metabolites	Komatsu et al. (2010)
See previous study	Strains SUKA 5 and SUKA17 reach higher final cell densities than the wild-type; SUKA 17 displays higher productivity of seven antibiotics	Komatsu et al. (2013)
Saccharomyces cerevisiae SH5209		
Up to 5% 15 chromosomal regions predicted to be non-essential	Aberrant expression patterns; reduced mitochondrial functions, reduced growth in liquid media; increase in ethanol production by 1.8 fold compared to parent strain	Murakami et al. (2007)
Schizosaccharomyces pombe		
5.2% four large deletions, one on each arm of chromosomes I and II	Slightly reduced specific glucose uptake rate and ethanol production, extended lag phase, higher ATP levels and protein production	Sasaki et al. (2013)

mechanism assisted by nuclease counter-selection. The final strain MDS12 lacked 8.1% of the wild-type genome. Its maximum growth rate was not statistically different from the wild-type, but it reached slightly higher final cell densities (Kolisnychenko et al. 2002). Further deletions of strain-specific regions, IS elements, and pseudogenes lead to strain MDS43 with a 15% reduced genome (Pósfai et al. 2006). The MDS strains did not show apparent growth defects in minimal or complex medium, but also did not have increased protein productivity in a fed-batch process. Their most outstanding properties reported were increased transformation efficiency and stable maintenance of plasmids due to the absence of IS elements (Pósfai et al. 2006).

Some strains of the "MDS" series were the subject of further investigations. The fermentation profile of strain MDS40 was examined (Sharma et al. 2007b). No striking differences between MDS40 and MG1655 could be observed. MDS41 and MDS42 showed an increased acetate formation and reduced biomass yield in fed-batch fermentations if recombinant chloramphenicol-acetyltransferase (CAT) was produced as a model protein (Sharma et al. 2007a). This was not the case with mutated inactive CAT. It remains unclear why this activity occurs in MDS41 and MDS42 but not in their wild-type parent (Sharma et al. 2007a). In another study MDS42 was engineered to produce L-threonine (Lee et al. 2009). In shaking flask cultivations, the engineered strain based on MDS42 reached almost twice the final concentration of L-threonine than an identically engineered strain based on MG1655. Transcriptional profiling indicated that the strain based on MDS42 had higher transcript levels of several process relevant genes. An exemplary jar fermentation was also conducted with the engineered strain reaching a final titer of 40.1 g/l threonine (Lee et al. 2009). Unfortunately, no control experiment with the engineered MG1655-based strain was conducted.

The absence of IS elements in MDS42 proved to be advantageous for research purposes in the stable maintenance of plasmids encoding toxic genes (Umenhoffer et al. 2010). While in MG1655 the toxic ORF examined was regularly inactivated by transposon insertion, this did not happen in MDS42. Only one in ten cultures of MDS42 carrying the toxic plasmid grew to high optical densities by mutational inactivation of the toxic ORF (Umenhoffer et al. 2010). Similar results were obtained with lentiviral expression vectors (Chakiath and Esposito 2007). These plasmids contain long direct repeat regions and are often unstable in *E. coli*. MDS42 outperformed all other examined *E. coli* strains in terms of plasmid stability, even strains that are commonly used for faithful cloning (Chakiath and Esposito 2007). To strengthen this trait Csörgo et al. (2012) further engineered MDS42 by removing stress-induced mutagenesis mechanisms. Mutation rates in the resulting strains were significantly reduced to less than half of those observed in MG1655 and almost two orders of magnitude lower than in *E. coli* BL21(DE3), a strain commonly used for protein production (Csörgo et al. 2012). This effect was even more pronounced under stress conditions. Notably, the engineered strains had growth rates identical to those of MG1655. The ability of MDS42 to reliably carry plasmid DNA initiated further comparative studies between MG1655 and MDS42 (Akeno et al. 2014). It was found that MDS42 suffered a higher burden than MG1655 when carrying plasmids of different sizes. Expression of foreign genes on plasmids was much stronger in MDS42 than in MG1655 at the expense of growth (Akeno et al. 2014).

A complementary approach aimed at investigating differences in transcriptional profiles between MDS42 and MG1655 (Ying et al. 2013). It was found that the transcription profile of MDS42 has a reduced chromosomal periodicity of six periods instead of seven under normal growth conditions. Given that MDS42 lacks about one seventh of its genome this might point at an altered chromosomal architecture. Upon heat shock treatment, the chromosomal periodicity of transcription was also reduced to six in MG1655, but it did not further decrease in MDS42. The overall transcriptional alterations in MDS42 showed partly overlapping patterns of heat shock effect and genome reduction effect in both antagonistic and synergistic ways (Ying et al. 2013). This study clearly shows that chromosomal periodicity must be taken into account when rationally designing large deletion series.

In continuation of the MDS deletion series Karcagi et al. (2016) created new strains by further reducing the K-12 genome. The final strain in this study was MDS69 with a genome reduced by about 20%. Over the course of the deletions, it became increasingly apparent that some deletions affected the fitness of the strains. To elucidate this, competition assays against MG1655 were performed. All tested strains from the MDS lineage, even early strain MDS14, were outcompeted by the wild-type, and this effect was more pronounced in the strains with more extensive reductions. Furthermore, the MDS strains showed reduced biomass yield in chemostat cultures with either glucose or ammonia as the limiting nutrient. The maintenance coefficient remained unchanged though (Karcagi et al. 2016). In summary, the basic fermentative parameters of the MDS series strains turned out not to be superior to those of their wild-type parent—a result that could have been available for several years if the early MDS strains had been tested more thoroughly in bioreactors.

What do the studies on the MDS series of strains teach us? First, proper choice of targets for genome reduction is essential. The removal of K-islands proved to be almost neutral towards growth, and the removal of IS elements provided the basis for an application of the strains. The increased genomic stability is a significant advantage in special applications. Second, thorough testing of the strains is mandatory. Characterization should take place under conditions as close to industrial application as possible. Unfortunately, while the potential for molecular biology applications was identified quickly, the economic potential of the MDS strains has hardly been investigated. Production of CAT as a model protein showed no advantages, but this might be based on the nature of the protein produced. The increase in L-threonine production reported is a most promising result, but even in this study the characterization in a bioreactor system was rudimentarily at best. When finally, after the construction of MDS69 fundamental parameters were accurately measured it became apparent that the hope for improvements in the basic physiology would not be fulfilled. Nevertheless, the strains from the MDS series have been commercialized by Scarab Genomics LLC, and patents have been filed for the production of recombinant non-toxic mutant derivative of diphtheria toxin CRM197 by MDS42 and MDS69 (Blattner et al. 2017). The company claims to produce CRM197 at productivities of 3—5 g/l/day in a continuous fermentation mode using a strain designated MDS69meta. The activities of Scarab Genomics LLC prove that genome-reduced strains with advantageous properties have industrial relevance—even if these advantages are mainly genomic and plasmid stability. Since we can expect that MDS42 and MDS69 will outperform their wild-type parent in other scenarios involving the maintenance of unstable DNA constructs, these strains can truly be regarded as chassis strains for this extraordinary application.

Based on the experiences collected with the MDS strains, Umenhoffer et al. (2017) attempted the construction of a genome-reduced chassis based on BL21 (DE3). After inactivation or deletion of prophages, IS elements and error-prone DNA polymerase the resulting strain BLK16 showed increased transformation efficiency and dramatically improved plasmid stability. Its genome is a mosaic of BL21(DE3) and K-12 sequences. The authors found minor alterations in growth, but clear disadvantages were not apparent (Umenhoffer et al. 2017). An in-depth characterization of its production properties is still pending, but it might quite well be that BLK16 emerges as an intelligently reduced chassis strain for protein production.

The "MGF" minimum genome factory deletion series initially followed a similar path like the "MDS" series (Anazawa 2014; Mizoguchi et al. 2007; Mizoguchi et al. 2008). Candidate regions for deletion were derived from a comparison of the *E. coli* K-12 genome with that of *Buchnera* species. These regions were deleted individually and those not affecting growth transferred into a single genome for a total genome reduction of about 22% in strain MGF-01. Note that strain MGF-01 is based on *E. coli* K-12 W3110. *E. coli* W3110 is a close relative of MG1655 but shows some alterations in central carbon metabolism (Vijayendran et al. 2007). MGF-01 reaches a 1.5-fold higher final cell density in M9 minimal medium than W3110. This is likely based on a reduced overflow metabolism as MGF-01 accumulated substantially less acetate than W3110 while displaying identical maximum specific growth rate. The effect of elevated final cell densities was gradual and increased with the number of deletions. Additionally, MGF-01 reached 2.4-fold higher threonine titers than MG1655 after a threonine production cassette was integrated in both strains (Mizoguchi et al. 2007, 2008).

In direct contrast to these findings later studies found a significant decrease in maximum specific growth rate and cell yield for the deletion strains leading to MGF-01 (Kurokawa et al. 2016). The effects were also gradual and more pronounced with an increase in the deletion size. The observed effects were not independent of the growth medium chosen and much more severe in minimal media. The data was collected from microplate cultivations without process control which limits their meaningfulness, but the findings are well in line with the observation that the MDS strains which carry deletions similar in size are overgrown by the wild-type in competition assays (Karcagi et al. 2016). Further investigations on MGF-01 and its parent strains revealed an altered mutational rate of these strains (Nishimura et al. 2017). Strains carrying a reduced genome showed higher mutational rates than their wild-type parent concomitant with a reduction in growth rate. Upon serial transfer of one deletion strain for approximately 400 generations its growth rate increased by 30% to wild-type levels and its mutational rate decreased by one order of magnitude also to wild-type levels (Nishimura et al. 2017). It appears likely that over the course of serial transfer beneficial mutations restoring the disturbed organization of cellular resources by genome reduction were fixated within the population rescuing initial flux maldistribution (Fig. 2).

Based on MGF-01 Hirokawa et al. (2013) constructed further deletion strains. The removal of IS elements and other non-essential regions was accompanied by rational reinsertion of some important sequences to maintain good growth as well as reversal of the well-known K-12 mutations in *ilvG* and *rph-1*. To reflect this change in strategy, the constructed series was renamed "DGF" for designed genome factory. The DGF strains have substantially reduced genomes only slightly larger than 3 Mb, and DGF-298 is even below this threshold (Hirokawa et al. 2013). They all show similar or better growth in minimal medium M9 and corn steep liquor like W3110, but it is difficult to pinpoint the "best" strain within their series. What remains to be shown is their putative superiority in the production of proteins or small molecules in a bioreactor fermentation scenario.

Apart from the MDS and MGF projects other groups have also attempted to construct superior deletion strains with similar design principles. The "MS" series reached a total genomic reduction content of 23% in strain MS56 based on MG1655 (Park et al. 2014). Target deletion regions were chosen similar to MDS42 and specifically included all IS elements. MS56 lacked IS element-based inactivation of plasmid genes and stably maintained a plasmid over serial cultivation for 20 days. Upon intentionally co-cultivating mixed populations of MS56 carrying active or

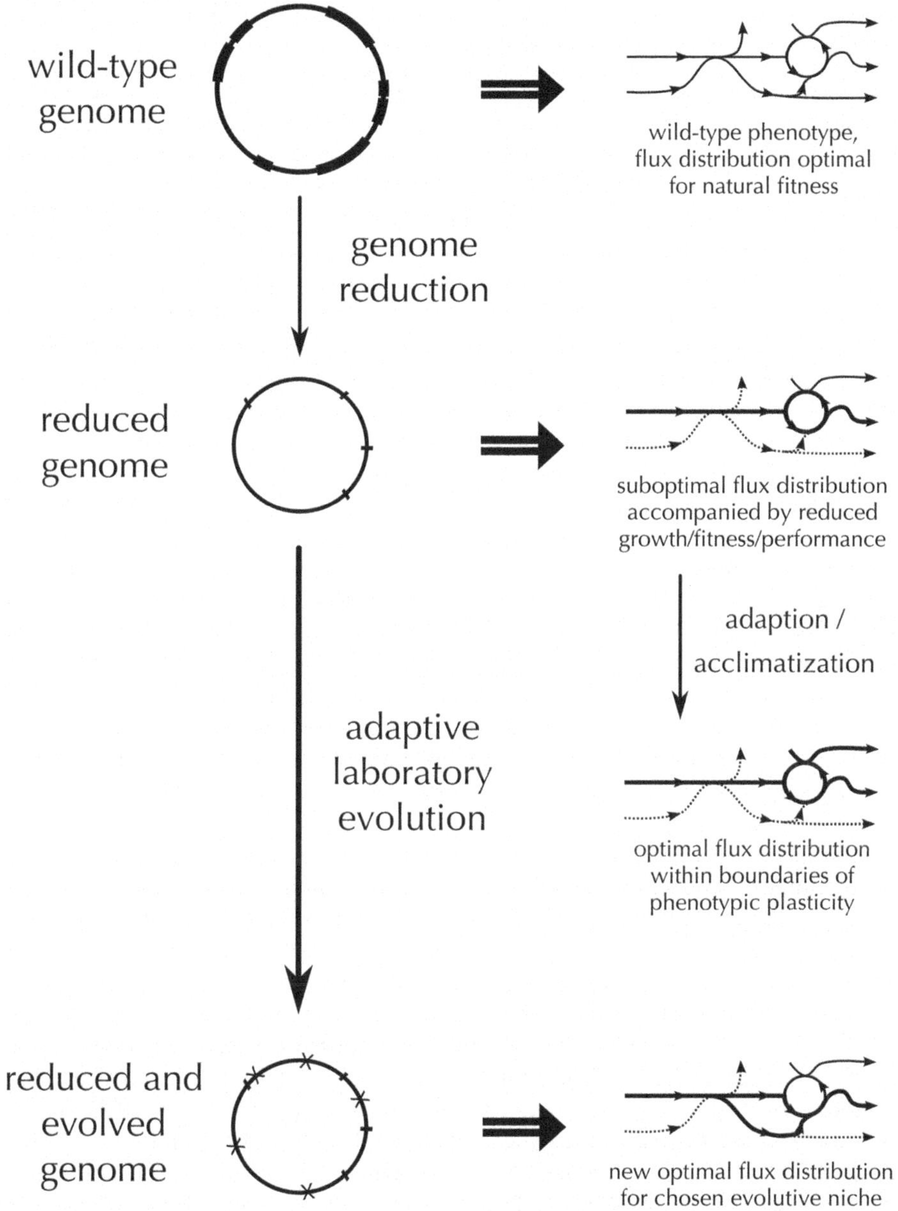

Fig. 2 Schematic depiction of genotypes and corresponding metabolic fluxes. Suboptimal flux distribution in genome-reduced strains can be improved on a short time-scale by adaption on transcriptional level. Adaptive laboratory evolution enables the fixation of beneficial mutations that rescue flux distribution deficits

inactive plasmids, the authors could demonstrate that inactive plasmid-carrying cells rapidly take over the population (Park et al. 2014).

Bacillus subtilis *Bacillus* species are the major source of technical enzymes (Schallmey et al. 2004). Their unrivaled ability to efficiently secrete native proteins and the surprising versatility of different species regarding robustness to external stress has placed them at the top of enzyme production (Liu et al. 2013). The Gram-positive model species *Bacillus subtilis* has been extensively studied for years and is well accessible through molecular biology methods (Earl et al. 2008; Kunst et al. 1997). Various research groups have put substantial effort into unraveling its "essential" set of genes which facilitates genome reduction studies (Commichau et al. 2013; Kobayashi et al. 2003; Tanaka et al. 2013). Moreover, it has been shown that a significant proportion of energy available to *B. subtilis* is directed towards preadaptive responses even under optimal growth conditions (Dauner et al. 2001). *Bacillus* species thus appear as promising species for genome reduction.

Bacillus subtilis was among the first organisms whose genome was reduced. In an initial study, Westers et al. (2003) deleted about 7.7% of its genome, freeing it of prophages and certain AT-rich islands. The deletions were well tolerated, and the resulting phenotype did not display major alterations proving the feasibility of this approach (Westers et al. 2003). Some years later, Reuß et al. (2017) established a new deletion series based on this strain. The final strains of this series, PS38 and PG10, displayed slightly reduced growth rates and a filamentous growth phenotype composed of cells not fully separated. Interestingly, PS38 and PG10 showed differences in amino acid metabolism. The authors did not investigate the production potential of these strains, but transcriptional and proteomic profiling of the deletion strains point towards more promising targets for successful deletions (Reuß et al. 2017).

The *Bacillus subtilis* "minimum genome factory" project aimed specifically at increasing the production capabilities of *Bacillus subtilis*. The involved groups consequently screened for deletions enhancing the production of secreted proteins (Ara et al. 2007). Sequential deletions of a total of 24% of the wild-type genome resulted in *B. subtilis* MG1M which displayed about 10% reduced growth rate in complex and minimal medium. The desired trait of increased enzyme production could not be shown though (Ara et al. 2007). In a subsequent study, Morimoto et al. (2008) noticed that MG1M's phenotype was unstable after successive culture and decided to establish a new deletion series. Strain MGB874 lacked 20.7% of its parent's genome. Its maximum growth rate in both complex LB and minimal SMM medium was substantially reduced. However, under the conditions of heterologous protein production in 2xL-Mal medium in a jar fermenter, its growth rate was similar to that of the wild-type. Substantial transcriptional changes could be observed as well as an increased sugar consumption rate towards the end of the cultivation. The authors observed a concomitant increase in heterologous protein activity in the culture supernatant (Morimoto et al. 2008). To further elucidate the underlying phenotypic changes, Manabe et al. (2011) constructed partial deletion strains and identified the *rocDEF-rocR* region as critical. Deletion of this region alone was sufficient to increase the specific cell yield in the parent *B. subtilis* strain 168 and compensated for the reduction in cell yield caused by other deletions in MGB874. Due to the absence of this region arginine catabolism was inactive. The presence of arginine in the later stages of the fermentation then repressed arginine biosynthesis from glutamate, leading to elevated intracellular glutamate levels (Manabe et al. 2011). Glutamate feeding also increased the cell yield for all strains but did not improve protein production on top. It was found that plasmid copy numbers and transcript per plasmid levels were increased in the deletion strain MGB874 during late fermentation stages which was likely a major cause for its higher protein productivity (Manabe et al. 2011). To further increase the production potential of strain MGB874, deletion of *rocG* was performed next (Manabe et al. 2013). While this deletion reduced the protein production in the jar fermenters used up to this point, the authors could demonstrate that this effect was caused by ammonia starvation and low external pH. Fermentation of MGB874 *ΔrocG* in an ammonia-controlled pH-auxostat not only rescued the production phenotype but also lead to an extended production phase in MGB874 *ΔrocG*. Ultimately, MGB874 *ΔrocG* surpassed the titers achieved by its parent strain for a final alkaline cellulase Egl-237 concentration of 5.5 g/l (Manabe et al. 2013). Once again deletion of *rocG* in the parent strain *B. subtilis* strain 168 also increased its protein production, but the high levels of MGB874 *ΔrocG* could not be reached.

What do we learn from the *Bacillus subtilis* minimum genome factory? It enabled the identification of the effect of the *rocDEF-rocR* deletion which is substantial in the medium used in the described studies. Its reintroduction in the parent strain is also quite beneficial in terms of cell yield. On the other hand, it is not surprising that heterologous protein production is higher in MGB874 since it displays higher transcriptional levels of the target gene. This effect should also be achievable in *B. subtilis* wild-type by proper choices of plasmid system and expression cassette. It is difficult to assess how the remaining global effects regarding delayed stationary phase and sporulation contribute to MGB874's enhanced protein productivity. Deletion of *rocG* further prolonged MGB874's production phase if fermented with ammonia surplus and pH control. It should be noted that the growth defects of MB874 observed in LB medium and minimal medium did not translate to the jar fermentations in 2xL-Mal medium. The reduced production of MGB874 *ΔrocG* in the same medium turned into an advantage once pH was externally controlled by ammonia addition which is easily implemented technically. These two findings teach us an important lesson: apparently disadvantageous features of chassis organism can be quite well acceptable as long as they are compensated by the applied production strategy. Chassis organisms should be tailored for the niche of a bioreactor with process control—and this is the only suitable environment to evaluate them.

In a more recent study, Li et al. (2016) capitalized on the experience collected in the *B. subtilis* minimum genome factory project. A new series of deletion strains based on *B. subtilis* strain 168 derivatives was created by deleting prophages, antibiotic gene clusters and several large non-essential regions. It culminated in strain BSK814G2 lacking 19.7% of the parent genome (Li et al. 2016). The strains were thoroughly characterized in bioreactors and shaking flaks. A substantial increase in biomass yield concomitant with reduced maintenance coefficients was observed. On the downside, maximum specific growth rate and specific glucose uptake rates were reduced. Interestingly, the effects were gradually more pronounced in strains carrying longer deletions. The strains along with their parent strain were then engineered to overproduce either guanosine or thymidine. In both cases the genome-reduced strains produced severalfold higher titers than the identically engineered control strain. Similar to the behavior of MGB874 from the minimum genome factory project, BSK814G2 had elevated transcript levels of the overexpressed enzymes which might partly explain its production advantages (Li et al. 2016). The case of BSK841G2 is particularly interesting because it performs well in small molecule production, thus showing that genome-reduced *Bacillus* species will also be useful for applications other than enzyme production.

Corynebacterium glutamicum The actinobacterium *Corynebacterium glutamicum* is the dominating microbial host for the production of amino acids and similar compounds (Lee et al. 2016). Production of glutamic acid and lysine is a billion dollar market, and *C. glutamicum* is increasingly used for other products as well (Becker and Wittmann 2012; Eberhardt et al. 2014). Due to its accessibility as a host and the availability of genomic information and functional annotation, its properties can be rapidly modified. Combined with the long experience in fermentation of *C. glutamicum* this allows rapid commercialization of new processes (Ikeda and Nakagawa 2003).

The first major deletion series in *C. glutamicum* was conducted to test for the essentiality of strain-specific islands (SSIs) identified by the comparison of *C. glutamicum* ATTC 13032 to *C. glutamicum* R (Suzuki et al. 2005a). The deletion of 11 SSIs from the genome of *C. glutamicum* R was performed. Remarkably, the strains displayed no negative traits compared to their parent wild-type. In fact doubling time in minimal medium was slightly improved in some mutants, although the observed effects were very small (Suzuki et al. 2005a). Nevertheless, the feasibility of this approach was demonstrated successfully. Next, 8 of 11 SSIs were combined in a single strain for a total genome reduction of 5.7% (Suzuki et al. 2005b). The genome-reduced strain had no significantly different growth in minimal medium. In-depth phenotyping of the strain was not conducted.

Corynebacterium glutamicum ATCC 13032 carries several prophages which can be induced under stress conditions. The observation that spontaneous induction is possible and in *dtxr* mutants the largest prophage CGP3 is frequently excised under stress conditions initiated further investigations (Frunzke et al. 2008; Nanda et al. 2014). By deletion of the CGP1, CGP2, and CGP3 regions from the *C. glutamicum* ATCC 13032, the strain MB001 was obtained (Baumgart et al. 2013). Its growth behavior was identical to that of the wild-type, and even in direct competition assays, no fitness difference could be detected. However, its growth in the presence of mitomyin C, which triggers CPG3 induction, was more stable and faster compared to its parent, but both strains reached identical final cell densities. Moreover, MB001 showed increased transformability and carried a higher plasmid copy number. These differences could be traced back to the deletion of the restriction-modification system within CPG3. As a consequence of the increased plasmid copy number, the production of heterologously expressed fluorescent protein was also increased (Baumgart et al. 2013). The creation of MB001 is a particularly interesting case because its advantages can be clearly assigned to genetic alterations within the deleted regions while its construction was planned simply on the premise of deleting all prophages as a whole. MB001 itself formed the base for a new genome reduction study targeted at the identification of irrelevant genomic regions based on conservation among *C. glutamicum* and related species (Unthan et al. 2015). Twenty-six irrelevant genomic regions which did not affect growth in minimal medium were identified. The authors could show that some combinatorial deletions of irrelevant regions were not compatible resulting in reduced growth rates, while other combinations were possible without fitness losses (Unthan et al. 2015). In a follow-up study Baumgart et al. (2018) constructed the chassis strain *C. glutamicum* C1∗ from MB001 by adding compatible deletions for a total genome reduction of 13.4% of the *C. glutamicum* ATCC 13032 genome. *C. glutamicum* C1∗ displays identical growth like its parent on minimal glucose and complex medium (Baumgart et al. 2018). Note that the proposed theoretical combination of all initially identified irrelevant genomic regions would have led to a genome reduction of 22% (Unthan et al. 2015). It is evident that for the construction of chassis organisms, trade-offs must be made. Some initially irrelevant genes complement each other, and deletion of one such gene converts the other one in a relevant gene. Deletion targets must be handpicked and, in the absence of phenotypic information on gene functions predicting the compatibility is not always possible. What remains to be shown is an actual advantage of *C. glutamicum* C1∗ over its wild-type parent in the development or final performance of a bioprocess.

Another genome reduction approach in *C. glutamicum* ATCC 13032 was targeted at insertion sequence (IS) elements (Choi et al. 2015). Insertion sequences are mobile genetic elements and well known for their ability to alter the expression of genes. Insertion into an overexpressed protein coding DNA abolishes target protein production. Moreover it often confers a growth advantage which can lead to non-producers overtaking the entire population. *C. glutamicum* lacking all IS elements of either family ISCg1 or ISCg2 had identical growth like their wild-type parent. They had higher content of polyhydroxybutyrate or produced more recombinant protein or γ-aminobutyric acid when the necessary genes for the biosynthesis of the products were provided on a plasmid. Additionally, the reduced strains had a higher transformation efficiency (Choi et al. 2015). Overall, the removal of IS elements is an attractive starting point for genome reduction: Plasmid gene inactivation is a relevant parameter in many biotechnological applications, and side effects of IS elements removal are typically not expected.

Pseudomonas putida The Gram-negative soil bacterium *Pseudomonas putida* is well known for its robustness towards external stressors and versatile catabolic capabilities (Loeschcke and Thies 2015). In recent years it has attracted increased attention for its potential to synthesize bioplastics and bioactive compounds (Poblete-Castro et al. 2012). Due to the complexity of its metabolism, *P. putida* is an attractive target for genome reduction studies.

Many *Pseudomonas* species are highly mobile (Sampedro et al. 2015), and first studies targeted the flagellar operon for deletions (Martínez-García et al. 2014b). Mixed results were obtained: The resulting strain lacking flagella displayed impaired growth on a variety of carbon sources. This data was collected in microtiter plates though where sufficient oxygen supply must be carefully evaluated (Wewetzer et al. 2015). Wild-type *Pseudomonas putida* KT2440 can use aerotaxis to move towards the surface of a culture broth which may be a major advantage under such conditions. Interestingly, the deletion strain showed a reduced lag phase, particularly if cultivated on fructose. It also displayed aberrant stress resistance with reduced resistance towards several antibiotics and an acidic environment. On the other hand, there was an increase in stress resistance against oxidative stressors concomitant with an increase in reducing power availability (Martínez-García et al. 2014b).

In a parallel study, Martínez-Garcia et al. (2015) deleted annotated prophage regions in *P. putida* leading to an increased tolerance against DNA damage by UV light or chemical DNA damaging agents. Finally the combination of both approaches yielded strain *P. putida* EM383 with a genome reduced by 4.3%. It produced a phenotype which retained all advantageous traits previously observed for the individual deletions (Martínez-García et al. 2014a). While its growth in complex media was slower than that of its wild-type parent, it reached a higher final cell density and displayed faster growth in glucose minimal medium.

Lieder et al. (2015) subsequently thoroughly tested *P. putida* EM383 generated in the aforementioned study in small-scale bioreactors. Strikingly, under the controlled conditions of a bioreactor, it outcompeted its parent strain in every single process relevant parameter measured: maximum specific growth rate in minimal medium, biomass yield in minimal medium, maintenance coefficient, organic acid byproduct formation, availability of energy as measured by adenylate energy charge (AEC), plasmid stability, and specific heterologous protein productivity (Lieder et al. 2015). This study stands out among the competition and for good reason: It is the only study in which a genome-reduced microorganism was actually thoroughly phenotyped in a realistic production scenario. It clearly shows that intelligent genome reduction can substantially impact the performance of strains in bioreactors. Note that the phenotypic advantages were much less apparent in the preceding studies where cultivations were not performed under sufficiently controlled conditions. Therefore we can learn an important lesson from this research: Deletion strains must be tested with the same effort as they are constructed, and this characterization must take place in the only industrially relevant niche—the controlled environment of a bioreactor.

Recently, an attempt at genome reduction has also been made in *Pseudomonas chloraphis* GP72, a strain from the green pepper rhizosphere which produces antifungal compounds (Liu et al. 2007). Major regions predicted to be non-essential were deleted from the chromosome (Shen et al. 2017). While growth of deletion strains was in general equal or worse than that of the wild-type, some deletion strains displayed improved production of phenazines, valuable secondary metabolites. On the downside, essentiality of genes was chosen with regard to growth in complex media only which putatively eliminates the chance of cultivating deletion strains in defined media. Nevertheless, it will be most interesting to follow the future path of this promising microbe.

Lactococcus lactis The Gram-positive bacterium *Lactococcus lactis* is traditionally used in the production of dairy products due to its ability to efficiently produce lactic acid from lactose (Song et al. 2017). In recent years *L. lactis* has increasingly attracted attention as a host for other industrial processes such as protein production (Mierau et al. 2005). To date, there is only one study dedicated to genome reduction of *L. lactis* (Zhu et al. 2017). Four large genomic regions containing prophage proteins were deleted for a total genomic reduction of 2.83%. The resulting strains displayed lower lag phases and increased maximum specific growth rates in microtiter plates. While differences in complex medium M17G were relatively small, final cell densities in defined medium SA containing glucose and amino acids were up to 40% higher than for the wild-type strain. The authors could further show that plasmid-based expression of *lecC* leads to higher mRNA levels and larger leucocin C inhibitory zones on agar plates. These findings were confirmed on mRNA and protein level using RFP as a reporter in microtiter plates (Zhu et al. 2017). We are eager to see whether the promising performance gains observed will hold in the controlled environment of a bioreactor.

***Streptomyces* Species** *Streptomyces* species are mycelial Gram-positive bacteria used extensively to produce antibiotics for human therapy (Bérdy 2005; Procópio et al. 2012). To date studies on genome reduction have been performed in *S. avermitilis* and *S. coelicolor*.

Gomez-Escribano and Bibb (2011) constructed *S. coelicolor* M1146 which is devoid of native antibiotic gene clusters. It displayed normal growth compared to its wild-type parent. HPLC analysis of its products was facilitated due to the absence of disturbing secondary metabolites. Fermentation data from shaking flasks indicates increased mRNA levels of heterologously expressed genes for the production of the antibiotics chloramphenicol and congocidine. A concomitant increase in antibiotic titers could also be measured although data scatters considerably. In another study as much as 14% of the parent genome was deleted in *S. coelicolor* (Zhou et al. 2012). The strains exhibited normal growth and showed differences in actinorhodin synthesis. No clear verdict on the benefit of this substantial genome reduction is possible though as the authors observed major fluctuations in their fermentation attempts (Zhou et al. 2012).

The situation is clearer in *Streptomyces avermitilis*. Deletion strain SUKA5 lacking a major 1.4 Mb subtelomeric region and oligomycin biosynthesis genes produced severalfold higher titers of streptomycin than its parent strain (Komatsu et al. 2010). The heterologously expressed streptomycin biosynthesis genes originated from *S. griseus*. Under the conditions of this study, *S. avermitilis* strain SUKA5 also easily surpassed *S. griseus* in terms of streptomycin productivity. It appears likely that the removal of endogenous secondary metabolism as demonstrated by HPLC measurements dramatically improved the availability of precursors for the new product. In a follow-up study, the authors could demonstrate that SUKA5 and its daughter strain SUKA17 reached higher final cell densities than

S. avermitilis wild-type in defined medium (Komatsu et al. 2013). Additionally, SUKA17 displayed higher productivity for seven heterologously produced antibiotics.

The data collected in this study exemplifies the gray area between genome reduction and metabolic engineering. From the perspective of a metabolic engineer, one could quite well argue that Komatsu and his colleagues in effect performed flux rerouting by increasing precursor supply. On the other hand, they successfully removed a total of 18.5% of *S. avermitilis* genomic DNA, much more than what would have been necessary for simple flux rerouting. Moreover, the methods applied by them are clearly not derived from the standard repertoire of metabolic engineers. Ultimately, if we consider the fate of every single carbon atom entering a cell, then any manipulation leads to flux rerouting. Genome reduction as defined by us then becomes a segment of metabolic engineering.

Yeasts: *Saccharomyces cerevisiae*, *Schizosaccharomyces pombe*, and Their Relatives The history of the domestication of yeast is almost as long as that of sessile humanity. Initially used for the production of bread and alcoholic beverages, yeast strains have found countless applications in modern biotechnology and biochemistry (Gallone et al. 2018; Pscheidt and Glieder 2008). Despite their importance in biotechnology, yeasts have so far scarcely been the subject of major genome reduction attempts. The *S. cerevisiae* genome of approximately 12 Mb contains substantial genetic redundancy but is also less accessible than that of most prokaryotic model organisms (Goffeau et al. 1996). Things are complicated by the fact that many different yeast strains (not restricted to the species *S. cerevisiae*) are circulating in laboratories all over the world and are actively used for research and production (Pscheidt and Glieder 2008).

While a substantial portion of the yeast genome is considered to be non-essential, competition experiments with single deletion mutants of *S. cerevisiae* BY4743 did not unveil potential benefits of such deletions (Sliwa and Korona 2005). Given that the vast majority of proteins is expressed at very low levels (Ghaemmaghami et al. 2003), it is quite possible that only major deletions will lead to appreciable benefits. In contrast, another study found the mating pathway of *S. cerevisiae* DBY15084 to be sufficiently energy intensive so as to be an identifiable potential deletion target (Lang et al. 2009). In the only study specifically targeted at reducing a yeast genome, about 5% of the genome of *S. cerevisiae* SH5209 were deleted (Murakami et al. 2007). The resulting strains showed impaired growth in liquid media and reduced mitochondrial functions but also increased ethanol formation.

Another model organism, the "fission yeast" *Schizosaccharomyces pombe*, has also been the subject of genome reduction (Giga-Hama et al. 2007). Deletions of single genes as well as up to 100 kb in a single experiment have been performed successfully (Hirashima et al. 2006). In a parallel study, strains with multiple protease knockouts were constructed to improve the production of heterologous proteins (Idiris et al. 2006). The resulting multiple deletion strains display reduced proteolytic capabilities and improved production of the human growth hormone hGH. Similar results were reported from other studies (Sasaki et al. 2013). *S. pombe* lacking 5.2% of its parent's genome had higher intracellular ATP concentration and showed improved protein production. On the other hand specific glucose uptake rate and growth rate were slightly reduced (Sasaki et al. 2013). *S. pombe* continues to be the subject of research. New genomic modification systems based on the CRISPR-Cas9 toolbox have been developed and will make this species more accessible (Zhao and Boeke 2018). Yet, much effort is still necessary to further promote this somewhat underdeveloped expression host and characterize its deletion mutants.

5 Future Prospects

Molecular biologists have shown that large genomic deletions are feasible in many species. While the majority of early genome reduction studies were conducted using *E. coli* or *B. subtilis*, the genome of other industrially relevant organisms is in principle also accessible (see Table 1). Some groups of organisms such as *Streptomyces* species appear to be particularly appealing targets of genome reduction studies. The reduction of their extensive secondary metabolism promises to increase precursor availability for synthesis of the target compounds. Moreover, *Streptomyces* species are well-established hosts in the pharmaceutical industry which facilitates strain testing and process development.

Besides, new or poorly characterized hosts are on the rise. Strains of *Komagataella phaffii* (formerly known as *Pichia pastoris*) produce a plethora of heterologously expressed proteins (Macauley-Patrick et al. 2005). Other yeasts like *Yarrowia lipolytica* or *Kluyeromyces lactis* have special niche applications (Rebello et al. 2018). *Geobacter* species might be the missing link towards electrification of biotechnological processes (Bond and Lovley 2003; Franks and Nevin 2010). *Vibrio natriegens* is waiting in the wings to accelerate strain development and bioprocesses (Hoffart et al. 2017; Weinstock et al. 2016). To date, none of these promising hosts has been the subject of genome reduction. Given the experience collected with other hosts we would expect to see fast and positive results by elimination of superfluous intrinsic pathways and genomic sequences.

The major challenge in all genomic reduction studies is the proper choice of targets. The genomic reduction series designed so far followed two strategies. The first approach used comparative genomics to predict non-essential regions. The initial targets in the MDS and MGF series of *E. coli* deletion mutants were identified by this method (Kolisnychenko et al. 2002; Mizoguchi et al. 2007). Early steps in *C. glutamicum* genome reduction were also based on the comparison of related strains (Suzuki et al. 2005a). If closely related species are compared, this strategy promises quick results coupled with good neutrality towards growth, but the number of identifiable targets is limited. Comparing more distantly related genomes yields more targets but at an increased risk of hitting indispensable functions. The second approach is knowledge-based and initially involves the removal of sequences acquired by orthogonal gene transfer such as prophages and insertion elements (Baumgart et al. 2018; Li et al. 2016; Morimoto et al. 2008; Pósfai et al. 2006).

Next, secondary metabolism and traits superfluous in the context of a bioreactor like the flagellar biosynthesis are targeted (Lieder et al. 2015; Martínez-García et al. 2014b). The major downsides of this strategy are that handpicking of genes is required and interactions between deletions are often unpredictable.

With an increasing availability of sequence and metabolic data, we expect knowledge-driven approaches to dominate genome reduction studies in the future. Many of the studies summarized in Table 1 have shown that benefits from genome reduction are frequently associated with undesired side effects (Hashimoto et al. 2005; Murakami et al. 2007; Reuß et al. 2017). The avoidance of side effects requires limiting genome reduction to impactful genes for the specific needs defined. The identification of such impactful genes can be based on transcriptomic or proteomic data. Löffler et al. (2016) found oscillating transcription profiles for many genes during repeated short-term passage of cells through a low-nutrient zone. Such zones exist in large-scale bioreactors, and a genome-reduced chassis for this environment could carry deletions in differentially expressed genes. In general, tailored approaches are attractive for another reason: The scope of genome reduction is clearly defined, which enables early support by bioinformatics analysis such as essentiality prediction or flux balance analysis (Daniels et al. 2016; Erdrich et al. 2015). After genome reduction and engineering, tailored microorganisms can be assayed in the niche they were designed for (Fig. 3).

Besides the few studies on yeasts, eukaryotic systems have not been the target of genome reduction so far. With the advent of CRISPR-Cas technology, their accessibility has dramatically increased. Tumor biology has demonstrated impressively that eukaryotic genomes are quite flexible despite their size. Cancer cells often show surprising chromosomal instability and rapid fixation of point mutations—a subject that has been thoroughly covered elsewhere (Greaves and Maley 2012; Meacham and Morrison 2013). It thus appears feasible to extensively engineer and select eukaryotic cell lines used for biopharmaceutical production in the future. The use of CRISPR-Cas editing for the generation of advantageous cell lines has been demonstrated by several research groups in Chinese Hamster Ovary (CHO) cell lines (Kellner et al. 2018; Shin et al. 2018; Wang et al. 2018). If it is possible to remove almost a fourth of *E. coli*'s 4.6 MB genome, how much genomic DNA can be removed from the 2359 MB of the CHO genome? Studies on genomic and chromosomal stability of immortalized CHO cell lines indicate high plasticity in all but seven or eight chromosomes (Cao et al. 2012; Derouazi et al. 2006). We believe that genome reduction studies have great potential in CHO cells in terms of functional annotation, increase in cell stability and predictability, and the reduction of superfluous side metabolism.

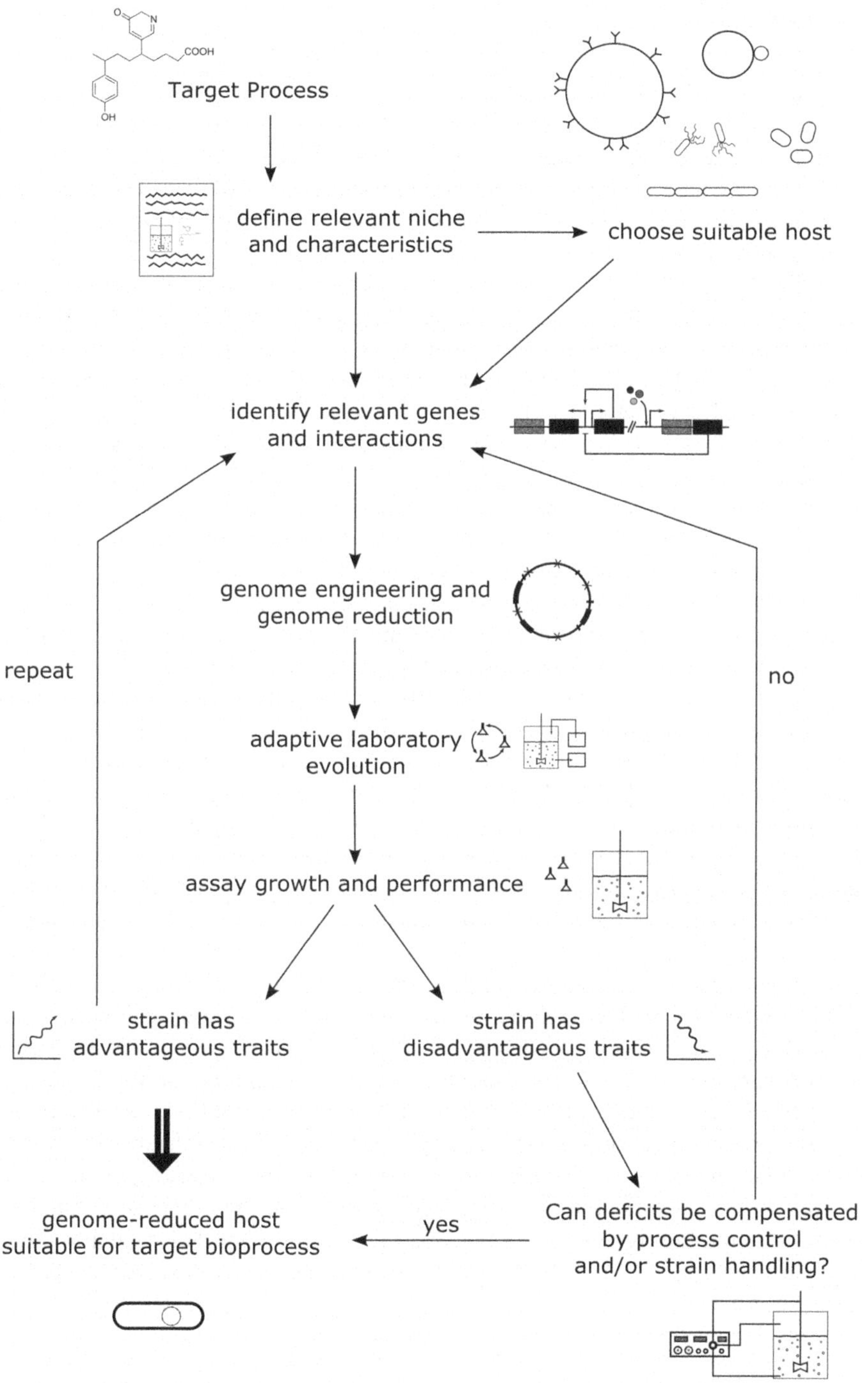

Fig. 3 Schematic workflow for the generation of genome-reduced hosts with increased process performance

6 Conclusion

As our understanding of the complex interactions in cellular metabolism increases, so does our ability to manipulate biological systems. Genome reduction is an approach to reduce them. Smaller, leaner systems tailored to the specific needs of a bioprocess are theorized to be less wasteful and thus more efficient. The wide array of organisms which have been the subject of successful genome reduction indicates general applicability.

A critical step in genome reduction is the proper evaluation of strains. Strains must be evaluated in the niche they were designed for as the case of *B. subtilis* MGB874 *ΔrocG* has demonstrated (Manabe et al. 2013). Generally speaking, a holistic view and clear process targets are necessary. Growth, substrate uptake rate, or other general parameters can be criteria for constructing useful deletion strains, but others may be just as important. The commercialization of *E. coli* strains from the MDS series, despite their slightly reduced process performance in standard cultivations, was based on their extreme genetic stability (Chakiath and Esposito 2007; Karcagi et al. 2016). These strains fill a niche: complex protein production in continuous cultivation enabled by excellent plasmid stability (Blattner et al. 2017). On the other hand, well-chosen targets for genome reduction in *P. putida* positively affected all process relevant parameters, indicating general superiority of the reduced strains in the niche of a controlled bioreactor (Lieder et al. 2015).

Studies on *E. coli* by Nishimura et al. (2017) showed that evolution after genome reduction can restore growth defects generated by large deletions. These results are well in line with observations made by evolving *E. coli* key enzyme knockout mutants (Long et al. 2017). Initially, knockout strains display suboptimal flux distribution and metabolite concentrations. Mutations in regulatory networks as well as relevant single enzymes then restore initial flux distributions or enable new optima to be found (McCloskey et al. 2018). Applying adaptive laboratory evolution after extensive genome reduction should thus be a standard practice if unexpected growth defects occur.

Evolution itself has demonstrated that organisms with very small genomes are feasible. Increases and decreases in cellular and organismal complexity are reoccurring phenomena over the history of life (Wolf and Koonin 2013). Reductive evolution has led to some species losing the majority of their genome as they adapt to a specialized niche (Andersson and Kurland 1998). In principle the controlled milieu of a bioreactor production process could reasonably be such a niche as well. It is up to molecular and systems biologists to find applicable design principles for small genomes. With the availability of sequence information, the powerful methods of modern systems biology and the extensive information collected on many model organisms, genome reduction is about to become a feasible tool for host optimization.

References

Adli M (2018) The CRISPR tool kit for genome editing and beyond. Nat Commun 9:1911. https://doi.org/10.1038/s41467-018-04252-2

Ajikumar PK, Xiao W-H, Tyo KEJ, Wang Y, Simeon F, Leonard E, Mucha O, Phon TH, Pfeifer B, Stephanopoulos G (2010) Isoprenoid pathway optimization for Taxol precursor overproduction in Escherichia coli. Science 330:70–74. https://doi.org/10.1126/science.1191652

Akeno Y, Ying B-W, Tsuru S, Yomo T (2014) A reduced genome decreases the host carrying capacity for foreign DNA. Microb Cell Factories 13:49. https://doi.org/10.1186/1475-2859-13-49

Alberts B, Miake-Lye R (1992) Unscrambling the puzzle of biological machines: the importance of the details. Cell 68:415–420. https://doi.org/10.1016/0092-8674(92)90179-G

Anazawa H (2014) The concept of the Escherichia coli minimum genome factory. In: Anazawa H, Shimizu S (eds) Microbial production: from genome design to cell engineering. Springer, Tokyo, pp 25–32

Andersson SG, Kurland CG (1998) Reductive evolution of resident genomes. Trends Microbiol 6:263–268

Ara K, Ozaki K, Nakamura K, Yamane K, Sekiguchi J, Ogasawara N (2007) Bacillus minimum genome factory: effective utilization of microbial genome information. Biotechnol Appl Biochem 46:169–178. https://doi.org/10.1042/BA20060111

Baba T, Ara T, Hasegawa M, Takai Y, Okumura Y, Baba M, Datsenko KA, Tomita M, Wanner BL, Mori H (2006) Construction of Escherichia coli K-12 in-frame, single-gene knockout mutants: the Keio collection. Mol Syst Biol 2:0008. https://doi.org/10.1038/msb4100050

Baeshen MN, Al-Hejin AM, Bora RS, Ahmed MMM, Ramadan HAI, Saini KS, Baeshen NA, Redwan EM (2015) Production of biopharmaceuticals in *E. coli*: current scenario and future perspectives. J Microbiol Biotechnol 25:953–962. https://doi.org/10.4014/jmb.1412.12079

Bailey JE (1991) Toward a science of metabolic engineering. Science 252:1668–1675

Bailey JE (1999) Lessons from metabolic engineering for functional genomics and drug discovery. Nat Biotechnol 17:616–618. https://doi.org/10.1038/10794

Bailey JE, Sburlati A, Hatzimanikatis V, Lee K, Renner WA, Tsai PS (2002) Inverse metabolic engineering: a strategy for directed genetic engineering of useful phenotypes. Biotechnol Bioeng 79:568–579. https://doi.org/10.1002/bit.10441

Barve A, Rodrigues JFM, Wagner A (2012) Superessential reactions in metabolic networks. Proc Natl Acad Sci U S A 109:E1121–E1130. https://doi.org/10.1073/pnas.1113065109

Baumgart M, Unthan S, Rückert C, Sivalingam J, Grünberger A, Kalinowski J, Bott M, Noack S, Frunzke J (2013) Construction of a prophage-free variant of Corynebacterium glutamicum ATCC 13032 for use as a platform strain for basic research and industrial biotechnology. Appl Environ Microbiol 79:6006–6015. https://doi.org/10.1128/AEM.01634-13

Baumgart M, Unthan S, Kloß R, Radek A, Polen T, Tenhaef N, Müller MF, Küberl A, Siebert D, Brühl N, Marin K, Hans S, Krämer R, Bott M, Kalinowski J, Wiechert W, Seibold G, Frunzke J, Rückert C, Wendisch VF, Noack S (2018) Corynebacterium glutamicum chassis C1∗: building and testing a novel platform host for synthetic biology and industrial biotechnology. ACS Synth Biol 7:132–144. https://doi.org/10.1021/acssynbio.7b00261

Becker J, Wittmann C (2012) Bio-based production of chemicals, materials and fuels – Corynebacterium glutamicum as versatile cell factory. Curr Opin Biotechnol 23:631–640. https://doi.org/10.1016/j.copbio.2011.11.012

Benson DA, Cavanaugh M, Clark K, Karsch-Mizrachi I, Lipman DJ, Ostell J, Sayers EW (2013) GenBank. Nucleic Acids Res 41:D36–D42. https://doi.org/10.1093/nar/gks1195

Bérdy J (2005) Bioactive microbial metabolites. J Antibiot 58:1–26. https://doi.org/10.1038/ja.2005.1

Berg HC (2003) The rotary motor of bacterial flagella. Annu Rev Biochem 72:19–54. https://doi.org/10.1146/annurev.biochem.72.121801.161737

Blattner FR, Plunkett G, Bloch CA, Perna NT, Burland V, Riley M, Collado-Vides J, Glasner JD, Rode CK, Mayhew GF, Gregor J, Davis NW, Kirkpatrick HA, Goeden MA, Rose DJ, Mau B, Shao Y (1997) The complete genome sequence of *Escherichia coli* K-12. Science 277:1453–1462

Blattner CR, Frisch D, Novy RE, Henker TM, Steffen EA, Blattner FR, Choi H, Posfai G, Landry CF (2017) Enhanced production of recombinant CRM197 in *E. coli*. (US patent 20170073379)

Bond DR, Lovley DR (2003) Electricity production by Geobacter sulfurreducens attached to electrodes. Appl Environ Microbiol 69:1548–1555. https://doi.org/10.1128/AEM.69.3.1548-1555.2003

Burgard AP, Vaidyaraman S, Maranas CD (2001) Minimal reaction sets for Escherichia coli metabolism under different growth requirements and uptake environments. Biotechnol Prog 17:791–797. https://doi.org/10.1021/bp0100880

Cao Y, Kimura S, Itoi T, Honda K, Ohtake H, Omasa T (2012) Construction of BAC-based physical map and analysis of chromosome rearrangement in Chinese hamster ovary cell lines. Biotechnol Bioeng 109:1357–1367. https://doi.org/10.1002/bit.24347

Carpentier A-S, Torrésani B, Grossmann A, Hénaut A (2005) Decoding the nucleoid organisation of *Bacillus subtilis* and Escherichia coli through gene expression data. BMC Genomics 6:84. https://doi.org/10.1186/1471-2164-6-84

Chakiath C, Esposito D (2007) Improved recombinational stability of lentiviral expression vectors using reduced-genome *Escherichia coli*. Biotech 43:466–470. https://doi.org/10.2144/000112585

Choe D, Cho S, Kim SC, Cho B-K (2016) Minimal genome: worthwhile or worthless efforts toward being smaller? Biotechnol J 11:199–211. https://doi.org/10.1002/biot.201400838

Choi JW, Yim SS, Kim MJ, Jeong KJ (2015) Enhanced production of recombinant proteins with Corynebacterium glutamicum by deletion of insertion sequences (IS elements). Microb Cell Factories 14:207. https://doi.org/10.1186/s12934-015-0401-7

Colin VL, Rodríguez A, Cristóbal HA (2011) The role of synthetic biology in the design of microbial cell factories for biofuel production. J Biomed Biotechnol 2011:601834. https://doi.org/10.1155/2011/601834

Commichau FM, Pietack N, Stülke J (2013) Essential genes in *Bacillus subtilis*: a re-evaluation after ten years. Mol BioSyst 9:1068–1075. https://doi.org/10.1039/c3mb25595f

Cooper VS, Schneider D, Blot M, Lenski RE (2001) Mechanisms causing rapid and parallel losses of ribose catabolism in evolving populations of *Escherichia coli* B. J Bacteriol 183:2834–2841. https://doi.org/10.1128/JB.183.9.2834-2841.2001

Csörgo B, Fehér T, Tímár E, Blattner FR, Pósfai G (2012) Low-mutation-rate, reduced-genome *Escherichia coli*: an improved host for faithful maintenance of engineered genetic constructs. Microb Cell Factories 11:11. https://doi.org/10.1186/1475-2859-11-11

Daniels W, Bouvin J, Busche T, Kalinowski J, Bernaerts K (2016) Finding targets for genome reduction in *Streptomyces lividans* TK24 using flux balance analysis. IFAC-PapersOnLine 49:252–257. https://doi.org/10.1016/j.ifacol.2016.12.134

Dauner M, Storni T, Sauer U (2001) *Bacillus subtilis* metabolism and energetics in carbon-limited and excess-carbon chemostat culture. J Bacteriol 183:7308–7317. https://doi.org/10.1128/JB.183.24.7308-7317.2001

Delvigne F, Boxus M, Ingels S, Thonart P (2009) Bioreactor mixing efficiency modulates the activity of a prpoS: GFP reporter gene in *E. coli*. Microb Cell Factories 8:15. https://doi.org/10.1186/1475-2859-8-15

Derouazi M, Martinet D, Besuchet Schmutz N, Flaction R, Wicht M, Bertschinger M, Hacker DL, Beckmann JS, Wurm FM (2006) Genetic characterization of CHO production host DG44 and derivative recombinant cell lines. Biochem Biophys Res Commun 340:1069–1077. https://doi.org/10.1016/j.bbrc.2005.12.111

Dong X, Quinn PJ, Wang X (2011) Metabolic engineering of Escherichia coli and Corynebacterium glutamicum for the production of L-threonine. Biotechnol Adv 29:11–23. https://doi.org/10.1016/j.biotechadv.2010.07.009

Du F-L, Yu H-L, Xu J-H, Li C-X (2014) Enhanced limonene production by optimizing the expression of limonene biosynthesis and MEP pathway genes in *E. coli*. Bioresour Bioprocess 1:10. https://doi.org/10.1186/s40643-014-0010-z

Dugar D, Stephanopoulos G (2011) Relative potential of biosynthetic pathways for biofuels and bio-based products. Nat Biotechnol 29:1074–1078. https://doi.org/10.1038/nbt.2055

Earl AM, Losick R, Kolter R (2008) Ecology and genomics of *Bacillus subtilis*. Trends Microbiol 16:269–275. https://doi.org/10.1016/j.tim.2008.03.004

Eberhardt D, Jensen JVK, Wendisch VF (2014) L-citrulline production by metabolically engineered Corynebacterium glutamicum from glucose and alternative carbon sources. AMB Express 4:85. https://doi.org/10.1186/s13568-014-0085-0

Emmerling M, Dauner M, Ponti A, Fiaux J, Hochuli M, Szyperski T, Wüthrich K, Bailey JE, Sauer U (2002) Metabolic flux responses to pyruvate kinase knockout in *Escherichia coli*. J Bacteriol 184:152–164. https://doi.org/10.1128/JB.184.1.152-164.2002

Erdrich P, Steuer R, Klamt S (2015) An algorithm for the reduction of genome-scale metabolic network models to meaningful core models. BMC Syst Biol 9:48. https://doi.org/10.1186/s12918-015-0191-x

Esnault E, Valens M, Espéli O, Boccard F (2007) Chromosome structuring limits genome plasticity in *Escherichia coli*. PLoS Genet 3:e226. https://doi.org/10.1371/journal.pgen.0030226

Esvelt KM, Wang HH (2013) Genome-scale engineering for systems and synthetic biology. Mol Syst Biol 9:641. https://doi.org/10.1038/msb.2012.66

Farmer IS, Jones CW (1976) The energetics of *Escherichia coli* during aerobic growth in continuous culture. Eur J Biochem 67:115–122. https://doi.org/10.1111/j.1432-1033.1976.tb10639.x

Ferenci T (2005) Maintaining a healthy SPANC balance through regulatory and mutational adaptation. Mol Microbiol 57:1–8. https://doi.org/10.1111/j.1365-2958.2005.04649.x

Ferrer-Miralles N, Domingo-Espín J, Corchero JL, Vázquez E, Villaverde A (2009) Microbial factories for recombinant pharmaceuticals. Microb Cell Factories 8:17. https://doi.org/10.1186/1475-2859-8-17

Fischer E, Sauer U (2005) Large-scale in vivo flux analysis shows rigidity and suboptimal performance of *Bacillus subtilis* metabolism. Nat Genet 37:636–640. https://doi.org/10.1038/ng1555

Fleischmann R, Adams M, White O, Clayton R, Kirkness E, Kerlavage A, Bult C, Tomb J, Dougherty B, Merrick J (1995) Whole-genome random sequencing and assembly of Haemophilus influenzae Rd. Science 269:496–512. https://doi.org/10.1126/science.7542800

Ford K, McDonald D, Mali P (2018) Functional genomics via CRISPR-Cas. J Mol Biol 43(1):48–65. https://doi.org/10.1016/j.jmb.2018.06.034

Fossum S, Crooke E, Skarstad K (2007) Organization of sister origins and replisomes during multifork DNA replication in *Escherichia coli*. EMBO J 26:4514–4522. https://doi.org/10.1038/sj.emboj.7601871

Franks AE, Nevin KP (2010) Microbial fuel cells, a current review. Energies 3:899–919. https://doi.org/10.3390/en3050899

Fraser CM, Gocayne JD, White O, Adams MD, Clayton RA, Fleischmann RD, Bult CJ, Kerlavage AR, Sutton G, Kelley JM, Fritchman JL, Weidman JF, Small KV, Sandusky M, Fuhrmann J, Nguyen D, Utterback TR, Saudek DM, Phillips CA, Merrick JM, Tomb J-F, Dougherty BA, Bott KF, Hu P-C, Lucier TS (1995) The minimal gene complement of *Mycoplasma genitalium*. Science 270:397–404. https://doi.org/10.1126/science.270.5235.397

Freed E, Fenster J, Smolinski SL, Walker J, Henard CA, Gill R, Eckert CA (2018) Building a genome engineering toolbox in nonmodel prokaryotic microbes. Biotechnol Bioeng 115(9):2120–2138. https://doi.org/10.1002/bit.26727

Frunzke J, Bramkamp M, Schweitzer J-E, Bott M (2008) Population heterogeneity in Corynebacterium glutamicum ATCC 13032 caused by prophage CGP3. J Bacteriol 190:5111–5119. https://doi.org/10.1128/JB.00310-08

Gallone B, Mertens S, Gordon JL, Maere S, Verstrepen KJ, Steensels J (2018) Origins, evolution, domestication and diversity of Saccharomyces beer yeasts. Curr Opin Biotechnol 49:148–155. https://doi.org/10.1016/j.copbio.2017.08.005

Gama-Castro S, Salgado H, Santos-Zavaleta A, Ledezma-Tejeida D, Muñiz-Rascado L, García-Sotelo JS, Alquicira-Hernández K, Martínez-Flores I, Pannier L, Castro-Mondragón JA, Medina-Rivera A, Solano-Lira H, Bonavides-Martínez C, Pérez-Rueda E, Alquicira-Hernández S, Porrón-Sotelo L, López-Fuentes A, Hernández-Koutoucheva A, Del Moral-Chávez V, Rinaldi F, Collado-Vides J (2016) RegulonDB version 9.0: high-level integration of gene regulation, coexpression, motif clustering and beyond. Nucleic Acids Res 44:D133–D143. https://doi.org/10.1093/nar/gkv1156

Gao H, Zhuo Y, Ashforth E, Zhang L (2010) Engineering of a genome-reduced host: practical application of synthetic biology in the overproduction of desired secondary metabolites. Protein Cell 1:621–626. https://doi.org/10.1007/s13238-010-0073-3

Ghaemmaghami S, Huh W-K, Bower K, Howson RW, Belle A, Dephoure N, O'Shea EK, Weissman JS (2003) Global analysis of protein expression in yeast. Nature 425:737–741. https://doi.org/10.1038/nature02046

Giga-Hama Y, Tohda H, Takegawa K, Kumagai H (2007) Schizosaccharomyces pombe minimum genome factory. Biotechnol Appl Biochem 46:147–155. https://doi.org/10.1042/BA20060106

Gil R, Silva FJ, Peretó J, Moya A (2004) Determination of the core of a minimal bacterial gene set. Microbiol Mol Biol Rev 68:518–537. https://doi.org/10.1128/MMBR.68.3.518-537.2004

Goffeau A, Barrell BG, Bussey H, Davis RW, Dujon B, Feldmann H, Galibert F, Hoheisel JD, Jacq C, Johnston M, Louis EJ, Mewes HW, Murakami Y, Philippsen P, Tettelin H, Oliver SG (1996) Life with 6000 genes. Science 274:546–567

Gomez-Escribano JP, Bibb MJ (2011) Engineering Streptomyces coelicolor for heterologous expression of secondary metabolite gene clusters. Microb Biotechnol 4:207–215. https://doi.org/10.1111/j.1751-7915.2010.00219.x

Gong Z, Nielsen J, Zhou YJ (2017) Engineering robustness of microbial cell factories. Biotechnol J 12:1700014. https://doi.org/10.1002/biot.201700014

Goryshin IY, Naumann TA, Apodaca J, Reznikoff WS (2003) Chromosomal deletion formation system based on Tn5 double transposition: use for making minimal genomes and essential gene analysis. Genome Res 13:644–653. https://doi.org/10.1101/gr.611403

Greaves M, Maley CC (2012) Clonal evolution in cancer. Nature 481:306–313. https://doi.org/10.1038/nature10762

Guha TK, Wai A, Hausner G (2017) Programmable genome editing tools and their regulation for efficient genome engineering. Comput Struct Biotechnol J 15:146–160. https://doi.org/10.1016/j.csbj.2016.12.006

Hashimoto M, Ichimura T, Mizoguchi H, Tanaka K, Fujimitsu K, Keyamura K, Ote T, Yamakawa T, Yamazaki Y, Mori H, Katayama T, J-i K (2005) Cell size and nucleoid organization of engineered *Escherichia coli* cells with a reduced genome. Mol Microbiol 55:137–149. https://doi.org/10.1111/j.1365-2958.2004.04386.x

Hénin J, Tajkhorshid E, Schulten K, Chipot C (2008) Diffusion of glycerol through *Escherichia coli* aquaglyceroporin GlpF. Biophys J 94:832–839. https://doi.org/10.1529/biophysj.107.115105

Hirashima K, Iwaki T, Takegawa K, Giga-Hama Y, Tohda H (2006) A simple and effective chromosome modification method for large-scale deletion of genome sequences and identification of essential genes in fission yeast. Nucleic Acids Res 34:e11. https://doi.org/10.1093/nar/gnj011

Hirokawa Y, Kawano H, Tanaka-Masuda K, Nakamura N, Nakagawa A, Ito M, Mori H, Oshima T, Ogasawara N (2013) Genetic manipulations restored the growth fitness of reduced-genome *Escherichia coli*. J Biosci Bioeng 116:52–58. https://doi.org/10.1016/j.jbiosc.2013.01.010

Hoffart E, Grenz S, Lange J, Nitschel R, Müller F, Schwentner A, Feith A, Lenfers-Lücker M, Takors R, Blombach B (2017) High substrate uptake rates empower Vibrio natriegens as production host for industrial biotechnology. Appl Environ Microbiol 83:e01614–e01617. https://doi.org/10.1128/AEM.01614-17

Hooven TA, Catomeris AJ, Akabas LH, Randis TM, Maskell DJ, Peters SE, Ott S, Santana-Cruz I, Tallon LJ, Tettelin H, Ratner AJ (2016) The essential genome of *Streptococcus agalactiae*. BMC Genomics 17:406. https://doi.org/10.1186/s12864-016-2741-z

Idiris A, Tohda H, K-w B, Isoai A, Kumagai H, Giga-Hama Y (2006) Enhanced productivity of protease-sensitive heterologous proteins by disruption of multiple protease genes in the fission yeast *Schizosaccharomyces pombe*. Appl Microbiol Biotechnol 73:404–420. https://doi.org/10.1007/s00253-006-0489-0

Ikeda M, Nakagawa S (2003) The Corynebacterium glutamicum genome: features and impacts on biotechnological processes. Appl Microbiol Biotechnol 62:99–109. https://doi.org/10.1007/s00253-003-1328-1

Jain R, Srivastava R (2009) Metabolic investigation of host/pathogen interaction using MS2-infected *Escherichia coli*. BMC Syst Biol 3:121. https://doi.org/10.1186/1752-0509-3-121

Juhas M, Reuß DR, Zhu B, Commichau FM (2014) *Bacillus subtilis* and *Escherichia coli* essential genes and minimal cell factories after one decade of genome engineering. Microbiology 160:2341–2351. https://doi.org/10.1099/mic.0.079376-0

Karcagi I, Draskovits G, Umenhoffer K, Fekete G, Kovács K, Méhi O, Balikó G, Szappanos B, Györfy Z, Fehér T, Bogos B, Blattner FR, Pál C, Pósfai G, Papp B (2016) Indispensability of horizontally transferred genes and its impact on bacterial genome streamlining. Mol Biol Evol 33:1257–1269. https://doi.org/10.1093/molbev/msw009

Kellner K, Solanki A, Amann T, Lao N, Barron N (2018) Targeting miRNAs with CRISPR/Cas9 to improve recombinant protein production of CHO cells. Methods Mol Biol 1850:221–235. https://doi.org/10.1007/978-1-4939-8730-6_15

Képès F (2004) Periodic transcriptional organization of the *E.coli* genome. J Mol Biol 340:957–964. https://doi.org/10.1016/j.jmb.2004.05.039

Kiviet DJ, Nghe P, Walker N, Boulineau S, Sunderlikova V, Tans SJ (2014) Stochasticity of metabolism and growth at the single-cell level. Nature 514:376–379. https://doi.org/10.1038/nature13582

Kobayashi K, Ehrlich SD, Albertini A, Amati G, Andersen KK, Arnaud M, Asai K, Ashikaga S, Aymerich S, Bessieres P, Boland F, Brignell SC, Bron S, Bunai K, Chapuis J, Christiansen LC, Danchin A, Débarbouille M, Dervyn E, Deuerling E, Devine K, Devine SK, Dreesen O, Errington J, Fillinger S, Foster SJ, Fujita Y, Galizzi A, Gardan R, Eschevins C, Fukushima T, Haga K, Harwood CR, Hecker M, Hosoya D, Hullo MF, Kakeshita H, Karamata D, Kasahara Y, Kawamura F, Koga K, Koski P, Kuwana R, Imamura D, Ishimaru M, Ishikawa S, Ishio I, Le Coq D, Masson A, Mauël C, Meima R, Mellado RP, Moir A, Moriya S, Nagakawa E, Nanamiya H, Nakai S, Nygaard P, Ogura M, Ohanan T, O'Reilly M, O'Rourke M, Pragai Z, Pooley HM, Rapoport G, Rawlins JP, Rivas LA, Rivolta C, Sadaie A, Sadaie Y, Sarvas M, Sato T, Saxild HH, Scanlan E, Schumann W, Seegers JFML, Sekiguchi J, Sekowska A, Séror SJ, Simon M, Stragier P, Studer R, Takamatsu H, Tanaka T, Takeuchi M, Thomaides HB, Vagner V, van Dijl JM, Watabe K, Wipat A, Yamamoto H, Yamamoto M, Yamamoto Y, Yamane K, Yata K, Yoshida K, Yoshikawa H, Zuber U, Ogasawara N (2003) Essential *Bacillus subtilis* genes. Proc Natl Acad Sci U S A 100:4678–4683. https://doi.org/10.1073/pnas.0730515100

Kolisnychenko V, Plunkett G, Herring CD, Fehér T, Pósfai J, Blattner FR, Pósfai G (2002) Engineering a reduced *Escherichia coli* genome. Genome Res 12:640–647. https://doi.org/10.1101/gr.217202

Komatsu M, Uchiyama T, Omura S, Cane DE, Ikeda H (2010) Genome-minimized Streptomyces host for the heterologous expression of secondary metabolism. Proc Natl Acad Sci U S A 107:2646–2651. https://doi.org/10.1073/pnas.0914833107

Komatsu M, Komatsu K, Koiwai H, Yamada Y, Kozone I, Izumikawa M, Hashimoto J, Takagi M, Omura S, Shin-ya K, Cane DE, Ikeda H (2013) Engineered Streptomyces avermitilis host for heterologous expression of biosynthetic gene cluster for secondary metabolites. ACS Synth Biol 2:384–396. https://doi.org/10.1021/sb3001003

Koob MD, SHAW AJ, CAMERON DC (1994) Minimizing the genome of *Escherichia coli*. Ann N Y Acad Sci 745:1–3. https://doi.org/10.1111/j.1749-6632.1994.tb44359.x

Koonin EV (2000) How many genes can make a cell: the minimal-gene-set concept. Annu Rev Genomics Hum Genet 1:99–116. https://doi.org/10.1146/annurev.genom.1.1.99

Kunst F, Ogasawara N, Moszer I, Albertini AM, Alloni G, Azevedo V, Bertero MG, Bessières P, Bolotin A, Borchert S, Borriss R, Boursier L, Brans A, Braun M, Brignell SC, Bron S, Brouillet S, Bruschi CV, Caldwell B, Capuano V, Carter NM, Choi SK, Cordani JJ, Connerton IF, Cummings NJ, Daniel RA, Denziot F, Devine KM, Düsterhöft A, Ehrlich SD, Emmerson PT, Entian KD, Errington J, Fabret C, Ferrari E, Foulger D, Fritz C, Fujita M, Fujita Y, Fuma S, Galizzi A, Galleron N, Ghim SY, Glaser P, Goffeau A, Golightly EJ, Grandi G, Guiseppi G, Guy BJ, Haga K, Haiech J, Harwood CR, Hènaut A, Hilbert H, Holsappel S, Hosono S, Hullo MF, Itaya M, Jones L, Joris B, Karamata D, Kasahara Y, Klaerr-Blanchard M, Klein C, Kobayashi Y, Koetter P, Koningstein G, Krogh S, Kumano M, Kurita K, Lapidus A, Lardinois S, Lauber J, Lazarevic V, Lee SM, Levine A, Liu H, Masuda S, Mauël C, Médigue C, Medina N, Mellado RP, Mizuno M, Moestl D, Nakai S, Noback M, Noone D, O'Reilly M, Ogawa K, Ogiwara A, Oudega B, Park SH, Parro V, Pohl TM, Portelle D, Porwollik S, Prescott AM, Presecan E, Pujic P, Purnelle B, Rapoport G, Rey M, Reynolds S, Rieger M, Rivolta C, Rocha E, Roche B, Rose M, Sadaie Y, Sato T, Scanlan E, Schleich S, Schroeter R, Scoffone F, Sekiguchi J, Sekowska A, Seror SJ, Serror P, Shin BS, Soldo B, Sorokin A, Tacconi E, Takagi T, Takahashi H, Takemaru K, Takeuchi M, Tamakoshi A, Tanaka T, Terpstra P, Togoni A, Tosato V, Uchiyama S, Vandebol M, Vannier F, Vassarotti A, Viari A, Wambutt R, Wedler H, Weitzenegger T, Winters P, Wipat A, Yamamoto H, Yamane K, Yasumoto K, Yata K, Yoshida K, Yoshikawa HF, Zumstein E, Yoshikawa H, Danchin A (1997) The complete genome sequence of the gram-positive bacterium *Bacillus subtilis*. Nature 390:249–256. https://doi.org/10.1038/36786

Kurakin A (2005) Stochastic cell. IUBMB Life 57:59–63. https://doi.org/10.1080/15216540400024314

Kurokawa M, Seno S, Matsuda H, Ying B-W (2016) Correlation between genome reduction and bacterial growth. DNA Res 23:517–525. https://doi.org/10.1093/dnares/dsw035

Lang GI, Murray AW, Botstein D (2009) The cost of gene expression underlies a fitness trade-off in yeast. Proc Natl Acad Sci U S A 106:5755–5760. https://doi.org/10.1073/pnas.0901620106

Lee SY, Kim HU (2015) Systems strategies for developing industrial microbial strains. Nat Biotechnol 33:1061–1072. https://doi.org/10.1038/nbt.3365

Lee JH, Sung BH, Kim MS, Blattner FR, Yoon BH, Kim JH, Kim SC (2009) Metabolic engineering of a reduced-genome strain of *Escherichia coli* for L-threonine production. Microb Cell Factories 8:2. https://doi.org/10.1186/1475-2859-8-2

Lee J-Y, Na Y-A, Kim E, Lee H-S, Kim P (2016) The Actinobacterium Corynebacterium glutamicum, an industrial workhorse. J Microbiol Biotechnol 26:807–822. https://doi.org/10.4014/jmb.1601.01053

Leprince A, van Passel MWJ, dos Santos VAPM (2012) Streamlining genomes: toward the generation of simplified and stabilized microbial systems. Curr Opin Biotechnol 23:651–658. https://doi.org/10.1016/j.copbio.2012.05.001

Li Y, Zhu X, Zhang X, Fu J, Wang Z, Chen T, Zhao X (2016) Characterization of genome-reduced *Bacillus subtilis* strains and their application for the production of guanosine and thymidine. Microb Cell Factories 15:94. https://doi.org/10.1186/s12934-016-0494-7

Lieder S, Nikel PI, de Lorenzo V, Takors R (2015) Genome reduction boosts heterologous gene expression in *Pseudomonas putida*. Microb Cell Factories 14:23. https://doi.org/10.1186/s12934-015-0207-7

Liu H, He Y, Jiang H, Peng H, Huang X, Zhang X, Thomashow LS, Xu Y (2007) Characterization of a phenazine-producing strain Pseudomonas chlororaphis GP72 with broad-spectrum antifungal activity from green pepper rhizosphere. Curr Microbiol 54:302–306. https://doi.org/10.1007/s00284-006-0444-4

Liu L, Liu Y, Shin H-D, Chen RR, Wang NS, Li J, Du G, Chen J (2013) Developing *Bacillus* spp. as a cell factory for production of microbial enzymes and industrially important biochemicals in the context of systems and synthetic biology. Appl Microbiol Biotechnol 97:6113–6127. https://doi.org/10.1007/s00253-013-4960-4

Loeschcke A, Thies S (2015) Pseudomonas putida-a versatile host for the production of natural products. Appl Microbiol Biotechnol 99:6197–6214. https://doi.org/10.1007/s00253-015-6745-4

Löffler M, Simen JD, Jäger G, Schäferhoff K, Freund A, Takors R (2016) Engineering *E. coli* for large-scale production – strategies considering ATP expenses and transcriptional responses. Metab Eng 38:73–85. https://doi.org/10.1016/j.ymben.2016.06.008

Long CP, Gonzalez JE, Feist AM, Palsson BO, Antoniewicz MR (2017) Dissecting the genetic and metabolic mechanisms of adaptation to the knockout of a major metabolic enzyme in *Escherichia coli*. Proc Natl Acad Sci U S A 115:222–227. https://doi.org/10.1073/pnas.1716056115

Lowe G, Meister M, Berg HC (1987) Rapid rotation of flagellar bundles in swimming bacteria. Nature 325:637–640. https://doi.org/10.1038/325637a0

Macauley-Patrick S, Fazenda ML, McNeil B, Harvey LM (2005) Heterologous protein production using the Pichia pastoris expression system. Yeast 22:249–270. https://doi.org/10.1002/yea.1208

Macnab RM (1992) Genetics and biogenesis of bacterial flagella. Annu Rev Genet 26:131–158. https://doi.org/10.1146/annurev.ge.26.120192.001023

Macnab RM (1996) Flagella and motility. In: Neidhardt FC, Curtiss R (eds) *Escherichia coli* and Salmonella: cellular and molecular biology, 2nd edn. ASM, Washington D.C, pp 123–145

Manabe K, Kageyama Y, Morimoto T, Ozawa T, Sawada K, Endo K, Tohata M, Ara K, Ozaki K, Ogasawara N (2011) Combined effect of improved cell yield and increased specific productivity enhances recombinant enzyme production in genome-reduced *Bacillus subtilis* strain MGB874. Appl Environ Microbiol 77:8370–8381. https://doi.org/10.1128/AEM.06136-11

Manabe K, Kageyama Y, Morimoto T, Shimizu E, Takahashi H, Kanaya S, Ara K, Ozaki K, Ogasawara N (2013) Improved production of secreted heterologous enzyme in *Bacillus subtilis* strain MGB874 via modification of glutamate metabolism and growth conditions. Microb Cell Factories 12:18. https://doi.org/10.1186/1475-2859-12-18

Martínez-García E, Nikel PI, Aparicio T, de Lorenzo V (2014a) Pseudomonas 2.0: genetic upgrading of *P. putida* KT2440 as an enhanced host for heterologous gene expression. Microb Cell Factories 13:159. https://doi.org/10.1186/s12934-014-0159-3

Martínez-García E, Nikel PI, Chavarría M, de Lorenzo V (2014b) The metabolic cost of flagellar motion in *Pseudomonas putida* KT2440. Environ Microbiol 16:291–303. https://doi.org/10.1111/1462-2920.12309

Martínez-García E, Jatsenko T, Kivisaar M, de Lorenzo V (2015) Freeing *Pseudomonas putida* KT2440 of its proviral load strengthens endurance to environmental stresses. Environ Microbiol 17:76–90. https://doi.org/10.1111/1462-2920.12492

McCloskey D, Xu S, Sandberg TE, Brunk E, Hefner Y, Szubin R, Feist AM, Palsson BO (2018) Evolution of gene knockout strains of *E. coli* reveal regulatory architectures governed by metabolism. Nat Commun 9:3796. https://doi.org/10.1038/s41467-018-06219-9

Meacham CE, Morrison SJ (2013) Tumor heterogeneity and cancer cell plasticity. Nature 501:328–337. https://doi.org/10.1038/nature12624

Mears PJ, Koirala S, Rao CV, Golding I, Chemla YR (2014) Escherichia coli swimming is robust against variations in flagellar number. elife 3:e01916. https://doi.org/10.7554/eLife.01916

Mierau I, Leij P, van Swam I, Blommestein B, Floris E, Mond J, Smid EJ (2005) Industrial-scale production and purification of a heterologous protein in *Lactococcus lactis* using the nisin-controlled gene expression system NICE: the case of lysostaphin. Microb Cell Factories 4:15. https://doi.org/10.1186/1475-2859-4-15

Milo R (2013) What is the total number of protein molecules per cell volume? A call to rethink some published values. BioEssays 35:1050–1055. https://doi.org/10.1002/bies.201300066

Mizoguchi H, Mori H, Fujio T (2007) Escherichia coli minimum genome factory. Biotechnol Appl Biochem 46:157–167. https://doi.org/10.1042/BA20060107

Mizoguchi H, Sawano Y, J-i K, Mori H (2008) Superpositioning of deletions promotes growth of *Escherichia coli* with a reduced genome. DNA Res 15:277–284. https://doi.org/10.1093/dnares/dsn019

Monod J (1949) The growth of bacterial cultures. Annu Rev Microbiol 3:371–394. https://doi.org/10.1146/annurev.mi.03.100149.002103

Morimoto T, Kadoya R, Endo K, Tohata M, Sawada K, Liu S, Ozawa T, Kodama T, Kakeshita H, Kageyama Y, Manabe K, Kanaya S, Ara K, Ozaki K, Ogasawara N (2008) Enhanced recombinant protein productivity by genome reduction in *Bacillus subtilis*. DNA Res 15:73–81. https://doi.org/10.1093/dnares/dsn002

Murakami K, Tao E, Ito Y, Sugiyama M, Kaneko Y, Harashima S, Sumiya T, Nakamura A, Nishizawa M (2007) Large scale deletions in the *Saccharomyces cerevisiae* genome create strains with altered regulation of carbon metabolism. Appl Microbiol Biotechnol 75:589–597. https://doi.org/10.1007/s00253-007-0859-2

Nakashima N, Miyazaki K (2014) Bacterial cellular engineering by genome editing and gene silencing. Int J Mol Sci 15:2773–2793. https://doi.org/10.3390/ijms15022773

Nanda AM, Heyer A, Krämer C, Grünberger A, Kohlheyer D, Frunzke J (2014) Analysis of SOS-induced spontaneous prophage induction in Corynebacterium glutamicum at the single-cell level. J Bacteriol 196:180–188. https://doi.org/10.1128/JB.01018-13

Nishimura I, Kurokawa M, Liu L, Ying B-W (2017) Coordinated changes in mutation and growth rates induced by genome reduction. MBio 8:e00676–e00617. https://doi.org/10.1128/mBio.00676-17

Noack S, Baumgart M (2018) Communities of niche-optimized strains: small-genome organism consortia in bioproduction. Trends Biotechnol 37(2):126–139. https://doi.org/10.1016/j.tibtech.2018.07.011

Noguchi Y, Nakai Y, Shimba N, Toyosaki H, Kawahara Y, Sugimoto S, Suzuki E-I (2004) The energetic conversion competence of *Escherichia coli* during aerobic respiration studied by 31P NMR using a circulating fermentation system. J Biochem 136:509–515. https://doi.org/10.1093/jb/mvh147

Notley-McRobb L, King T, Ferenci T (2002) rpoS mutations and loss of general stress resistance in *Escherichia coli* populations as a consequence of conflict between competing stress responses. J Bacteriol 184:806–811. https://doi.org/10.1128/JB.184.3.806-811.2002

O'Brien EJ, Utrilla J, Palsson BO (2016) Quantification and classification of *E. coli* proteome utilization and unused protein costs across environments. PLoS Comput Biol 12:e1004998. https://doi.org/10.1371/journal.pcbi.1004998

Oehler S, Eismann ER, Krämer H, Müller-Hill B (1990) The three operators of the lac operon cooperate in repression. EMBO J 9:973–979

Oesterle S, Wuethrich I, Panke S (2017) Toward genome-based metabolic engineering in bacteria. Adv Appl Microbiol 101:49–82. https://doi.org/10.1016/bs.aambs.2017.07.001

Park MK, Lee SH, Yang KS, Jung S-C, Lee JH, Kim SC (2014) Enhancing recombinant protein production with an *Escherichia coli* host strain lacking insertion sequences. Appl Microbiol Biotechnol 98:6701–6713. https://doi.org/10.1007/s00253-014-5739-y

Poblete-Castro I, Becker J, Dohnt K, dos Santos VM, Wittmann C (2012) Industrial biotechnology of *Pseudomonas putida* and related species. Appl Microbiol Biotechnol 93:2279–2290. https://doi.org/10.1007/s00253-012-3928-0

Pósfai G, Plunkett G, Fehér T, Frisch D, Keil GM, Umenhoffer K, Kolisnychenko V, Stahl B, Sharma SS, de Arruda M, Burland V, Harcum SW, Blattner FR (2006) Emergent properties of reduced-genome *Escherichia coli*. Science 312:1044–1046. https://doi.org/10.1126/science.1126439

Price MN, Wetmore KM, Deutschbauer AM, Arkin AP (2016) A comparison of the costs and benefits of bacterial gene expression. PLoS One 11:e0164314. https://doi.org/10.1371/journal.pone.0164314

Procópio REL, Silva IR, Martins MK, Azevedo JL, Araújo JM (2012) Antibiotics produced by Streptomyces. Braz J Infect Dis 16:466–471. https://doi.org/10.1016/j.bjid.2012.08.014

Pscheidt B, Glieder A (2008) Yeast cell factories for fine chemical and API production. Microb Cell Factories 7:25. https://doi.org/10.1186/1475-2859-7-25

Raj A, van Oudenaarden A (2008) Nature, nurture, or chance: stochastic gene expression and its consequences. Cell 135:216–226. https://doi.org/10.1016/j.cell.2008.09.050

Rath D, Jawali N (2006) Loss of expression of cspC, a cold shock family gene, confers a gain of fitness in *Escherichia coli* K-12 strains. J Bacteriol 188:6780–6785. https://doi.org/10.1128/JB.00471-06

Rebello S, Abraham A, Madhavan A, Sindhu R, Binod P, Babu AK, Aneesh EM, Pandey A (2018) Non-conventional yeast cell factories for sustainable bioprocesses. FEMS Microbiol Lett 365 (21):fny222. https://doi.org/10.1093/femsle/fny222

Reuß DR, Altenbuchner J, Mäder U, Rath H, Ischebeck T, Sappa PK, Thürmer A, Guérin C, Nicolas P, Steil L, Zhu B, Feussner I, Klumpp S, Daniel R, Commichau FM, Völker U, Stülke J (2017) Large-scale reduction of the *Bacillus subtilis* genome: consequences for the transcriptional network, resource allocation, and metabolism. Genome Res 27:289–299. https://doi.org/10.1101/gr.215293.116

Salsman J, Dellaire G (2017) Precision genome editing in the CRISPR era. Biochem Cell Biol 95:187–201. https://doi.org/10.1139/bcb-2016-0137

Sampedro I, Parales RE, Krell T, Hill JE (2015) *Pseudomonas* chemotaxis. FEMS Microbiol Rev 39:17–46. https://doi.org/10.1111/1574-6976.12081

Sanger F, Coulson AR, Hong GF, Hill DF, Petersen GB (1982) Nucleotide sequence of bacteriophage λ DNA. J Mol Biol 162:729–773. https://doi.org/10.1016/0022-2836(82)90546-0

Sasaki M, Kumagai H, Takegawa K, Tohda H (2013) Characterization of genome-reduced fission yeast strains. Nucleic Acids Res 41:5382–5399. https://doi.org/10.1093/nar/gkt233

Sauer M, Mattanovich D (2012) Construction of microbial cell factories for industrial bioprocesses. J Chem Technol Biotechnol 87:445–450. https://doi.org/10.1002/jctb.3711

Schallmey M, Singh A, Ward OP (2004) Developments in the use of *Bacillus* species for industrial production. Can J Microbiol 50:1–17. https://doi.org/10.1139/w03-076

Schempp FM, Drummond L, Buchhaupt M, Schrader J (2018) Microbial cell factories for the production of Terpenoid flavor and fragrance compounds. J Agric Food Chem 66:2247–2258. https://doi.org/10.1021/acs.jafc.7b00473

Schmidt A, Kochanowski K, Vedelaar S, Ahrné E, Volkmer B, Callipo L, Knoops K, Bauer M, Aebersold R, Heinemann M (2016) The quantitative and condition-dependent *Escherichia coli* proteome. Nat Biotechnol 34:104–110. https://doi.org/10.1038/nbt.3418

Schuster S, Hilgetag C (1994) On elementary flux modes in biochemical reaction systems at steady state. J Biol Syst 02:165–182. https://doi.org/10.1142/S0218339094000131

Segrè D, Vitkup D, Church GM (2002) Analysis of optimality in natural and perturbed metabolic networks. Proc Natl Acad Sci U S A 99:15112–15117. https://doi.org/10.1073/pnas.232349399

Sharma SS, Campbell JW, Frisch D, Blattner FR, Harcum SW (2007a) Expression of two recombinant chloramphenicol acetyltransferase variants in highly reduced genome *Escherichia coli* strains. Biotechnol Bioeng 98:1056–1070. https://doi.org/10.1002/bit.21491

Sharma SS, Blattner FR, Harcum SW (2007b) Recombinant protein production in an *Escherichia coli* reduced genome strain. Metab Eng 9:133–141. https://doi.org/10.1016/j.ymben.2006.10.002

Shen X, Wang Z, Huang X, Hu H, Wang W, Zhang X (2017) Developing genome-reduced *Pseudomonas chlororaphis* strains for the production of secondary metabolites. BMC Genomics 18:715. https://doi.org/10.1186/s12864-017-4127-2

Shin J, Lee N, Cho S, Cho B-K (2018) Targeted genome editing using DNA-free RNA-guided Cas9 ribonucleoprotein for CHO cell engineering. Methods Mol Biol 1772:151–169. https://doi.org/10.1007/978-1-4939-7795-6_8

Simen JD, Löffler M, Jäger G, Schäferhoff K, Freund A, Matthes J, Müller J, Takors R (2017) Transcriptional response of *Escherichia coli* to ammonia and glucose fluctuations. Microb Biotechnol 10:858–872. https://doi.org/10.1111/1751-7915.12713

Sliwa P, Korona R (2005) Loss of dispensable genes is not adaptive in yeast. Proc Natl Acad Sci U S A 102:17670–17674. https://doi.org/10.1073/pnas.0505517102

Smalley DJ, Whiteley M, Conway T (2003) In search of the minimal *Escherichia coli* genome. Trends Microbiol 11:6–8. https://doi.org/10.1016/S0966-842X(02)00008-2

Song AA-L, In LLA, Lim SHE, Rahim RA (2017) A review on *Lactococcus lactis*: from food to factory. Microb Cell Factories 16:55. https://doi.org/10.1186/s12934-017-0669-x

Stephanopoulos G (1999) Metabolic fluxes and metabolic engineering. Metab Eng 1:1–11. https://doi.org/10.1006/mben.1998.0101

Stoebel DM, Dean AM, Dykhuizen DE (2008) The cost of expression of *Escherichia coli* lac operon proteins is in the process, not in the products. Genetics 178:1653–1660. https://doi.org/10.1534/genetics.107.085399

Stouthamer AH (1973) A theoretical study on the amount of ATP required for synthesis of microbial cell material. Antonie Van Leeuwenhoek 39:545–565. https://doi.org/10.1007/BF02578899

Suzuki N, Okayama S, Nonaka H, Tsuge Y, Inui M, Yukawa H (2005a) Large-scale engineering of the Corynebacterium glutamicum genome. Appl Environ Microbiol 71:3369–3372. https://doi.org/10.1128/AEM.71.6.3369-3372.2005

Suzuki N, Nonaka H, Tsuge Y, Inui M, Yukawa H (2005b) New multiple-deletion method for the Corynebacterium glutamicum genome, using a mutant lox sequence. Appl Environ Microbiol 71:8472–8480. https://doi.org/10.1128/AEM.71.12.8472-8480.2005

Szathmáry E (2005) Life: in search of the simplest cell. Nature 433:469–470. https://doi.org/10.1038/433469a

Takors R, Bathe B, Rieping M, Hans S, Kelle R, Huthmacher K (2007) Systems biology for industrial strains and fermentation processes–example: amino acids. J Biotechnol 129:181–190. https://doi.org/10.1016/j.jbiotec.2007.01.031

Tanaka K, Henry CS, Zinner JF, Jolivet E, Cohoon MP, Xia F, Bidnenko V, Ehrlich SD, Stevens RL, Noirot P (2013) Building the repertoire of dispensable chromosome regions in *Bacillus subtilis* entails major refinement of cognate large-scale metabolic model. Nucleic Acids Res 41:687–699. https://doi.org/10.1093/nar/gks963

Tao H, Bausch C, Richmond C, Blattner FR, Conway T (1999) Functional genomics: expression analysis of *Escherichia coli* growing on minimal and rich media. J Bacteriol 181:6425–6440

Taymaz-Nikerel H, Borujeni AE, Verheijen PJT, Heijnen JJ, van Gulik WM (2010) Genome-derived minimal metabolic models for *Escherichia coli* MG1655 with estimated in vivo respiratory ATP stoichiometry. Biotechnol Bioeng 107:369–381. https://doi.org/10.1002/bit.22802

Terpe K (2006) Overview of bacterial expression systems for heterologous protein production: from molecular and biochemical fundamentals to commercial systems. Appl Microbiol Biotechnol 72:211–222. https://doi.org/10.1007/s00253-006-0465-8

Thomas P, Terradot G, Danos V, Weiße AY (2018) Sources, propagation and consequences of stochasticity in cellular growth. Nat Commun 9:4528. https://doi.org/10.1038/s41467-018-06912-9

Trinh CT, Wlaschin A, Srienc F (2008) Elementary mode analysis: a useful metabolic pathway analysis tool for characterizing cellular metabolism. Appl Microbiol Biotechnol 81:813–826. https://doi.org/10.1007/s00253-008-1770-1

Turina P, Samoray D, Gräber P (2003) H+/ATP ratio of proton transport-coupled ATP synthesis and hydrolysis catalysed by CF0F1-liposomes. EMBO J 22:418–426. https://doi.org/10.1093/emboj/cdg073

Umenhoffer K, Fehér T, Balikó G, Ayaydin F, Pósfai J, Blattner FR, Pósfai G (2010) Reduced evolvability of *Escherichia coli* MDS42, an IS-less cellular chassis for molecular and synthetic biology applications. Microb Cell Factories 9:38. https://doi.org/10.1186/1475-2859-9-38

Umenhoffer K, Draskovits G, Nyerges Á, Karcagi I, Bogos B, Tímár E, Csörgő B, Herczeg R, Nagy I, Fehér T, Pál C, Pósfai G (2017) Genome-wide abolishment of mobile genetic elements using genome shuffling and CRISPR/Cas-assisted MAGE allows the efficient stabilization of a bacterial chassis. ACS Synth Biol 6:1471–1483. https://doi.org/10.1021/acssynbio.6b00378

Unthan S, Baumgart M, Radek A, Herbst M, Siebert D, Brühl N, Bartsch A, Bott M, Wiechert W, Marin K, Hans S, Krämer R, Seibold G, Frunzke J, Kalinowski J, Rückert C, Wendisch VF, Noack S (2015) Chassis organism from Corynebacterium glutamicum–a top-down approach to identify and delete irrelevant gene clusters. Biotechnol J 10:290–301. https://doi.org/10.1002/biot.201400041

Valgepea K, Peebo K, Adamberg K, Vilu R (2015) Lean-proteome strains – next step in metabolic engineering. Front Bioeng Biotechnol 3:11. https://doi.org/10.3389/fbioe.2015.00011

Venter JC, Adams MD, Myers EW, Li PW, Mural RJ, Sutton GG, Smith HO, Yandell M, Evans CA, Holt RA, Gocayne JD, Amanatides P, Ballew RM, Huson DH, Wortman JR, Zhang Q, Kodira CD, Zheng XH, Chen L, Skupski M, Subramanian G, Thomas PD, Zhang J, Gabor Miklos GL, Nelson C, Broder S, Clark AG, Nadeau J, McKusick VA, Zinder N, Levine AJ, Roberts RJ, Simon M, Slayman C, Hunkapiller M, Bolanos R, Delcher A, Dew I, Fasulo D, Flanigan M, Florea L, Halpern A, Hannenhalli S, Kravitz S, Levy S, Mobarry C, Reinert K, Remington K, Abu-Threideh J, Beasley E, Biddick K, Bonazzi V, Brandon R, Cargill M, Chandramouliswaran I, Charlab R, Chaturvedi K, Deng Z, Di Francesco V, Dunn P, Eilbeck K, Evangelista C, Gabrielian AE, Gan W, Ge W, Gong F, Gu Z, Guan P, Heiman TJ, Higgins ME, Ji RR, Ke Z, Ketchum KA, Lai Z, Lei Y, Li Z, Li J, Liang Y, Lin X, Lu F, Merkulov GV, Milshina N, Moore HM, Naik AK, Narayan VA, Neelam B, Nusskern D, Rusch DB, Salzberg S, Shao W, Shue B, Sun J, Wang Z, Wang A, Wang X, Wang J, Wei M, Wides R, Xiao C, Yan C, Yao A, Ye J, Zhan M, Zhang W, Zhang H, Zhao Q, Zheng L, Zhong F, Zhong W, Zhu S, Zhao S, Gilbert D, Baumhueter S, Spier G, Carter C, Cravchik A, Woodage T, Ali F, An H, Awe A, Baldwin D, Baden H, Barnstead M, Barrow I, Beeson K, Busam D, Carver A, Center A, Cheng ML, Curry L, Danaher S, Davenport L, Desilets R, Dietz S, Dodson K, Doup L, Ferriera S, Garg N, Gluecksmann A, Hart B, Haynes J, Haynes C, Heiner C, Hladun S, Hostin D, Houck J, Howland T, Ibegwam C, Johnson J, Kalush F, Kline L, Koduru S, Love A, Mann F, May D, McCawley S, McIntosh T, McMullen I, Moy M, Moy L, Murphy B, Nelson K, Pfannkoch C, Pratts E, Puri V, Qureshi H, Reardon M, Rodriguez R, Rogers YH, Romblad D, Ruhfel B, Scott R, Sitter C, Smallwood M, Stewart E, Strong R, Suh E, Thomas R, Tint NN, Tse S, Vech C, Wang G, Wetter J, Williams S, Williams M, Windsor S, Winn-Deen E, Wolfe K, Zaveri J, Zaveri K, Abril JF, Guigó R, Campbell MJ, Sjolander KV, Karlak B, Kejariwal A, Mi H, Lazareva B, Hatton T, Narechania A, Diemer K, Muruganujan A, Guo N, Sato S, Bafna V, Istrail S, Lippert R, Schwartz R, Walenz B, Yooseph S, Allen D, Basu A, Baxendale J, Blick L, Caminha M, Carnes-Stine J, Caulk P, Chiang YH, Coyne M, Dahlke C, Mays A, Dombroski M, Donnelly M, Ely D, Esparham S, Fosler C, Gire H, Glanowski S, Glasser K, Glodek A, Gorokhov M, Graham K, Gropman B, Harris M, Heil J, Henderson S, Hoover J, Jennings D, Jordan C, Jordan J, Kasha J, Kagan L, Kraft C, Levitsky A, Lewis M, Liu X, Lopez J, Ma D, Majoros W, McDaniel J, Murphy S, Newman M, Nguyen T, Nguyen N, Nodell M, Pan S, Peck J, Peterson M, Rowe W, Sanders R, Scott J, Simpson M, Smith T, Sprague A, Stockwell T, Turner R, Venter E, Wang M, Wen M, Wu D, Wu M, Xia A, Zandieh A, Zhu X (2001) The sequence of the human genome. Science 291:1304–1351. https://doi.org/10.1126/science.1058040

Vickers CE, Blank LM, Krömer JO (2010) Grand challenge commentary: chassis cells for industrial biochemical production. Nat Chem Biol 6:875–877. https://doi.org/10.1038/nchembio.484

Vijayendran C, Polen T, Wendisch VF, Friehs K, Niehaus K, Flaschel E (2007) The plasticity of global proteome and genome expression analyzed in closely related W3110 and MG1655 strains of a well-studied model organism, *Escherichia coli*-K12. J Biotechnol 128:747–761. https://doi.org/10.1016/j.jbiotec.2006.12.026

Villaverde A (2010) Nanotechnology, bionanotechnology and microbial cell factories. Microb Cell Factories 9:53. https://doi.org/10.1186/1475-2859-9-53

Wang Q, Chung C-Y, Rosenberg JN, Yu G, Betenbaugh MJ (2018) Application of the CRISPR/Cas9 gene editing method for modulating antibody fucosylation in CHO cells. Methods Mol Biol 1850:237–257. https://doi.org/10.1007/978-1-4939-8730-6_16

Weinstock MT, Hesek ED, Wilson CM, Gibson DG (2016) Vibrio natriegens as a fast-growing host for molecular biology. Nat Methods 13:849–851. https://doi.org/10.1038/nmeth.3970

Westers H, Dorenbos R, van Dijl JM, Kabel J, Flanagan T, Devine KM, Jude F, Seror SJ, Beekman AC, Darmon E, Eschevins C, de Jong A, Bron S, Kuipers OP, Albertini AM, Antelmann H, Hecker M, Zamboni N, Sauer U, Bruand C, Ehrlich DS, Alonso JC, Salas M, Quax WJ (2003) Genome engineering reveals large dispensable regions in *Bacillus subtilis*. Mol Biol Evol 20:2076–2090. https://doi.org/10.1093/molbev/msg219

Weuster-Botz D, Takors R (2018) Wachstumskinetik. In: Chmiel H, Takors R, Weuster-Botz D (eds) Bioprozesstechnik, 4. [überbearbeitete und aktualisierte] Auflage, vol. 40. Springer, Berlin, pp 45–70

Wewetzer SJ, Kunze M, Ladner T, Luchterhand B, Roth S, Rahmen N, Kloß R, Costa E, Silva A, Regestein L, Büchs J (2015) Parallel use of shake flask and microtiter plate online measuring devices (RAMOS and BioLector) reduces the number of experiments in laboratory-scale stirred tank bioreactors. J Biol Eng 9:9. https://doi.org/10.1186/s13036-015-0005-0

Willrodt C, David C, Cornelissen S, Bühler B, Julsing MK, Schmid A (2014) Engineering the productivity of recombinant *Escherichia coli* for limonene formation from glycerol in minimal media. Biotechnol J 9:1000–1012. https://doi.org/10.1002/biot.201400023

Wolf YI, Koonin EV (2013) Genome reduction as the dominant mode of evolution. BioEssays 35:829–837. https://doi.org/10.1002/bies.201300037

Ye Y-N, Ma B-G, Dong C, Zhang H, Chen L-L, Guo F-B (2016) A novel proposal of a simplified bacterial gene set and the neo-construction of a general minimized metabolic network. Sci Rep 6:35082. https://doi.org/10.1038/srep35082

Ying B-W, Seno S, Kaneko F, Matsuda H, Yomo T (2013) Multilevel comparative analysis of the contributions of genome reduction and heat shock to the *Escherichia coli* transcriptome. BMC Genomics 14:25. https://doi.org/10.1186/1471-2164-14-25

Yu BJ, Sung BH, Koob MD, Lee CH, Lee JH, Lee WS, Kim MS, Kim SC (2002) Minimization of the *Escherichia coli* genome using a Tn5-targeted Cre/loxP excision system. Nat Biotechnol 20:1018–1023. https://doi.org/10.1038/nbt740

Zhao Y, Boeke JD (2018) Construction of designer selectable marker deletions with a CRISPR-Cas9 toolbox in *Schizosaccharomyces pombe* and new design of common entry vectors. G3: genes. G3 (Bethesda) 8:789–796. https://doi.org/10.1534/g3.117.300363

Zhou M, Jing X, Xie P, Chen W, Wang T, Xia H, Qin Z (2012) Sequential deletion of all the polyketide synthase and nonribosomal peptide synthetase biosynthetic gene clusters and a 900-kb subtelomeric sequence of the linear chromosome of *Streptomyces coelicolor*. FEMS Microbiol Lett 333:169–179. https://doi.org/10.1111/j.1574-6968.2012.02609.x

Zhu D, Fu Y, Liu F, Xu H, Saris PEJ, Qiao M (2017) Enhanced heterologous protein productivity by genome reduction in *Lactococcus lactis* NZ9000. Microb Cell Factories 16:1. https://doi.org/10.1186/s12934-016-0616-2

Construction of Minimal Genomes and Synthetic Cells

Donghui Choe, Sun Chang Kim, Bernhard O. Palsson, and Byung-Kwan Cho

Abstract A minimal genome strain containing only genes necessary for maintaining self-replicable life was proposed as a potential platform having various advantages in chemical and pharmaceutical industries. With recent advances in high-throughput DNA sequencing and synthesis technology, many reduced genomes have now been constructed. In this chapter, we will review previously constructed artificially reduced genomes to confirm the potential of their industrial utility. Some of them exhibit growth rates similar to those of their parental wild-type strains while offering higher genetic stability and productivity. Furthermore, we will discuss some technological hurdles and limitations encountered during the design and construction of reduced genomes.

Keywords Genome reduction · Minimal genome · Genome synthesis · Synthetic biology · Adaptive laboratory evolution

---

D. Choe
Department of Biological Science, Korea Advanced Institute of Science and Technology, Daejeon, South Korea

KI for the BioCentury, Korea Advanced Institute of Science and Technology, Daejeon, South Korea

S. C. Kim · B.-K. Cho (✉)
Department of Biological Science, Korea Advanced Institute of Science and Technology, Daejeon, South Korea

KI for the BioCentury, Korea Advanced Institute of Science and Technology, Daejeon, South Korea

Intelligent Synthetic Biology Center, Daejeon, South Korea
e-mail: bcho@kaist.ac.kr

B. O. Palsson
Department of Bioengineering, University of California, San Diego, CA, USA

Bioinformatics and Systems Biology Program, University of California, San Diego, CA, USA

Department of Pediatrics, University of California, San Diego, CA, USA

Novo Nordisk Foundation Center for Biosustainability, Technical University of Denmark, Lyngby, Denmark

A. R. Lara, G. Gosset (eds.), *Minimal Cells: Design, Construction, Biotechnological Applications*, https://doi.org/10.1007/978-3-030-31897-0_2

Abbreviations

ALE	Adaptive laboratory evolution
asRNA	Antisense RNA
CDS	Coding sequence
COG	Clusters of orthologous genes
CRISPR	Clustered regularly interspaced short palindromic repeats
IS element	Insertion sequence element
kbp	Kilo base pair
LUCA	Last universal common ancestor
Mbp	Mega base pair
ORF	Open reading frame
RNAi	RNA interference
sgRNA	Single guide RNA

1 Cell Factories and Small Genomes

1.1 Cell Factories

With the emergence of recombinant DNA technology, biological systems have been widely used in a variety of fields, such as the chemical and pharmaceutical industries. Bio-based production has many advantages compared to chemical production: (1) it does not require precious (sometimes toxic) catalysts, (2) it is carried out under mild conditions, (3) protein catalysts (enzymes) have high stereoselectivity, and (4) it is capable of producing macromolecules (e.g., protein drugs) and complex bio-active compounds (e.g., antibiotics). Cells specialized to manufacture these products are called *cell factories*, and they have been built from bacteria, yeast, and even mammalian cells. Cell factories have been built largely dependent on their innate metabolic capability, and also optimized successfully through metabolic engineering. However, cell factories often encounter systematic failures, such as low yield or productivity limited by endogenous metabolic capability, a complex process for removing biological components, low predictability of complex cellular processes, and low stability originating from clonal variations and mutations.

Synthetic biology and systems biology can address these problems. With recent advances in DNA sequencing and synthesis technology, synthetic biology can be used to design and build the new high-potential biological systems based on synthetic genetic circuits and pathways. Systems biology provides collective information concerning complex biological processes for the design step in the design-build-test-learn cycle of synthetic biology. Thus, systems and synthetic biology together offer the possibility of building designed cells for predictable, efficient, and streamlined production through minimization of a genome to contain only those genes necessary and sufficient for cell replication and product production.

1.2 Small Genomes in Nature

Minimal genomes exist in nature. Living organisms from the three domains of life have diverse genome sizes ranging from a few hundred thousand base pairs to more than a hundred billion base pairs. Among them, *Buchnera* spp., an intracellular symbiont split from a common ancestor with *E. coli*, lost 75% of its ancestral genome down to 250 kbp (McCutcheon and Moran 2011; Moran and Mira 2001). Considering that wild-type *E. coli* genomes are generally over 5 Mbp, *Buchnera* is an amazing example of massive natural genome reduction. Minimized genomes are found not only in *Buchnera* spp., but also many other symbiotic bacteria. *Nasuia deltocephalinicola*, an insect symbiont, has the smallest self-replicable genome with a size of 112 kbp (Bennett and Moran 2013). These examples demonstrate an evolutionary trend of *niche*-adapted genomes. Symbionts do not require genes related to environmental response in diverse conditions, because their host provides a stable nutrient supply and protects against harsh environmental changes. These unnecessary genes have been removed from their genomes over long-term evolutionary periods.

Conversely, bacteria have genomes of several million base pairs. For example, the most widely studied model organism, *E. coli*, has a far larger genome (> 4 Mbp) than the symbionts, genomes that contain more than 4000 genes, including over a thousand of unknown function (Ghatak et al. 2019; Riley et al. 2006). *E. coli* can propagate under a variety of conditions, including in aerobic and anaerobic environments and with a wide range of nutrients, pH, and temperature. Many genes are responsible for handling environmental stresses and utilizing a diversity of nutrients. However, in laboratory conditions, where the environment is defined, and many stress responses are not needed many *E. coli* genes can be deleted without a negative effect on cell growth (Baba et al. 2006). Thus, bacterial genomes encode many genes that are not needed for laboratory use and industrial fermentation. These gene functions can result in wastage of energy and biomass precursors, replication of unnecessary genome sections, and synthesis of functionally redundant or useless transcripts, proteins, and metabolites. Thus, it was hypothesized that an organism without these unnecessary genes could be a completely novel platform for the production of valued products under laboratory conditions (Moya et al. 2009).

1.3 Small Genome Advantages

Biological systems are extremely sophisticated; we cannot fully predict the functioning of even the simplest of life forms, the bacteria. However, it is obvious that genomes with fewer genes should have more predictable phenotypes. Consider a computational model that predicts industrially relevant features such as growth rate, biomass yield, and productivity of *E. coli* cells versus those of Chinese hamster ovary (CHO) cells. It is relatively easier to reconstruct the regulatory and metabolic networks for modeling the cellular features of *E. coli* than CHO cells. As a result, there are hundreds of computational bacterial models available, while only a handful have been reported for CHO cells (Hartleb et al. 2016; Hefzi et al. 2016; Nolan and Lee 2011; Orth et al. 2011; Weaver et al. 2014).

A minimal genome is convenient for genetic manipulation. For example, the genetically recoded *E. coli* for codon expansion has been constructed, in which the rarest stop codons, 321 UAG, were completely replaced with the UAA stop codon (Lajoie et al. 2013). Although the UAG codon is rare, genome editing of over 300 loci remains challenging. A minimal genome with fewer genes will reduce this effort substantially. Apart from the symbiont examples such as *Buchnera* spp. and *N. deltocephalinicola*, previously constructed artificial minimal genomes showed advantages such as high genetic stability (Park et al. 2014), chemical productivity (Mizoguchi et al. 2008), and transformation efficiency (Pósfai et al. 2006).

2 Elucidating Essential Parts of a Genome

2.1 Essential Gene Set

To construct a minimal genome, genes essential for maintaining life need to be elucidated. First, a minimal gene set was computationally deduced by comparing the first three fully sequenced genomes. In the 1990s, when few small bacterial genome sequences were available, Mushegian and Koonin compared *M. genitalium* and *Haemophilus influenza* (Mushegian and Koonin 1996). These bacteria both have relatively small genomes and exhibit distinct evolutionary tracks. By the logic of comparative genomics, genes conserved in multiple organisms are likely to have an essential function. A total of 240 orthologous genes were found in *M. genitalium* and *H. influenza*; however, several genes encoding essential cellular functions, such as phosphoglycerate mutase and nucleoside diphosphate kinase, were missed in the analysis. Interestingly, the two species have different phosphoglycerate mutases unrelated to each other, so clearly non-orthologous genes can occasionally replace an ancient gene, disrupting conservation. Including non-orthologous gene displacement, a total of 262 genes were predicted to constitute the core biological functions of the two species. Considering 1.5 billion years of evolution between the two bacteria and their common ancestor, it is striking that 50% of genes remain conserved.

With advances in high-throughput sequencing technology, tens of thousands of genome sequences have become available since 2000. Scientists compared hundreds of genomes to determine genes that are ubiquitous across species; interestingly, few genes were. Brown and colleagues compared 45 genomes and found only 23 conserved genes (Brown et al. 2001). Similarly, Harris and colleagues reported 80 core genes from the Clusters of Orthologous Groups (COG) database (Harris et al. 2003), and Koonin reported 63 ubiquitous genes from 100 genomes (Koonin 2003).

Charlebois and Doolittle compared 147 prokaryotic genomes among 14 phyla and found only 34 universal genes (Charlebois and Doolittle 2004). Although parameters and conservation measures differed among the reports, they all indicated that there are only a handful of conserved genes, which are definitely insufficient for maintaining life. The number of universal genes tends to decrease when more genomes are considered. This decrease is related to the distance of the last universal common ancestor (LUCA). The earlier the LUCA differentiated, the greater the chance that non-orthologous displacement occurred, resulting in a smaller number of universal genes. Thus, the theoretical prediction of essential genes using comparative genomics is limited concerning genes of unknown function, non-orthologous displacements, and billions of years of evolutionary history.

Besides computational predictions, investigators used experimental methods to probe essential genes. The simplest method is to remove a specific locus from the genome and see if life is sustained. *Mycoplasma genitalium* has the smallest genome (580 kbp) of any free-living organism known to date. Even though it has the smallest genome (517 genes encoded in this genome), many genes in *M. genitalium* could be disrupted by transposons (Hutchison et al. 1999). In *Bacillus subtilis*, 79 random genomic regions had been mutated (Itaya 1995). Only six loci were indispensable, comprising an estimated 318–562 kbp of the B. *subtilis* genome, similar to the size of the *M. genitalium* genome. Thus, various methods were devised to determine gene essentiality in bacteria. Direct inactivation of individual genes via recombination, transposon insertion, and antisense RNA (asRNA) has been conducted as well. By disrupting individual genes in *B. subtilis* with the insertion of a non-replicating plasmid, Kobayashi and colleagues found that only 271 of 4101 genes are essential (Kobayashi et al. 2003). In *E. coli*, 3985 of 4288 open reading frames (ORFs) were knocked out, indicating that the remaining 303 ORFs are essential for life (Baba et al. 2006). The number of essential genes in *B. subtilis* and *E. coli* is comparable to that of *M. genitalium*. Although the targeted gene knockout study provides direct evidence of gene essentiality, the method is time-consuming and requires intensive works to generate thousands of deletion experiments.

To overcome these limitations, high-throughput methods employing the disruptive characteristics of transposon mutagenesis have been widely used. The transposon is a genetic element that can randomly move within a genome, disrupting the gene where the transposon DNA is inserted. When a transposon is introduced to a genome, a mutant with insertion in an essential gene cannot survive. Using these characteristics, one can discriminate essential and non-essential genes by identifying the transposon insertion site in surviving mutants. Genome-wide transposon insertion maps have been elucidated in *M. genitalium* using 1300 and 3000 mutants, showing that 265–382 coding sequences (CDSs) are essential (Glass et al. 2006; Hutchison et al. 1999). A transposon insertion map composed of 3000 unique insertion sites has a resolution of approximately one insertion per 200 bp on average. With this resolution, the essentiality of small genetic elements such as functional RNAs (e.g., tRNAs and ncRNAs) could not be examined. Furthermore, some essential genes were resistant to transposon insertion at the 3′ end because short truncation or elongation did not affect their functionality. To circumvent this limitation, a method to inactivate genes by asRNAs was invented (Ji et al. 2001). Using asRNA, a certain gene can be conditionally knocked down, unlike irreversible gene disruption by transposons. Thus, once an asRNA-containing library is constructed, investigators can assess gene essentiality or fitness in multiple environmental conditions and iterations without repeatedly constructing knockout strains or transposon mutants.

Identification of transposon insertion sites is dependent on isolation of single clones and Sanger sequencing. When combined with high-throughput sequencing techniques, numerous insertion sites can be identified in parallel. Thus, transposon mutagenesis coupled with next generation sequencing (Tn-Seq) enables elucidation of essential genes using transposon mutagenesis at an unprecedented resolution. From a 2×10^5 *E. coli* transposon mutant library, 620 of 4291 protein-coding genes were estimated to be essential according to statistical analysis (Gerdes et al. 2003). The discrepancy between Tn-Seq and individual knockout studies might originate during cell propagation. Even though a gene is not strictly essential, inactivation of an important gene can result in severe growth defects, which might be underrepresented or diminished during cell propagation. Also, statistical cutoffs and varying experimental conditions may cause this discrepancy.

Finally, with the recent revolution of clustered regularly interspaced short palindromic repeats (CRISPR) technology, catalytically inactive Cas9 (dCas9) can transcriptionally repress a target gene expression (Qi et al. 2013). Since the CRISPR system requires only a 20-nt proto-spacer sequence in chimeric single-guide RNA (sgRNA) for its specificity, a massive genome-wide sgRNA library can be easily constructed via DNA synthesis. Genome-wide CRISPR interference (CRISPRi) elucidated 379 essential genes in *E. coli* by inhibiting all genes with ~59,000 sgRNAs (Rousset et al. 2018). The number of essential genes estimated by the CRISPRi technique is slightly larger than those estimated by other methods. In bacteria, many genes are transcribed polycistronically. That is, disruption of a leader polycistron inactivates downstream genes contained in the operon. Thus, even if the leader gene is inessential, transcriptional inhibition can lead to lethal effects in an essential downstream gene. The polycistron structure induces an overestimation of essential genes, resulting in a higher estimate of essential genes by CRISPRi compared to other methods. With current high-throughput DNA synthesis technology, large sgRNA libraries could be easily synthesized; thus, gene essentiality of organisms with far larger genomes, such as humans, could be elucidated (Shalem et al. 2014; Wang et al. 2014).

Overall, multiple efforts have been made to elucidate essential genes in various organisms. Although the exact number of essential genes varies among methods, 300–500 genes are considered to be sufficient for maintaining life. Lastly, the direct examination of individual gene essentiality using single-gene knockout studies, Tn-Seq, and CRISPRi have inherent limitations. These methods rely on removal or inactivation of a single-gene, while inactivation of more than two genes simultaneously has never been tested. For example, assume that two genes have the same essential function in *E. coli*. These genes can be deleted individually because the other gene can rescue the function; however, simultaneous inactivation of both will be lethal. Construction of all double knockout strains in even the simplest bacterium, *M. genitalium,* requires a quarter million strains. Considering the time and expense of constructing ~4000 *E. coli* single gene knockout strains are huge (Baba et al. 2006), technological innovation is required to unveil higher-order combinatorial gene essentiality.

2.2 *Essentiality of Sub- and Non-genic Elements*

With a high-resolution essentiality scan using Tn-Seq, the essentiality of genetic elements smaller than a gene, such as promoters, terminators, and RNA genes, was reported (Christen et al. 2011; Lluch-Senar et al. 2015). Strikingly, different protein domains in one gene can differ in essentiality. For example, the C-terminal ATPase domain of the essential tyrosine kinase DivL, involved in cell cycle regulation in *Caulobacter crescentus*, was tolerant of disruptive transposon insertion, while other domains were not (Christen et al. 2011). Furthermore, genomic regions that encode virtually no functional product (e.g., protein or functional RNA) can still be pivotal for maintaining life. For instance, the origin of replication is an essential part of a genome despite encoding no functional products. In addition, bacterial genomes seem to have a more complex ordered structure than previously believed due to nucleoid-associated proteins (NAPs) (Srinivasan et al. 2015). It has been reported that NAP binding regions are important for replication (Lin and Grossman 1998). Likewise, there are many factors that must be considered when designing a minimal genome. A thorough examination and understanding of the genome are required during design.

3 Reduced Genomes

3.1 *Reduced* E. coli *Genomes*

E. coli is the most extensively studied model organism, and many reduced genomes have been constructed with deletion sizes ranging from 300 kbp to 1.38 Mbp (Table 1 and Fig. 1). Considering that laboratory *E. coli* K-12 strains have genomes of approximately 4.64 Mbp, deletions span from 6.8 to 29.7% of the original genome.

3.1.1 *E. coli* CDΔ3456

Two reduced-genome *E. coli* strains were reported in 2002. One, CDΔ3456, was reported by Yu and colleagues with a deletion size of over 300 kbp (Yu et al. 2002). The strain was constructed by deleting a large genomic region between two *loxP*

Table 1 Summary of reduced genomes

Ancestor	Name	Original genome size	Reduction (proportion)	Note
Escherichia coli str. K-12 substr. MG1655	CDΔ3456	4.64 Mbp	313 kbp (6.8%)	Normal growth
Escherichia coli str. K-12 substr. MG1655	MDS12	4.64 Mbp	376 kbp (8.1%)	Normal growth, 10% higher cell density
Escherichia coli str. K-12 substr. MG1655	MDS42	4.64 Mbp	708 kbp (15.3%)	Normal growth, increased transformation efficiency
Escherichia coli str. K-12 substr. MG1655	MS56	4.64 Mbp	1068 kbp (23.0%)	Normal growth, increased genetic stability
Escherichia coli str. K-12 substr. MG1655	Δ16	4.64 Mbp	1377 kbp (29.7%)	Lower growth rate, aberrant nucleoid structure
Escherichia coli str. K-12 substr. W3110	MGF-01	4.65 Mbp	1030 kbp (22.2%)	50% higher cell density, higher threonine production
Bacillus subtilis str. 168	Δ6	4.22 Mbp	320 kbp (7.7%)	Normal growth
Bacillus subtilis str. 168	PG10	4.22 Mbp	1456 kbp (34.5%)	Lower growth rate
Bacillus subtilis str. 168	PG38	4.22 Mbp	1535 kbp (36.4%)	Lower growth rate
Bacillus subtilis str. 168	MGB469	4.22 Mbp	469 kbp (11.1%)	Normal growth
Bacillus subtilis str. 168	MG1M	4.22 Mbp	991 kbp (23.5%)	Normal growth
Bacillus subtilis str. 168	MBG874	4.22 Mbp	874 kbp (20.7%)	Lower growth rate, higher protein production
Streptomyces avermitilis	SUKA17	9.03 Mbp	1674 kbp (18.5%)	Higher antibiotics production
Schizosaccharomyces pombe	A8	12.6 Mbp	657 kb (5.2%)	Higher protein production

sites recognized by Cre site-specific recombinase (Fig. 2). The *loxP* sites were pre-inserted at a random position in the genome using transposons. Specifically, two different transposon mutant libraries were made with two different antibiotic-resistance gene cassettes, and all transposon insertion sites were identified. Genomes of two mutants (containing two different marker genes) with transposons positioned each at one end of the region targeted for deletion were fused together via P1 transduction. Two different antibiotic-resistance gene markers were used to screen for successfully transduced cells. Then, two *loxP* sites in the fused genome were recombined by Cre recombinase, generating a large targeted deletion. Six deletion strains were built, and multiple deletions were cumulated into one genome using additional P1 transductions. During cumulative deletion, some deletions could not be combined together. This phenomenon occurs frequently during genome reduction. Some genomic regions can be deleted from a genome individually; however, specific combinations cannot be deleted simultaneously. This kind of relationship is

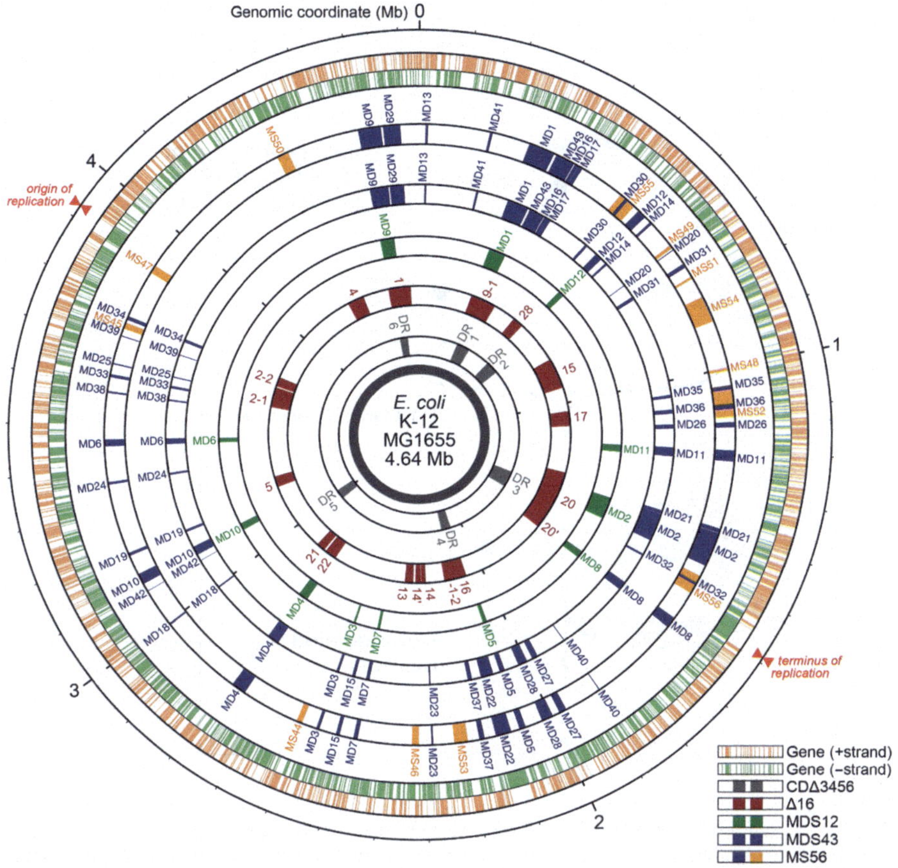

Fig. 1 Genomic map of reduced *E. coli* genomes. The outermost track contains the position of all genes in *E. coli* MG1655 ("Gene" in both strands). Colored bars indicate the positions of deleted regions in each reduced genome strain. Red arrowheads indicate origin or terminus of DNA replication

called synthetic lethality, which may be due to gene duplication or orthologs (Nijman 2011). When one copy of two genes is deleted, the other copy can maintain biological function, while a double mutant cannot. Avoiding synthetically lethal combinations, four regions were combined in CDΔ3456, which lacks 287 open reading frames (ORFs) containing 179 unknown genes and genes related to histidine biosynthesis, fimbriae, and several transporters. The final clone had a growth rate equivalent to the parental *E. coli*.

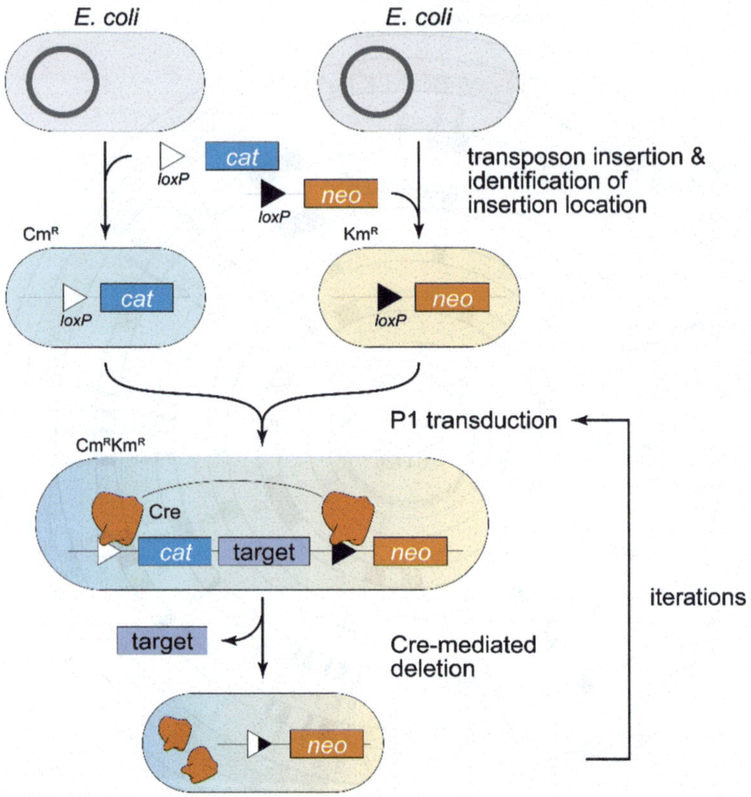

Fig. 2 Genome reduction method for *E. coli* CDΔ3456. *loxP* recognition sequence of the site-specific DNA recombinase Cre, *cat* chloramphenicol resistance gene, *Neo* kanamycin resistance gene, Cm^R *and* Km^R chloramphenicol- and kanamycin-resistant phenotypes

3.1.2 *E. coli* MS56 and Its Relatives

The reduced-genome *E. coli* MDS12 was reported in 2002, which lacks 12 K-islands from the *E. coli* K-12 strain (Kolisnychenko et al. 2002). The K-islands are genomic regions that K-12 acquired via horizontal gene transfer. K-islands contain unnecessary genes, such as prophages and transposons, making the genome unstable. The 12 K-islands were sequentially deleted by the scarless deletion method using I-*Sce*I meganuclease and the double-strand break repair system (Fig. 3). Specifically, a DNA cassette containing a chloramphenicol resistance gene was introduced into *E. coli* MG1655, and the cassette replaced a homologous target region. Although the target was deleted, removal of the resistance gene was required for the next round of deletion; thus, the gene was excised by I-*Sce*I. After deletion, the double-strand break was repaired by RecA, resulting in a scarless deletion strain. After 12 iterative deletions and P1 transductions, the combined length of deletions was 376 kbp, containing 409 ORFs. Since the deleted genes have no essential functions, MDS12

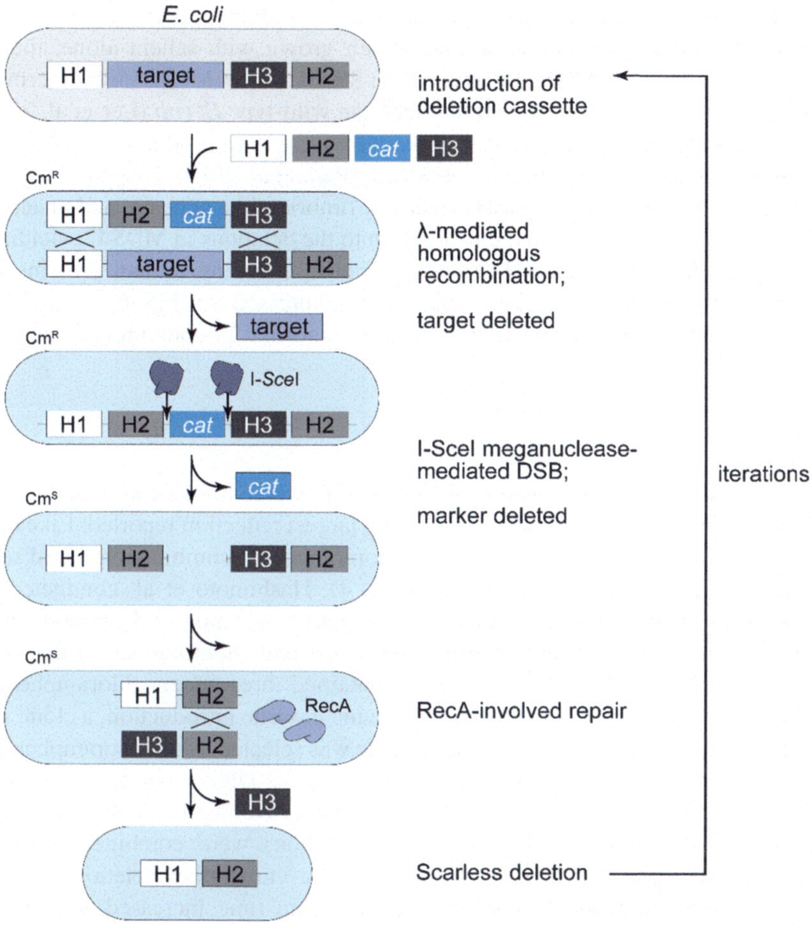

Fig. 3 Genome reduction method for *E. coli* MS56 and its relatives. *Cat* chloramphenicol resistance gene, Cm^S *and* Cm^R chloramphenicol-sensitive and chloramphenicol-resistant phenotypes, I-*Sce*I intron-encoded meganuclease

showed no difference in growth rate and DNA transformation efficiency. Interestingly, MDS12 exhibited an approximately 10% higher final cell density than wild-type *E. coli*. In MDS12, conserved energy and material may be converted to biomass, demonstrating an advantage of a reduced genome.

Four years later, Pósfai and colleagues reported *E. coli* MDS41, 42, and 43, descendants of MDS12 containing additional deletions (a total deletion of 663–708 kbp from wild-type) (Pósfai et al. 2006). MDS42 is free of all transposable and insertion sequence (IS) elements. In *E. coli*, it has been reported that 20–25% of all mutations are related to IS elements (Pósfai et al. 2006). MDS42 showed no IS-mediated gene inactivation. With this property, MDS42 stably propagated plasmid DNA carrying toxic ORFs, and genomic components were also stable. *E. coli* has a silent *bgl* operon that enables salicin utilization; thus, *E. coli* cannot normally utilize salicin as its sole carbon source. When grown with salicin alone, the *bgl* operon activation rate in MDS41 was less than 8% of that for MG1655. Furthermore, MDS42 produced over 80% more threonine than wild-type *E. coli* (Lee et al. 2009).

With the success of MDS strains, Park and colleagues introduced 14 additional deletions to MDS42 to construct *E. coli* MS56 (Park et al. 2014). Dispensable genes and genetic elements including hydrogenases, fimbriae-like adhesin, and anaerobic respiratory enzymes were removed in addition to the deletions in MDS42, yielding a 1.07 Mbp genome reduction. Heterologous genes encoding human proteins that arrest *E. coli* growth could be successfully expressed in MS56 without any IS-mediated inactivation, which occurs rapidly in normal IS-containing *E. coli*.

3.1.3 Δ16

Hashimoto and colleagues constructed *E. coli* Δ16, which has a genome reduced by 1.377 Mbp (29.7% of its original genome), the largest reduction reported. Like other reduced genomes, Δ16 was constructed by homologous recombination-based serial deletions combined by P1 transduction (Fig. 4). Hashimoto et al. conducted the deletion using two methods: positive and negative selections (Hashimoto et al. 2005). First, a DNA cassette containing sequence ends homologous to the target region was constructed. The cassette also contained three genes: chloramphenicol acetyltransferase (*cat*), *rpsL*, and *sacB*. Following cassette introduction, a clone with a target genomic region replaced by the cassette was selected using chloramphenicol. Then, the cassette was replaced by a new homologous DNA cassette, followed by negative selection with the *rpsL* and *sacB* genes, making the cell sensitive to streptomycin and sucrose. Sixteen scarless deletions were combined using P1 transduction to make the final strain Δ16 with a cumulative deletion of 1.377 Mbp. During the sequential deletions, the doubling time increased sequentially from 26.2 min in MG1655 to 45.4 min in Δ16. The shape of the cell became longer and wider, and the nucleoid showed irregular distribution and asynchronous replication with respect to cell division.

3.1.4 MGF-01

Based on non-essential gene information obtained by comparing *E. coli* and *Buchnera* spp. (a symbiont discussed earlier), Mizoguchi and colleagues constructed *E. coli* MGF-01 by removing *E. coli*-specific genes from *E. coli* W3110 (Mizoguchi et al. 2008). As with other deletion methods, MGF-01 was constructed using sequential deletions combined with repetitive P1 transductions (Fig. 5). Briefly, the target DNA sequence was removed by homologous replacement with a DNA cassette containing *sacB* and *cat* genes. Then, selection markers were replaced by another replacement DNA cassette, and a markerless clone was selected by negative selection with sucrose. Finally, 53 deletions were accumulated in 28 P1 transduction

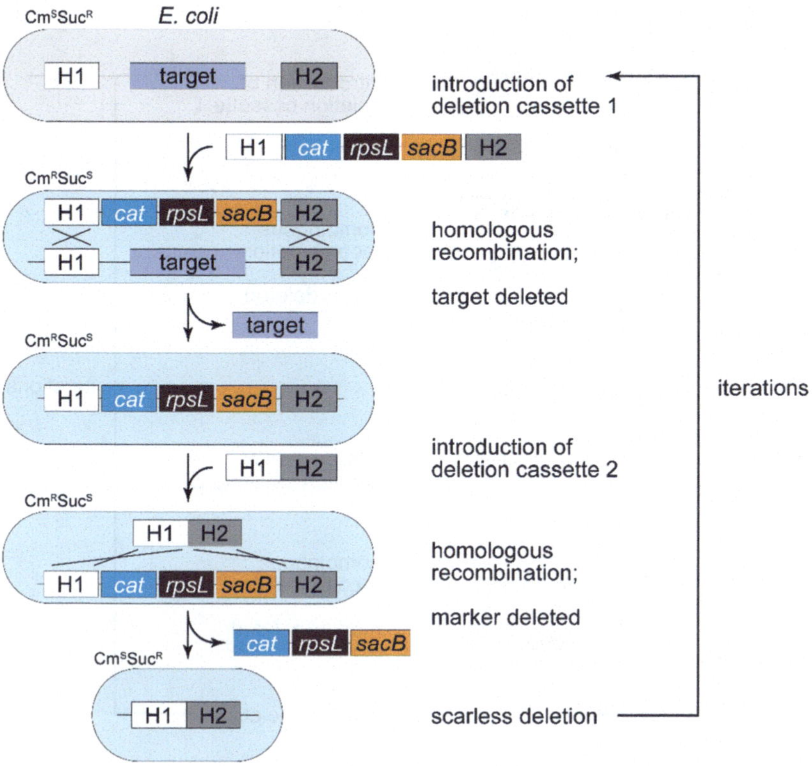

Fig. 4 Genome reduction method for *E. coli* Δ16. *sacB* levansucrase, which produces the toxic fructose polymer levan from sucrose, *Cat* chloramphenicol resistance gene, Cm^S *and* Cm^R chloramphenicol-sensitive and chloramphenicol-resistant phenotypes, Suc^S *and* Suc^R sucrose-sensitive and sucrose-resistant phenotypes

cycles, resulting in the combined deletion length of 1.03 Mbp. Interestingly, MGF-01 yielded a 1.5 times higher final biomass than the wild-type parent. It was claimed that MGF-01 utilizes glucose more efficiently than the parental strain due to a significant reduction in overabundant acetate accumulation (from 1.37 to 0.50 g/L in *E. coli* W3110 and MGF-01, respectively). Moreover, MGF-01 showed 2.44 and 1.69 times higher threonine titer and yield, respectively, with a significantly lower acetate byproduct compared to the wild-type. These unexpected beneficial properties of MGF-01 are distinct from those of other reduced-genome *E. coli* that have phenotypes similar to their wild-type ancestors. However, further study on the differences in MGF-01 is required.

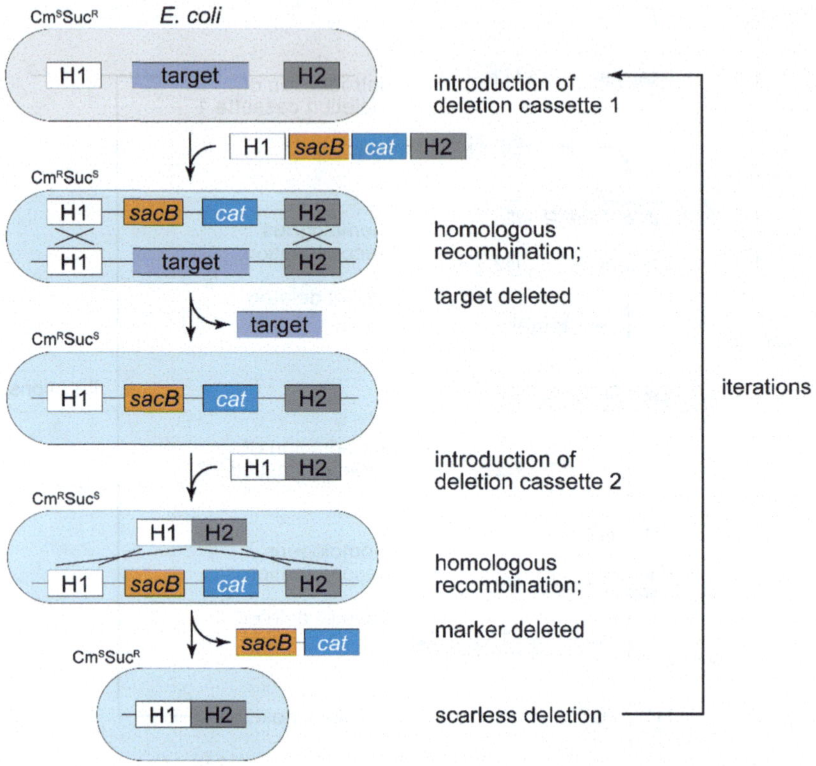

Fig. 5 Genome reduction method for *E. coli* MGF-01. *sacB* levansucrase, which produces toxic fructose polymer levan from sucrose, *Cat* chloramphenicol resistance gene, Cm^S *and* Cm^R chloramphenicol-sensitive and chloramphenicol-resistant phenotypes. Suc^S *and* Suc^R sucrose-sensitive and sucrose-resistant phenotypes

3.2 *Reduced Genomes in Other Species*

3.2.1 *B. subtilis* PG38 and Its Relatives

Bacillus subtilis is one of the most widely studied gram-positive bacteria, and it is a good production host for various proteins through its secretion system. Westers and colleagues constructed genome-reduced *B. subtilis* Δ6 by removing six genomic loci containing genes responsible for polyketides, protein antibiotic biosynthesis, prophages, and prophage-like elements (Westers et al. 2003). These loci contain 332 genes in total, spanning 320 kbp. Deletions were carried out by integrating and excising the selection marker using the integration plasmid pG+ host4 (Biswas et al. 1993) (Fig. 6). Genome-reduced *B. subtilis* Δ6 had no apparent changes in cell physiology. The strain had the same growth rate, glucose/acetate metabolic fluxes, heterologous protein secretion, and biomass yield compared to its parental strain. Unexpectedly, Δ6 showed increased cell motility on an agarose plate, although no

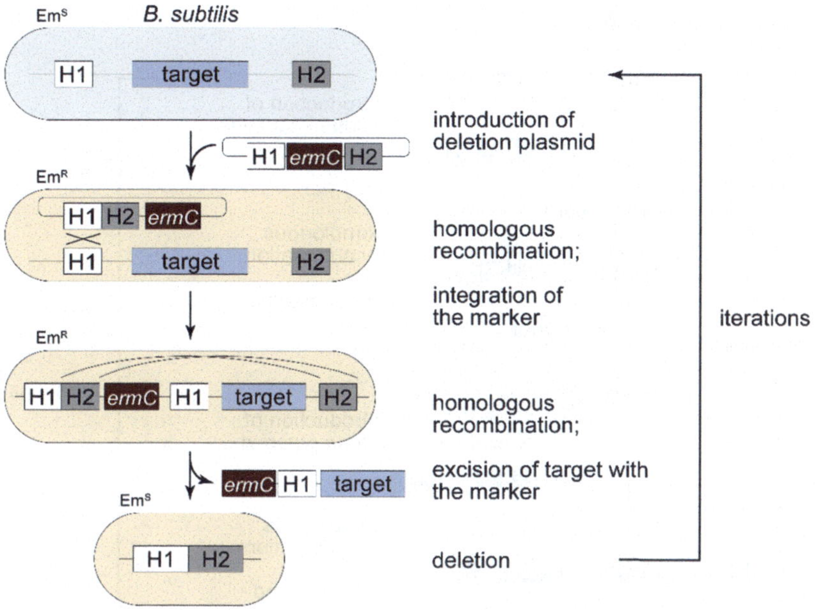

Fig. 6 Genome reduction method for *B. subtilis* Δ6 and its relatives. *H1 and H2* homology regions. *ermC* erythromycin resistance gene, Em^S *and* Em^R erythromycin-sensitive and erythromycin-resistant phenotypes

deleted genes were related to cell motility. Genome-reduced strains occasionally show unexpected phenotypes such as this, illustrating our limited understanding of even small bacterial genomes (Choe et al. 2019).

The genome of *B. subtilis* Δ6 was further reduced to produce PG10 and PG38 (Reuß et al. 2017). Reuß and colleagues constructed two independent genome-reduced strains, PG10 and PG38, containing 88 and 94 iterative deletions, respectively. The deletions contain non-essential genes related to sporulation, motility, antibiotic production, and secondary metabolism. During the serial deletions, intermediate strains lost DNA competence progressively. Thus, they introduced additional competence proteins (ComK and ComS) into the genome to effectively transfer DNA required for deletion. The two derivative strains showed a slower growth rate (~40% longer doubling time) and long filamentous cell morphology. Although growth rate of the two strains was reduced, they had the largest proportion of genome reduction by far (1.46 and 1.54 Mbp; over one-third of their original genomes) from that required for viable cells.

3.2.2 *B. subtilis* MGB874 and Its Relatives

Ara and colleagues constructed a reduced-genome *B. subtilis* MGB469 from *B. subtilis* strain 168 (Ara et al. 2007). The reduced strain lacks nine genomic loci

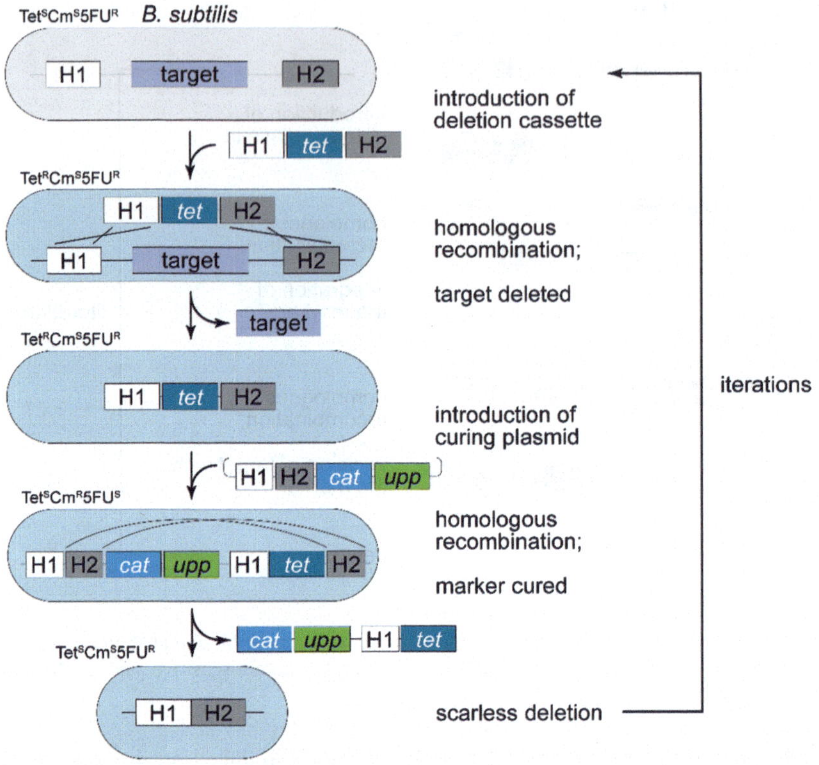

Fig. 7 Genome reduction method for *B. subtilis* MGB874 and its relatives. *Tet* tetracycline resistance gene, *Cat* chloramphenicol resistance gene, *Upp* uracil phosphoribosyl transferase that make cell sensitive to 5-fluorouracil, Tet^S *and* Tet^R tetracycline-sensitive and tetracycline-resistant phenotypes, Cm^S *and* Cm^R chloramphenicol-sensitive and chloramphenicol-resistant phenotypes, $5FU^S$ *and* $5FU^R$ 5-fluorouracil-sensitive and 5-fluorouracil-resistant phenotypes

related to prophages or prophage-like elements and two antibiotic synthesis genes (plipastatin and polyketide). The loci were first removed from the *B. subtilis* genome individually to confirm that they contained no essential genes. Then, the deletions were combined sequentially, so that the removed regions spanned 469 kbp altogether (Fig. 7). The resulting strain, MGB469, showed no difference in growth rate and protein productivity, which led to further genome reduction. They sought a genomic locus which could be reduced to increase protein productivity to finally construct a genome-reduced strain with higher productivity than its ancestor. Six genomic loci that enhance cellulase productivity when deleted individually were further removed from MGB469, creating strain MG1M with 991 kbp total genome reduction. However, cellulase and protease production in MG1M showed no noticeable change.

As *B. subtilis* strains MGB469 and MG1M had phenotypes comparable to their wild-type counterpart, and there was no benefit in further genome reduction,

Morimoto and colleagues revisited intermediate strain MGB469 to construct a new genome-reduced *B. subtilis* with advantageous characteristics (Morimoto et al. 2008). They tested the deletion of 74 genomic regions, including prophages and secondary metabolic genes. Of 63 regions where deletion was possible, 11 sequential deletions were introduced into MGB469, resulting in *B. subtilis* strain MGB874 with a deletion length of 874 kbp. Although MGB874 had normal cell morphology and nucleoid structure, its growth rate was reduced to 70% that of wild-type *B. subtilis*. Unlike other reduced genomes of *B. subtilis*, MGB874 managed to produce 1.7- and 2.5-fold more cellulase and protease than *B. subtilis* 168. An investigation into the strain's transcriptome indicated that a number of transcriptomic changes, including sporulation, degradative enzyme secretion, and sigma factors, were responsible for its increased productivity. The strain MGB874 demonstrates a reduced genome as a potent host for industrial protein production.

3.2.3 *Streptomyces avermitilis* SUKA17

Streptomyces is an industrially and clinically important Actinobacteria genus known for producing most biologically active secondary metabolites such as antibiotics. *Streptomyces* have relatively large genomes among bacteria, and their chromosomal DNA is linear rather than circular. Komatsu and colleagues constructed SUKA17 from *Streptomyces avermitilis*, one of the most widely used industrial species (Komatsu et al. 2010). *S. avermitilis* harbors more than 20 secondary metabolite biosynthetic gene clusters, and these clusters are located mainly at the end of the genome, called the subtelomeric region. Comparative genomics studies on *Streptomyces griseus*, *Streptomyces coelicolor*, and *S. avermitilis* revealed essential core genes located at the center of the genome. Thus, genes related to major secondary metabolism in the subtelomeric region, including terpene metabolism, were removed from *S. avermitilis* using homologous recombination and Cre-mediated recombination (Fig. 8). *S. avermitilis* is not recombinogenic like *E. coli* and *B. subtilis*. Thus, two *loxP* sites were first introduced into the genome from a circular DNA template via homologous recombination. Then, the large genomic locus target was removed by Cre-mediated recombination. The deletion consisted of genomic regions containing 1272 ORFs with a collective size of 1.67 Mbp. With its simpler genome lacking the metabolic burden of secondary metabolite synthesis, SUKA17 containing the heterologous gene cluster produced more streptomycin and cephamycin C than native producers *S. griseus* and *S. clavuligerus*.

3.2.4 *Schizosaccharomyces pombe* A8

While we have discussed several prokaryotes, there is also a genome-reduced eukaryote. Giga-Hama and colleagues reported genome-reduced fission yeast *Schizosaccharomyces pombe* A8 (Giga-Hama et al. 2007). They obtained information about essential genes by comparing the *S. pombe* genome with closely related

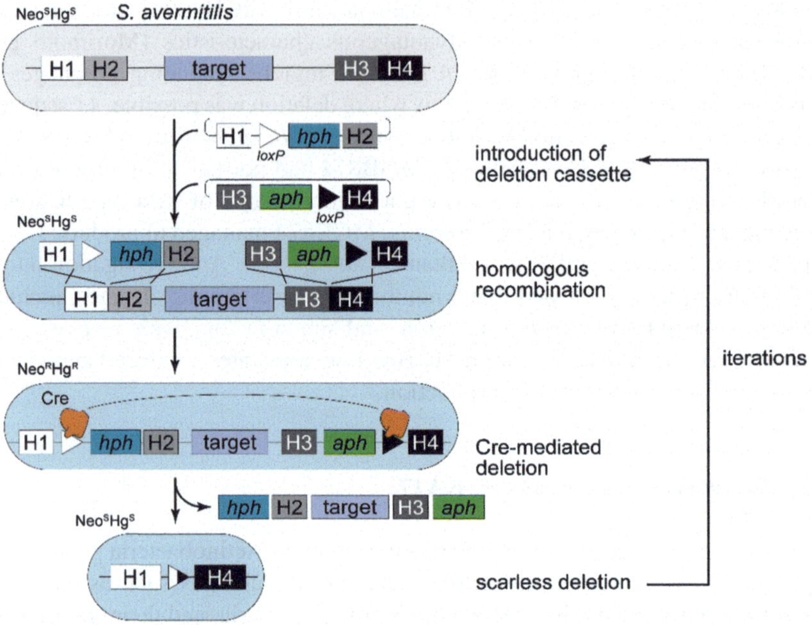

Fig. 8 Genome reduction method for *S. avermitilis* SUKA17. *loxP* recognition sequence of site-specific DNA recombinase Cre, *hph* hygromycin B phosphotransferase, *Aph* neomycin phosphotransferase, Neo^S *and* Neo^R neomycin-sensitive and neomycin-resistant phenotypes, Hg^S *and* Hg^R hygromycin B-sensitive and hygromycin B-resistant phenotypes

budding yeast *Saccharomyces cerevisiae*. Using the latency to universal rescue (LATOUR) method (Fig. 9) (Hirashima et al. 2006), a 657.3 kbp region in chromosomes I and II was deleted from the 12.6 Mbp genome. *S. pombe* A8, lacking 223 of 5100 genes, showed growth comparable to wild-type. Intracellular energy levels (ATP and GTP) in A8 were higher than wild-type; as a result, the strain showed 30 times higher production of human growth hormone, likely due to an increased GTP level, a limiting factor of translation. A eukaryotic platform strain is important for producing protein products (such as protein therapeutics) that cannot be produced by prokaryotic hosts because of post-translational modification and folding. The eukaryotic reduced-genome factory A8 is a milestone in this field.

As described earlier, the minimal genome is a fascinating concept, and many efforts have been made to construct artificial minimal genomes from well-known organisms. With a variety of practical advantages such as increased biomass yield, productivity, and genomic stability, the reduced genome will be a great tool for industry and biotechnology.

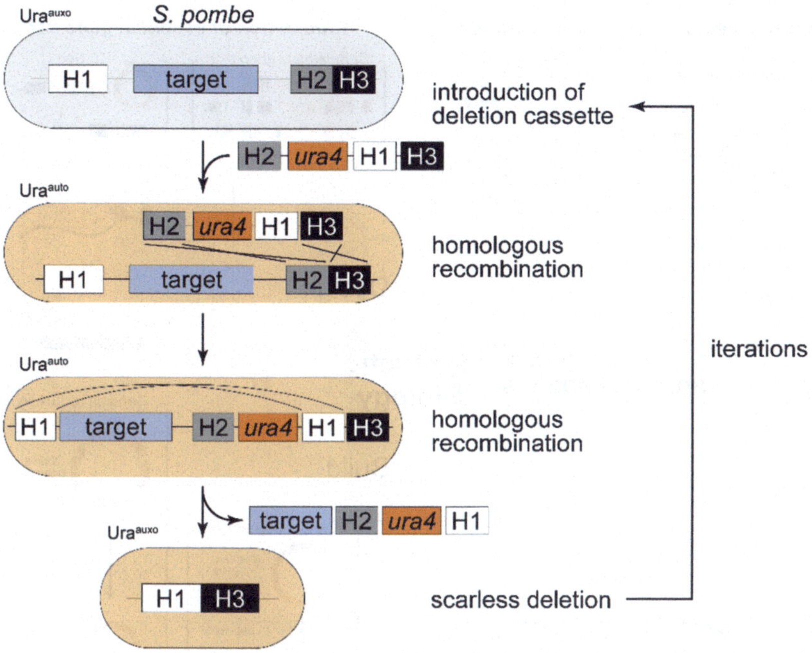

Fig. 9 Genome reduction method for *S. pombe* A8. *ura4* orotidine 5′-phosphate decarboxylase, *Ura*^{auxo} *and Ura*^{auto} uracil auxotroph and autotroph phenotypes

4 Synthetic Genomes and Cells

Reduced genomes discussed previously were all constructed using a top-down genome reduction as indicated by their names. However, investigators have synthesized genomes of viruses, phages, and bacteria from scratch. *Mycoplasma* JCVI-syn1.0 constructed by Gibson and colleagues harbors a fully chemically synthesized genome, a replica of the 1.08 Mbp *Mycoplasma mycoides* genome with slight modifications (Gibson et al. 2010). After elucidating essential genes in *M. mycoides* via transposon mutagenesis, the minimal genome *M. mycoides* JCVI-syn3.0 was synthesized (Hutchison et al. 2016). The JCVI-syn3.0 genome contains only 473 genes with a collective length of 531 kbp. Although the doubling time of JCVI-syn3.0 was three times longer (~180 min) than that of JCVI-syn1.0, it was much shorter than that of the natural minimal genome of *M. genitalium* (~16 hr doubling time). Recently, artificial yeast chromosomes were designed and are being synthesized through a worldwide consortium (Richardson et al. 2017). Although full genome synthesis requires astronomical time and expense (Sleator 2010), there is no doubt that it is an attractive alternative for genome reduction.

In this chapter, we described the design, build, and test cycle (DBT cycle) of a minimal genome (Fig. 10). We first looked into the designing step of a minimal (or reduced) genome based on an essential gene set elucidated by various methods.

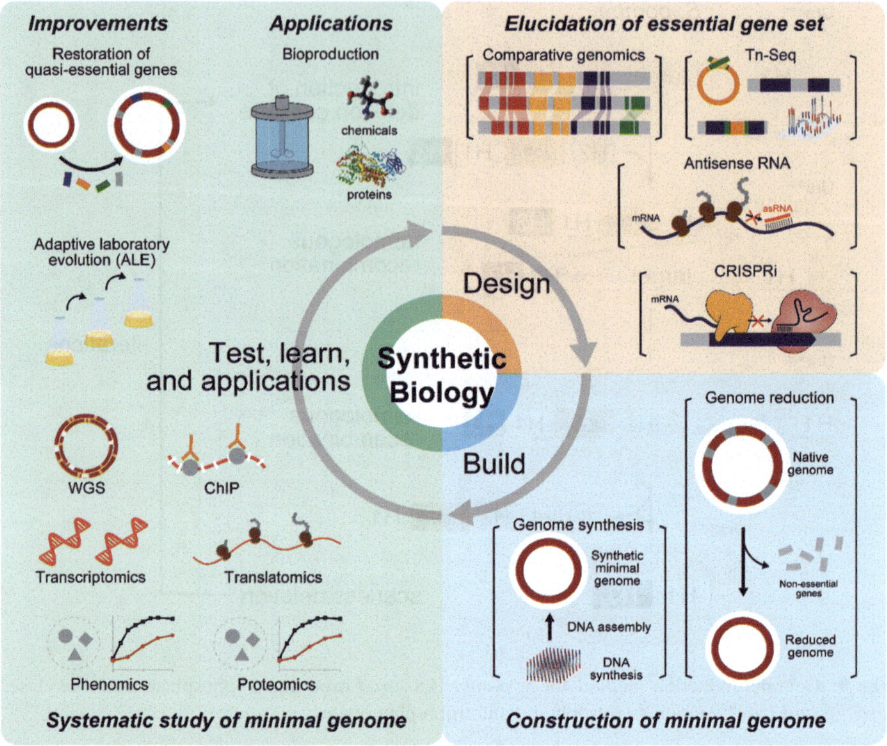

Fig. 10 Design, build, and test cycle of minimal genome research. *WGS* whole genome sequencing. *ChIP* chromatin immunoprecipitation *Tn-Seq* transposon mutagenesis coupled with next generation sequencing

Computational investigation of essential genes by comparing multiple genomes indicated that less than three hundred genes are essential. However, the comparison underestimates the number of essential genes because function of genes can be replaced by other genes. Experimental assessment using transposon mutagenesis, knockout study, asRNA, and CRISPR technique showed direct and precise information of gene essentiality. However, because the listed experimental methods used gene disruption or inactivation, combination of gene inactivation (e.g., synthetic gene lethality) could not be examined. To elucidate interconnected epistatic interactions between genes, a more delicate method is required to be invented.

Then we described characteristics of previously constructed reduced genomes with their construction methods. Even though high-throughput DNA synthesis technology has been improved, current technology level is far behind from full genome synthesis. Thus, scientists have tried to reduce pre-existing genomes to make minimal genomes. Genome reductions were accomplished by iterative deletion of the genome by homologous recombination. Because deletions needed to be accumulated multiple times, all the deletion methods include a removal step of the selective marker. Therefore, multiple deletions were accumulated in a sequential manner or sometimes combined by P1 transduction.

Most of the reduced genomes have growth rates comparable to their ancestors. Some of them have growth rate and final biomass yield even higher than their parental strains. Furthermore, the reduced genomes had advantageous characteristics such as increased transformation efficiency (Pósfai et al. 2006) and biochemical production (Lee et al. 2009). Rarely, reduced genome showed unexpected phenotypes such as growth retardation and aberrant cell morphology (Choe et al. 2019; Hashimoto et al. 2005) although genes related to growth, cell cycle, and morphology were not modified. A recent study proposed an adaptive laboratory evolution (ALE) technique to improve the growth phenotype of a reduced-genome *E. coli* (Choe et al. 2019). Multi-omics analyses of the evolved strain indicated that unbalanced metabolism induced growth retardation and transcriptome and translatomic remodeling via ALE rewired metabolic perturbation. This illustrates our incomplete understanding of gene functions, metabolism, and genomes, and thus a more thorough study on bacterial genomes is a required task to fill our knowledge gaps.

Other than genome reduction, Hutchison et al. fully synthesized the Mycoplasma minimal genome from scratch (Hutchison et al. 2016). Even though a minimal genome was synthesized, it could not support self-replicating life. Thus, authors restored genes that are not essential but necessary for robust growth, called quasi-essential genes. The synthetic minimal genome is a milestone in the field of synthetic biology and high-throughput gene synthesis technology is on its way of development (Plesa et al. 2018; Quan et al. 2011). With proven advantages, minimal genomes are a fascinating biological platform for science, industry, and many other applications. Despite that there are many hurdles remained, ever-improving cutting-edge technologies in systems and synthetic biology will overcome those challenges.

References

Ara K, Ozaki K, Nakamura K et al (2007) *Bacillus* minimum genome factory: effective utilization of microbial genome information. Biotechnol Appl Biochem 46:169–178

Baba T, Ara T, Hasegawa M et al (2006) Construction of *Escherichia coli* K-12 in-frame, single-gene knockout mutants: the Keio collection. Mol Syst Biol 2(2006):0008

Bennett GM, Moran NA (2013) Small, smaller, smallest: the origins and evolution of ancient dual symbioses in a phloem-feeding insect. Genome Biol Evol 5:1675–1688

Biswas I, Gruss A, Ehrlich SD, Maguin E (1993) High-efficiency gene inactivation and replacement system for gram-positive bacteria. J Bacteriol 175:3628–3635

Brown JR, Douady CJ, Italia MJ, Marshall WE, Stanhope MJ (2001) Universal trees based on large combined protein sequence data sets. Nat Genet 28:281–285

Charlebois RL, Doolittle WF (2004) Computing prokaryotic gene ubiquity: rescuing the core from extinction. Genome Res 14:2469–2477

Choe D, Lee JH, Yoo M et al (2019) Adaptive laboratory evolution of a genome-reduced *Escherichia coli*. Nat Commun 10:935

Christen B, Abeliuk E, Collier JM et al (2011) The essential genome of a bacterium. Mol Syst Biol 7:528

Gerdes SY, Scholle MD, Campbell JW et al (2003) Experimental determination and system level analysis of essential genes in *Escherichia coli* MG1655. J Bacteriol 185:5673–5684

Ghatak S, King ZA, Sastry A, Palsson BO (2019) The y-ome defines the 35% of *Escherichia coli* genes that lack experimental evidence of function. Nucleic Acids Res 47:2446–2454

Gibson DG, Glass JI, Lartigue C et al (2010) Creation of a bacterial cell controlled by a chemically synthesized genome. Science 329:52–56

Giga-Hama Y, Tohda H, Takegawa K, Kumagai H (2007) *Schizosaccharomyces pombe* minimum genome factory. Biotechnol Appl Biochem 46:147–155

Glass JI, Assad-Garcia N, Alperovich N et al (2006) Essential genes of a minimal bacterium. Proc Natl Acad Sci U S A 103:425–430

Harris JK, Kelley ST, Spiegelman GB, Pace NR (2003) The genetic core of the universal ancestor. Genome Res 13:407–412

Hartleb D, Jarre F, Lercher MJ (2016) Improved metabolic models for *E. coli* and *Mycoplasma genitalium* from GlobalFit, an algorithm yhat simultaneously matches growth and non-growth data sets. PLoS Comput Biol 12:e1005036

Hashimoto M, Ichimura T, Mizoguchi H et al (2005) Cell size and nucleoid organization of engineered *Escherichia coli* cells with a reduced genome. Mol Microbiol 55:137–149

Hefzi H, Ang KS, Hanscho M et al (2016) A consensus genome-scale reconstruction of chinese hamster ovary cell metabolism. Cell Syst 3:434–443 e438

Hirashima K, Iwaki T, Takegawa K, Giga-Hama Y, Tohda H (2006) A simple and effective chromosome modification method for large-scale deletion of genome sequences and identification of essential genes in fission yeast. Nucleic Acids Res 34:e11

Hutchison CA 3rd, Chuang RY, Noskov VN et al (2016) Design and synthesis of a minimal bacterial genome. Science 351:aad6253

Hutchison CA, Peterson SN, Gill SR et al (1999) Global transposon mutagenesis and a minimal *Mycoplasma genome*. Science 286:2165–2169

Itaya M (1995) An estimation of minimal genome size required for life. FEBS Lett 362:257–260

Ji Y, Zhang B, Van SF et al (2001) Identification of critical staphylococcal genes using conditional phenotypes generated by antisense RNA. Science 293:2266–2269

Kobayashi K, Ehrlich SD, Albertini A et al (2003) Essential *Bacillus subtilis* genes. Proc Natl Acad Sci U S A 100:4678–4683

Kolisnychenko V, Plunkett G 3rd, Herring CD et al (2002) Engineering a reduced *Escherichia coli* genome. Genome Res 12:640–647

Komatsu M, Uchiyama T, Omura S, Cane DE, Ikeda H (2010) Genome-minimized *Streptomyces* host for the heterologous expression of secondary metabolism. Proc Natl Acad Sci U S A 107:2646–2651

Koonin EV (2003) Comparative genomics, minimal gene-sets and the last universal common ancestor. Nat Rev Microbiol 1:127–136

Lajoie MJ, Rovner AJ, Goodman DB et al (2013) Genomically recoded organisms expand biological functions. Science 342:357–360

Lee JH, Sung BH, Kim MS et al (2009) Metabolic engineering of a reduced-genome strain of *Escherichia coli* for L-threonine production. Microb Cell Factories 8:2

Lin DC, Grossman AD (1998) Identification and characterization of a bacterial chromosome partitioning site. Cell 92:675–685

Lluch-Senar M, Delgado J, Chen WH et al (2015) Defining a minimal cell: essentiality of small ORFs and ncRNAs in a genome-reduced bacterium. Mol Syst Biol 11:780

McCutcheon JP, Moran NA (2011) Extreme genome reduction in symbiotic bacteria. Nat Rev Microbiol 10:13–26

Mizoguchi H, Sawano Y, Kato J, Mori H (2008) Superpositioning of deletions promotes growth of *Escherichia coli* with a reduced genome. DNA Res 15:277–284

Moran NA, Mira A (2001) The process of genome shrinkage in the obligate symbiont *Buchnera aphidicola*. Genome Biol 2:RESEARCH0054

Morimoto T, Kadoya R, Endo K et al (2008) Enhanced recombinant protein productivity by genome reduction in *Bacillus subtilis*. DNA Res 15:73–81

Moya A, Gil R, Latorre A et al (2009) Toward minimal bacterial cells: evolution vs. design. FEMS Microbiol Rev 33:225–235

Mushegian AR, Koonin EV (1996) A minimal gene set for cellular life derived by comparison of complete bacterial genomes. Proc Natl Acad Sci U S A 93:10268–10273

Nijman SM (2011) Synthetic lethality: general principles, utility and detection using genetic screens in human cells. FEBS Lett 585:1–6

Nolan RP, Lee K (2011) Dynamic model of CHO cell metabolism. Metab Eng 13:108–124

Orth JD, Conrad TM, Na J et al (2011) A comprehensive genome-scale reconstruction of *Escherichia coli* metabolism--2011. Mol Syst Biol 7:535

Park MK, Lee SH, Yang KS et al (2014) Enhancing recombinant protein production with an *Escherichia coli* host strain lacking insertion sequences. Appl Microbiol Biotechnol 98:6701–6713

Plesa C, Sidore AM, Lubock NB, Zhang D, Kosuri S (2018) Multiplexed gene synthesis in emulsions for exploring protein functional landscapes. Science 359:343–347

Pósfai G, Plunkett G 3rd, Feher T et al (2006) Emergent properties of reduced-genome *Escherichia coli*. Science 312:1044–1046

Qi LS, Larson MH, Gilbert LA et al (2013) Repurposing CRISPR as an RNA-guided platform for sequence-specific control of gene expression. Cell 152:1173–1183

Quan J, Saaem I, Tang N et al (2011) Parallel on-chip gene synthesis and application to optimization of protein expression. Nat Biotechnol 29:449–452

Reuß DR, Altenbuchner J, Mader U et al (2017) Large-scale reduction of the *Bacillus subtilis* genome: consequences for the transcriptional network, resource allocation, and metabolism. Genome Res 27:289–299

Richardson SM, Mitchell LA, Stracquadanio G et al (2017) Design of a synthetic yeast genome. Science 355:1040–1044

Riley M, Abe T, Arnaud MB et al (2006) *Escherichia coli* K-12: a cooperatively developed annotation snapshot--2005. Nucleic Acids Res 34:1–9

Rousset F, Cui L, Siouve E et al (2018) Genome-wide CRISPR-dCas9 screens in *E. coli* identify essential genes and phage host factors. PLoS Genet 14:e1007749

Shalem O, Sanjana NE, Hartenian E et al (2014) Genome-scale CRISPR-Cas9 knockout screening in human cells. Science 343:84–87

Sleator RD (2010) The story of *Mycoplasma mycoides* JCVI-syn1.0: the forty million dollar microbe. Bioeng Bugs 1:229–230

Srinivasan R, Scolari VF, Lagomarsino MC, Seshasayee AS (2015) The genome-scale interplay amongst xenogene silencing, stress response and chromosome architecture in *Escherichia coli*. Nucleic Acids Res 43:295–308

Wang T, Wei JJ, Sabatini DM, Lander ES (2014) Genetic screens in human cells using the CRISPR-Cas9 system. Science 343:80–84

Weaver DS, Keseler IM, Mackie A, Paulsen IT, Karp PD (2014) A genome-scale metabolic flux model of *Escherichia coli* K-12 derived from the EcoCyc database. BMC Syst Biol 8:79

Westers H, Dorenbos R, van Dijl JM et al (2003) Genome engineering reveals large dispensable regions in *Bacillus subtilis*. Mol Biol Evol 20:2076–2090

Yu BJ, Sung BH, Koob MD et al (2002) Minimization of the *Escherichia coli* genome using a Tn5-targeted Cre/loxP excision system. Nat Biotechnol 20:1018–1023

Engineering Reduced-Genome Strains of *Pseudomonas putida* for Product Valorization

Nicolas T. Wirth and Pablo I. Nikel

Abstract Environmental bacteria, such as strains of the genus *Pseudomonas*, constitute ideal starting points for the design of robust cell factories. These microorganisms are pre-endowed with a number of metabolic and stress-endurance traits that make them optimal for the needs of contemporary biotechnology. Significant technological advances in recent times opened new avenues for metabolic engineering of *Pseudomonas* species. Against this background, in this chapter we discuss the current engineering efforts aimed at launching the Gram-negative soil bacterium *P. putida* as a *chassis* for product valorization and refinement. We focus on the use of reduced-genome strains of *P. putida*, endowed with enhanced physiological characteristics (e.g., increased availability of ATP and NADPH, the energy and redox currencies of the cell), for the construction of bacterial cell factories that can be used across a range of operating conditions. Cutting-edge synthetic biology approaches for genome engineering, which significantly reduced the time needed for the construction of such reduced-genome variants of *P. putida*, are likewise discussed. We conclude the chapter by discussing future trends and bottlenecks toward the establishment of a minimal-genome chassis based on *P. putida*.

Keywords Synthetic biology · Metabolic engineering · Minimal genome · *Pseudomonas putida* · Biotechnology · *Chassis*

N. T. Wirth · P. I. Nikel (✉)
The Novo Nordisk Foundation Center for Biosustainability, Technical University of Denmark, Lyngby, Denmark
e-mail: pabnik@biosustain.dtu.dk

A. R. Lara, G. Gosset (eds.), *Minimal Cells: Design, Construction, Biotechnological Applications*, https://doi.org/10.1007/978-3-030-31897-0_3

1 The Potential of *Pseudomonas putida* as a *Chassis* for Biotechnology

1.1 A Historical Perspective on Microbial Chassis *for Biotechnology Applications*

The microorganisms most frequently used as cell factories in microbial fermentation processes are species that have historically been adopted as model organisms in the laboratory (Calero and Nikel 2019), such as the Gram-negative bacterium *Escherichia coli* or the archetypal baker's yeast *Saccharomyces cerevisiae*—together with a few other microbial species that had been exploited for their natural capacity to produce a defined set of compounds (Beites and Mendes 2015). Examples of this sort include amino acids [e.g., *Corynebacterium glutamicum* (Baritugo et al. 2018)], antibiotics and other antimicrobials [e.g., Actinomycetes (Palazzotto et al. 2019)], proteins [e.g., *Bacillus subtilis* (Gu et al. 2018)], and the methylotrophic yeast *Pichia pastoris* (Yang and Zhang 2018; Peña et al. 2018). Before the inception of technological advances in biological sciences (and, in particular, genetic and genome engineering approaches), most of the strain engineering efforts were heavily reliant on the limited knowledge (i.e., physiology, genetics, and biochemistry) that had been gathered laboriously over the second half of the twentieth century for a modest number of organisms. Success stories of metabolic engineering have thus often focused on the overproduction of natively synthesized metabolites or molecules that are easily accessible from the extant metabolic network [i.e., *cis*-metabolism (Nikel and de Lorenzo 2018)] by the addition of simple biochemical pathways.

More recently, the field of metabolic engineering had witnessed a significant increase in the number and nature of the potential microbial *chassis* that can be used for biotechnological purposes (Danchin 2012; Kim et al. 2016). This occurrence stems from the fact that not only has the knowledge on alternative microbial species expanded enormously, but also the technology bottlenecks for engineering their core properties have been (for the most part) solved. Achievements within the "omics" fields enabled us to gain the required knowledge about essentially any microbial platform on a much shorter time-scale—and synthetic biology has provided us with tools to manipulate an unprecedented number of living cells at will (Abram and Udaondo 2019). When organisms were previously chosen due to the availability of genetic engineering tools or their innate ability to produce a desired compound, we are now experiencing a shift in paradigm: one first tries to establish the characteristics that an *ideal* biological platform should display for the desired application, e.g., the production of a given target product (including adverse metabolic effects of genetic implants and a suitability for a specific set of production conditions). Then, a suitable *chassis* can be chosen that *naturally* meets the defined criteria in the best possible way and is then engineered to fulfill its task. This led to an increased interest in species that are true generalists in their nature and that are often found as free-living organisms in diverse environments. As indicated in Sect. 1.2, many of such environmental microorganisms are found within the genus *Pseudomonas* and herein, particularly *Pseudomonas putida* has made a name for itself as a promising cell factory.

1.2 Bacteria from the Genus Pseudomonas *as* Chassis

Representatives of the species *Pseudomonas putida* are classified as Gram-negative, obligate aerobic γ-proteobacteria with one or several polar flagella that are found in most soil and water habitats where oxygen is available (Timmis 2002; Palleroni 2010; Palleroni et al. 1973). Members of these bacteria can grow at temperatures between 4 and 40 °C (although they grow optimally between 25 and 30 °C; Poblete-Castro et al. 2017), and they can form biofilms (Volke and Nikel 2018; Benedetti et al. 2016), through a process coordinated by cell-to-cell-communication systems (i.e., quorum sensing) in a cell density-dependent manner (Moore et al. 2006). Due to the ever-changing nature of their natural habitat, *P. putida* is endowed with a versatile metabolic network that provides it with the ability to adapt to many different physicochemical conditions, some of which are of special interest for industrial fermentation setups (e.g., extreme pH values, temperature gradients, and high concentrations of toxic substances) (Poblete-Castro et al. 2012, 2019; Jiménez et al. 2002). In particular, one of its most remarkable physiological traits was the very reason why its most prominent representative, *P. putida* strain mt-2, was discovered in the first place: the ability to degrade and consume recalcitrant xenobiotic compounds, such as toluene and xylenes (Nakazawa 2002)—and this trait goes hand in hand with a remarkably high tolerance to organic solvents (Segura et al. 2012). This outstanding feature results from the presence of, inter alia, very effective efflux systems that make *P. putida* an ideal producer of such compounds or a suitable biocatalyst for production processes in two-phase systems (Heipieper et al. 2007; Simon et al. 2014; Blank et al. 2008). The ability to tolerate high concentrations of organic solvents and aromatic compounds furthermore allows for the use of complex alternative feedstocks that are not accessible for other bacteria (Nikel and de Lorenzo 2018). An outstanding example in this sense is the wealth of substrates that can be derived from plant biomass, a large fraction of which is represented by lignin. Because bioproduction processes established thus far exploited lignocellulose predominantly as a source of sugars, the lignin fraction remains to be a largely unused waste product that is often used for the production of heat and electric energy via combustion. *Pseudomonas*, however, is able to consume a large variety of the aromatic monomers that are derived from lignin after hydrolysis—even enabling the non-hierarchical co-utilization of these substrates (Beckham et al. 2016). Moreover, although *P. putida* is not able to efficiently consume the most abundant sugars present in lignocellulose besides glucose, xylose, and arabinose, it can be easily engineered to do so (Wang et al. 2019; Dvořák and de Lorenzo 2018). As a consequence, it is now possible to utilize a large proportion of the carbon that is stored in plant biomass for the production of value-added compounds in fermentation approaches. Another possibility for the valorization of complex feedstocks containing arenes is to make direct use of the assimilatory pathways that are present within the *cis*-metabolism of *P. putida*. The well-studied *meta*- and *ortho*-cleavage routes to degrade a range of aromatic compounds (Marqués and Ramos 1993) converge into the central metabolic intermediate catechol, which is subsequently converted into *cis,cis*-muconic acid [(2*E*,4*E*)2,4-hexanedioic acid] (Chua and Hsieh 1990; Vardon et al. 2015; Johnson et al. 2016; Kohlstedt et al. 2018). The latter molecule has recently gained significant attention in the chemical industry as a platform chemical, representing a precursor for the synthesis of terephthalic acid, 3-hexenedioic acid, 2-hexenedioic acid, 1,6-hexanediol, ε-caprolactam, and ε-caprolactone; all of which serve as building blocks for value-added commercial polymers [e.g., polyethylene terephthalate (PET) from terephthalic acid or Nylon-6,6 from adipic acid]. Recently, the substrate spectrum that is suitable to produce *cis,cis*-muconic acid from aromatic precursors in *P. putida* was further expanded by introducing heterologous functions that feed the catechol branch with intermediates from degradation routes that produce protocatechuic acid as central metabolic intermediate. These operations resulted in a complex converging metabolic network that converts a diverse mix of aromatic substrates into catechol and consequently *cis, cis*-muconic acid (Linger et al. 2014). As discussed in Sect. 1.3, engineering the synthesis of this type of compounds (among many others) in *Pseudomonas* species is possible largely due to a specific metabolic wiring characteristic of this species.

1.3 Central Carbon Metabolism in Pseudomonas putida

Pseudomonas putida represents a striking example of a bowtie framework (Sudarsan et al. 2014), where a large variety of different nutrients are assimilated through a *catabolic funnel* into a core metabolic network to produce activated carriers and precursor metabolites for the synthesis of larger building blocks that constitute biomass (Fig. 1). The central carbon metabolism of *P. putida* exhibits a rather special architecture when it is exposed to the sugar substrate that is widely preferred in industry, glucose (Fig. 1). Fructose is the only carbohydrate known to be transported into the cytoplasm through a phosphoenolpyruvate-carbohydrate phosphotransferase system (PTS) system (Chavarría et al. 2016). All other sugars use active non-PTS transport systems, often at the costs of a higher energy demand. As many other (aerobic) prokaryotes, most pseudomonads lack a functional 6-phosphofructo-1-kinase (Pfk), a key enzyme of the Embden–Meyerhof–Parnas (EMP) glycolytic pathway, and thus rely on the Entner–Doudoroff (ED) route for the consumption of sugars and sugar acids (Flamholz et al. 2013; Conway 1992). The initial steps of glucose consumption occur through a set of three pathways that converge in 6-phosphogluconate (6PG) as a central node, and each includes substrate oxidation (generating reduced cofactors) and ATP-dependent phosphorylation steps. The activated intermediate 6PG is then subjected to dehydration (catalyzed by Edd) and aldol cleavage (catalyzed by Eda) that yield the two C3 intermediates pyruvate and glyceraldehyde-3-*P* (GA3P). GA3P can then be further processed to pyruvate through the lower block of glycolysis (yielding one ATP and one NADH) or be recycled through a cyclic metabolic route termed the *EDEMP cycle* [a metabolic architecture involving enzymes belonging to the ED, EMP, and pentose phosphate (PP) pathways]. This recycling of metabolites enables *Pseudomonas* to

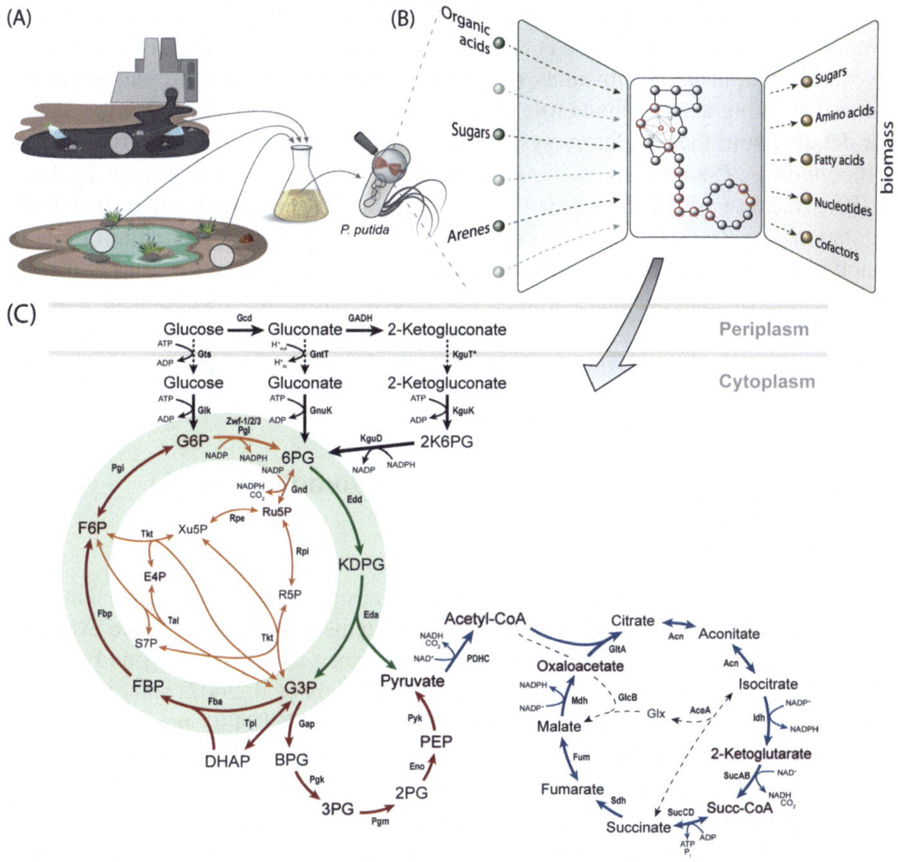

Fig. 1 Relevant metabolic characteristics of the Gram-negative environmental bacterium *Pseudomonas putida* as a *chassis* for biotechnology. (**a**) Strains of *P. putida* can be isolated in most soil and water habitats where oxygen is available, including wastewater reservoirs from chemical plants. (**b**) The metabolic network of *P. putida* KT2440 represents a striking example of a *bowtie framework*, where a large variety of different substrates are funneled through catabolic pathways into a set of central biochemical reactions that produce activated energy/redox carriers and precursor metabolites (the 12 elemental metabolites in this network (Noor et al. 2010) are highlighted in red) for the formation of biomass. (**c**) Biochemical pathways involved in glucose catabolism in *P. putida* KT2440. The metabolic flux revolves around the *EDEMP cycle* (highlighted in green) that includes activities from the Entner–Doudoroff pathway (reactions shown in dark green), the pentose phosphate pathway (reactions shown in orange), and the Embden–Meyerhof–Parnas route (reactions shown in red) (Nikel et al. 2015). Reactions of the tricarboxylic acid (TCA) cycle are indicated in blue. Transport reactions across the plasma membrane are shown with dashed lines. Note that the exact transport mechanism for 2-ketogluconate is unknown (shown with a dashed arrow). Abbreviations of metabolites are as follows: *G6P* glucose-6-*P*, *6PG* 6-phosphogluconate, *F6P* fructose-6-*P*, *6PG* 6-phosphogluconate, *FBP* fructose-1,6-P_2, *DHAP* dihydroxyacetone-*P*, *KDPG* 2-keto-3-deoxy-6-phosphogluconate, *Ru5P* ribulose-5-*P*, *R5P* ribose-5-*P*, *S7P* sedoheptulose-7-*P*, *E4P* erythrose-4-*P*, *G3P* glyceraldehyde-3-*P*, *3PG* glycerate-3-*P*, *2PG* glycerate-2-P, *PEP* phosphoenolpyruvate and *acetyl-CoA* acetyl-coenzyme A. Enzymes in this scheme are abbreviated as follows: *Gcd* glucose dehydrogenase, *GADH* gluconate dehydrogenase, *GtsABCD* D-glucose ABC-transporter, *GntT* gluconate/H^+ symporter, *KguT* putative

generate additional NADPH reducing power and is a key mechanism to counter high levels of oxidative stress (Nikel et al. 2015). A complex regulatory network leads to a sequential uptake of substrates if several carbon sources are present, whereby glucose is taking a subordinate role in favor of some organic acids and amino acids. For details about the regulatory systems that orchestrate the use of different carbon compounds in *Pseudomonas*, we recommend a previous review written by Rojo (2010). Because of the complexity and idiosyncrasies that characterize metabolic regulation in *Pseudomonas* species, a lack of knowledge (particularly about genotype–phenotype relationships) often becomes a bottleneck if one tries to predict the outcome of genetic interventions. As argued below, the inception of a set of dedicated tools for the engineering of *Pseudomonas* have emerged and considerably accelerated the design of whole-cell biocatalysts based on this species.

1.4 Toward the Domestication of **Pseudomonas putida** *as a* **Chassis**

Since the beginning of the twenty-first century, various modern techniques that were developed in the field of synthetic as well as systems biology were made available for *P. putida* and contributed to its broad acceptance as a promising and reliable biotechnological platform. The fully sequenced (Nelson et al. 2002) and extensively annotated (Belda et al. 2016) genome of *P. putida* KT2440 provided a solid fundament for the integration of several systems biology applications and enabled the defined manipulation of phenotypical traits useful for biotechnological purposes. It furthermore formed the basis to a set of genome-scale metabolic models that were regularly used to guide metabolic engineering strategies and helped elucidating the physiologic response of *P. putida* to different environments and physicochemical conditions (Nogales et al. 2008; Puchałka et al. 2008; Sohn et al. 2010; Oberhardt et al. 2011). Along the same lines, genetic amenability facilitated the adaptation and development of increasingly refined synthetic biology tools developed for specific gene and genome manipulations. A milestone for establishing a standardized

---

Fig. 1 (continued) 2-ketogluconate transporter, *Glk* glucose kinase, *GnuK* gluconate kinase, *KguK* 2-ketogluconate kinase, *KguD* 2K6PG reductase, *Pgi* phosphohexose isomerase, *Zwf* glucose-6-*P* dehydrogenase, *Pgl* phosphogluconolactonase, *Edd* 6-phosphogluconate dehydratase, *Eda* 2-dehydro-3-deoxy-6-phosphogluconate aldolase, *Fba* fructose-1,6-P_2 aldolase, *Fbp* fructose-1,6-P_2 phosphatase, *Gnd* phosphogluconate dehydrogenase, *Rpi* ribose-5-*P* isomerase, *Rpe* ribulose-5-*P* 3-epimerase, *Tkt* transketolase, *Tal* transaldolase, *Tpi* triosephosphate isomerase, *Gap* glyceraldehyde-3-*P* dehydrogenase, *Pgk* phosphoglycerate kinase, *Pgm* phosphoglycerate mutase, *Eno* enolase, *Pyk* pyruvate kinase, *PDHC* pyruvate dehydrogenase complex, *GltA* citrate synthase, *Acn* aconitase, *Idh* isocitrate dehydrogenase, *SucAB* 2-ketoglutarate dehydrogenase, *SucCD* succinyl-CoA synthetase, *Sdh* succinate dehydrogenase, *Fum* fumarate hydratase, *Mdh* malate dehydrogenase, *AceA* isocitrate lyase, and *GlcB* malate synthase

toolbox of well-characterized molecular tools has been the creation of the *Standard European Vector Architecture* (SEVA) platform (Martínez-García et al. 2014a; Silva-Rocha et al. 2013), containing various genetic elements (e.g., promoters, antibiotic resistance determinants, origins of plasmid replication, and reporter functions) built in a modular and easily interchangeable fashion. The inducible promoter systems within the SEVA collection were further complemented with synthetic libraries of constitutive promoters that allow for the predictable adjustment of the transcription strength of any desired gene and gene clusters (Elmore et al. 2017; Zobel et al. 2015). However, the strength of a promoter alone does not determine the overall expression strength of a gene, since translation heavily depends on the sequence context within the messenger RNA around the ribosome-binding site. This issue can be solved by exploiting the mechanism of translational coupling, which uses standardized, non-coding 5′-untranslated regions and leader sequences that are placed in front of a gene of interest, as firstly established for *E. coli* (Mutalik et al. 2013) and later demonstrated for *P. putida* as well (Zobel et al. 2015).

The plethora of tools and approaches for gene and genome manipulation of this species paved the way toward genetic domestication of *P. putida* as a *chassis* for biotechnology. However, despite considerable progress in our ability to access genetic manipulations of *Pseudomonas*, this bacterium still shows deficits when it comes to high-throughput genome engineering. This state of affairs can be explained to some extent by present knowledge gaps compared to the microbial forerunners of synthetic biology. More importantly, *P. putida* lacks the high capacity for homologous recombination (HR) displayed by other well-established *chassis*, e.g., *S. cerevisiae* or *B. subtilis* (Davy et al. 2017), or an inventory of well-characterized bacteriophages, e.g., phages characterized for *E. coli* (Li et al. 2019), that, together, provide the molecular tools for an efficient recombineering system. Since the availability of such systems would be instrumental for the development of veritable reduced-genome variants of *P. putida* (leading to the long-sought-after goal of constructing minimal-genome strains), several alternatives have emerged over the last few years to overcome the technical limitations in genome engineering. In the next section of the chapter (Sect. 2), we review state-of-the-art techniques for genome manipulation in *Pseudomonas* species, highlighting the advantages and disadvantages of these procedures.

2 Strategies for Genome Reduction of *Pseudomonas* Species

The knowledge that was accumulated over the time with a growing body of research helped to identify cellular functions within *P. putida* that are either neutral or disadvantageous for the construction of a robust cell factory. Examples of such functions will be addressed in further detail in the next section (Sect. 3). Genetic determinants of such "dispensable" functions—although the concept of dispensability here would largely depend on the intended application of the resulting strains—can be found clustered together or highly dispersed over the bacterial chromosome. The method of choice to erase several stretches of DNA at different locations should thus either enable the editing of multiple regions at once or require as little time as possible for each iteration in a stepwise approach—while at the same time being insensitive to the length of the removed sequence. Driven by the revolution of synthetic biology, different approaches have been designed, attempting to provide reliable, quick, and easy ways to implement defined genetic modifications in *P. putida*. Originally, these methods exclusively relied on native HR functions mediated by *recA*, which enables the allelic exchange of a plasmid-encoded sequence with a homologous counterpart on the bacterial chromosome (as discussed in Sect. 2.1). However, one disadvantage inherent to the HR-mediated mechanism is its impartiality regarding the result of the recombination. In some cases, this occurrence can increase the screening efforts to identify the desired mutant cells in a (sometimes, larger) population of wild-type cells with an increased fitness. This technical weakness could be overcome with the advent of the CRISPR/Cas9 technologies that enables the specific counter-selection against the unwanted (i.e., wild-type) genotype (as explained in Sect. 2.2).

2.1 *Genome Engineering of* Pseudomonas *Using Suicide Plasmids*

In the most established genome engineering methods, the template DNA sequences needed for efficient HR are delivered into *Pseudomonas* encoded on suicide plasmids carrying selectable markers (e.g., antibiotic resistance determinants). These template DNA sequences include the desired genetic changes flanked on each side by DNA segments (typically ca. 500 bp) that are homologous to the target site of the genome. In this context, a (conditional) suicide vector is a plasmid carrying an origin of vegetative replication (*oriV*) which is natively not functional in its target host. The most prominent and widely used example of a suicide *ori* in *P. putida* represents that of the plasmid R6K, whose functionality relies on the presence of its cognate replication factor π, encoded by the *pir* gene (Rakowski and Filutowicz 2013). The advantage of such a vector is that bacterial strains harboring *pir* can be used without limitations for the cloning and propagation of the plasmid, while survival of *Pseudomonas* under selection pressure depends on the integration of antibiotic resistance determinants into the chromosome. This is mediated by HR of the plasmid-encoded inserts and the target genomic sequence (Fig. 2). This allelic exchange can be performed within a single step (double crossover), where the two homologous arms (HAs) flanking a mutagenesis region recombine at the same time, or in two consecutive steps (i.e., insertion followed by excision). Hereby, the whole plasmid sequence is co-integrated into the chromosome through a single crossover event at either of the two individual homologous regions (HA1 and HA2). A second HR leads to the resolution of the plasmid and can leave behind the desired genetic modification. For systems relying on a double-crossover mechanism, a selectable marker is firstly introduced between the two homology regions that can later be

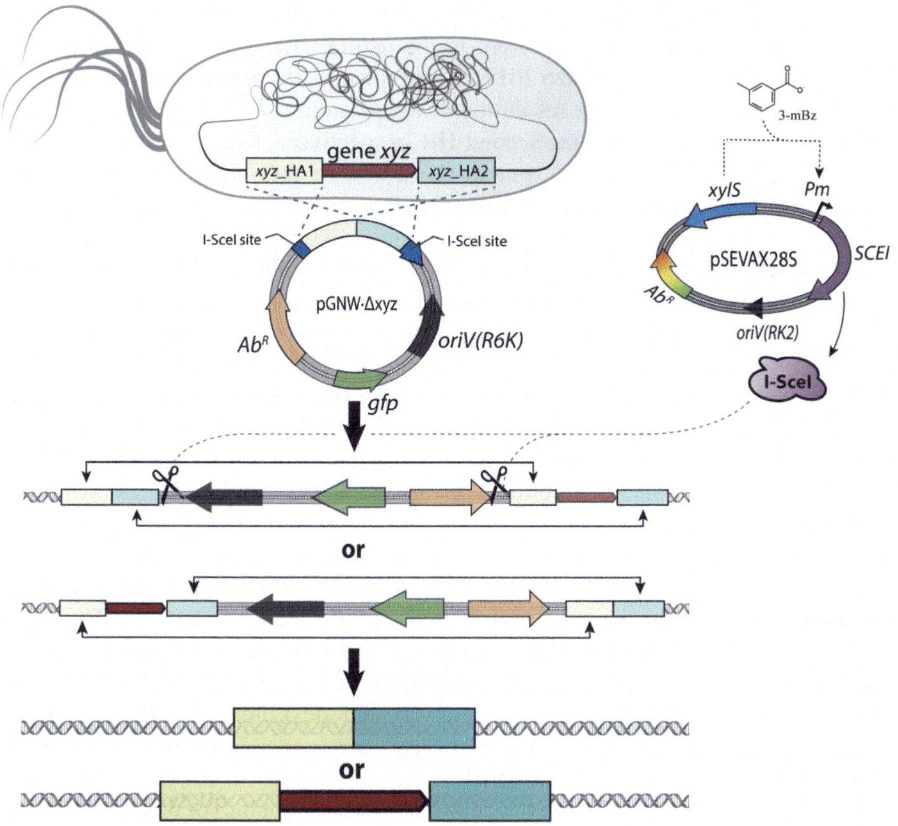

Fig. 2 Genome engineering of *Pseudomonas* using suicide plasmids. The molecular mechanisms that lead to the desired chromosomal modification are shown for the deletion of gene *xyz* with a recently published method (Wirth et al. 2019). Two homology arms (HA) that flank the target mutagenesis site are cloned into the plasmid pGNW, which acts as a suicide vector in *P. putida*. The plasmid is co-integrated into the chromosome through RecA-mediated homologous recombination at either of the two HAs, thereby conferring a selectable antibiotic resistance as well as a fluorescence signal to identify transformed cells. Introduction of a replicable plasmid [e.g., vector pSEVA628S, harboring the gene encoding the homing endonuclease I-*Sce*I, the expression of which is inducible upon addition of 3-methylbenzoate (3-mBz)] results in the meganuclease-dependent cuts on the two I-*Sce*I target recognition sites located within pGNW thereby introducing DNA double strand breaks in the genome. To avoid the lethal cutting mediated by I-*Sce*I, the cells undergo a second homologous recombination, which yields either the desired mutant or a revertant genotype, depending on which of the two HRs recombine

removed by means of the site-specific recombination systems Flp/*FRT* from yeast (Hoang et al. 1998) or Cre/*lox* from phage P1 (Ayres et al. 1993; Marx and Lidstrom 2002). These systems have the inherent disadvantage of leaving behind scars within the chromosome at the sites of recombination, which, after repeated use of the system for multiple genome interventions, are prone to recombine with each other, potentially leading to the deletion or inversion of large genomic segments.

For two-step based genome engineering approaches, one predominant problem was the low efficiency that is inherent in the natural HR system of *Pseudomonas* (Martínez-García and de Lorenzo 2017). Consequently, different methods have been exploited either to enforce the resolution of the plasmid or to counter-select against cells, which did not undergo the second HR, respectively. One of the first counter-selection strategies adapted to *P. putida* was based on the *sacB* gene derived from *Bacillus subtilis*, encoding for a levansucrase that acts on sucrose to produce levan polysaccharides, which accumulate in the periplasm of Gram-negative bacteria where they exert a toxic effect (Steinmetz et al. 1983). The resolution of a suicide plasmid encoding *sacB* can therefore be selected for by growth in a sucrose-containing medium. A similar approach is based on the *upp* gene whose natural function in *P. putida* is to phosphoribosylate uracil to yield uridine monophosphate (UMP, 5′-uridylic acid), which is a required step in the pyrimidine salvage pathway. Upp furthermore acts on the antimetabolite 5-fluorouracil (5-FU) that is thereby converted into 5-fluoro-UMP. After metabolic transformation of 5-FU into 5-fluoro-UMP, this compound acts as a suicide inhibitor of the essential enzyme thymidylate synthase, resulting in cell death. Cells deficient in the uracil phosphoribosyltransferase function show no physiological effect after exposure to 5-FU. After deletion of the non-essential *upp* in *P. putida*, the gene can be used as a counter-selectable marker for molecular tools by merely adding 5-FU to the culture medium (Graf and Altenbuchner 2011). Equally working on the biosynthesis of pyrimidine nucleobases is a dual-selection system based on a deletion of the gene *pyrF*, combined with the antimetabolite 5-fluoroorotidine-5′-*P* (Galvão and de Lorenzo 2005). The gene product of *pyrF* (orotidine-5′-*P* decarboxylase) is responsible for the formation of UMP from orotidine-5′-*P* (OMP), which makes it essential for the de novo biosynthesis of uracil. With a *pyrF* deletion, cells become auxotrophic for uracil and can be "rescued" via the ectopic expression of *pyrF* from a plasmid (positive selection). At the same time, OMP decarboxylase acts on 5-fluoroorotidine-5′-*P* (5-FOMP) and converts it into 5-fluoro-UMP with the same toxic effect as described above, allowing for a negative selection of genetic elements carrying *pyrF*. However, all counter-selection methods have shown to display shortcomings. The compound that is externally added for negative selection often does not suppress growth completely, making it difficult to identify cells that performed the second HR, especially when trying to eliminate (conditionally) essential functions whose deletion results in a growth deficiency. The *upp*/5-FU and the *pyrF*/5-FOMP methods furthermore require the deletion of relevant metabolic functions in the host, which is not desired in a robust cell factory. Finally, all the counter-selection methods that are based on the action of a single gene are prone to mutate, rendering them ineffective over repeated use of the markers and key genetic elements. In principle, all these drawbacks could be solved by establishing counter-selection techniques based on highly specific endonucleases that are produced endogenously. A widely adopted method uses a suicide plasmid harboring the conditional origin of replication R6K, an antibiotic resistance determinant and a polylinker region (i.e., multiple cloning site) flanked by two recognition sequences for the homing nuclease I-*Sce*I from *S. cerevisiae* (Martínez-García and de Lorenzo 2011; Wirth et al. 2019). Upon co-integration of the plasmid via homologous recombination, the *SCEI* gene is expressed from an additional helper plasmid and the meganuclease produced thereof introduces double-stranded breaks at both target sites encoded on the integrative plasmid. Cells can escape this lethal cleavage within the chromosome if they get rid of the I-*Sce*I recognition sites by a second crossover event (Fig. 2). The redundancies of the target sequences, as well as the *SCEI* gene, which is encoded on plasmids that are maintained at multiple copy numbers, make this system highly efficient and robust. A remaining disadvantage of all counter-selection systems discussed in this section is their limited selectiveness regarding the resulting DNA sequence in the chromosome. In order to introduce a change in the genetic sequence, the modified target region is flanked by two homologous arms (HA1 and HA2, Fig. 2). Each of the two crossover events that first lead to the co-integration of the suicide plasmid and then to its resolution can be performed by either of the two HAs. Only if the site of recombination differs between the first and the second crossover, a mutant genotype is created (Fig. 2). If both HR events involve the same HA, the wild-type sequence is restored.

2.2 *CRISPR/Cas9 Technologies and Recombineering Approaches for* Pseudomonas

Until recently, established genome engineering strategies were not able to discriminate between mutant and wild-type cells unless a genetic manipulation resulted in a clear selectable phenotype. On the contrary, the introduction of new biochemical functions or the deletion of genes with a relevant physiological role often results in an evolutionary advantage for the wild-type cells in such a way that often significant screening efforts had to be made for selecting the mutants. This problem has been largely solved when customized CRISPR/Cas9 systems were made accessible after their first introduction in 2012 (Jinek et al. 2012; Gasiunas et al. 2012; Jakočiūnas et al. 2017). Cas9 serves as an endonuclease that can be guided to a specific DNA sequence through its association with a synthetic guide RNA (sgRNA) and is therefore suitable to select *against* particular genotypes. Shortly after its emergence, the new technology was combined with traditional molecular tools for genetic engineering to create new, powerful gene edition systems that further facilitated engineering approaches in several cell factories (Fernández-Cabezón et al. 2019).

Different approaches for *P. putida* intended to perform chromosomal modifications within a single step (Fig. 3), by using CRISPR/Cas9 counter-selection to stimulate the allelic exchange of a sequence from either a plasmid via a double-crossover (Cook et al. 2018) or with a synthetic DNA fragment (either single or double stranded) in a process called *recombineering* (Choi et al. 2018; Aparicio et al. 2018). Indeed, it has been established in *E. coli* that one can simply insert modifications into any chromosomal target site using linear, double-stranded (ds) DNA (e.g., PCR products), by flanking fragments with short (typically 40–50 bp) homologous arms and delivering them into cells that overexpress the three genes of the Red

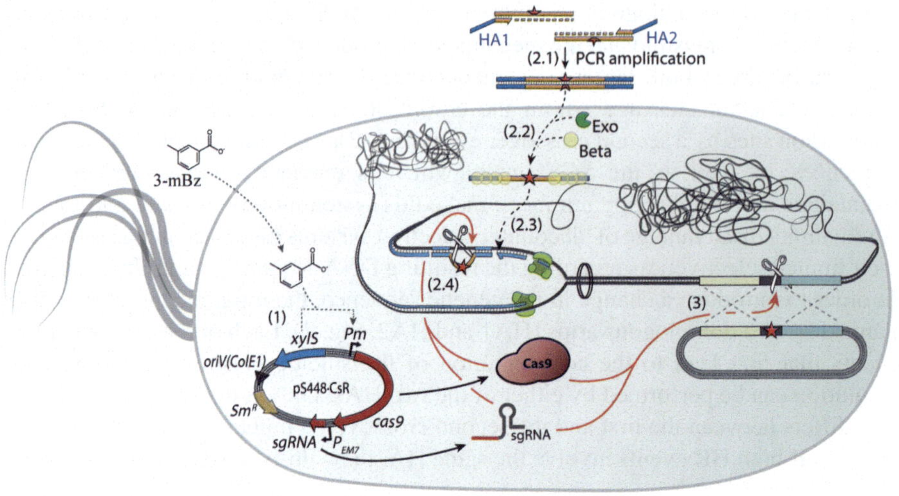

Fig. 3 Genome engineering applications in *Pseudomonas* species using CRISPR/Cas9 as counter-selection system. A plasmid-based CRISPR/Cas9 expression system is shown in the scheme with vector pS448·CsR as an example, as described by Wirth et al. (2019). This plasmid encodes a streptomycin-resistance gene (Sm^R), the *cas9* gene under the control of the inducible XylS/*Pm* expression system as well as a constitutively-expressed (with the P_{EM7} promoter), customizable cassette including a specific synthetic guide RNA (sgRNA). Autonomous replication of this vector is controlled by the ColE1 origin of vegetative replication [*oriV*(ColE1)]. The steps needed for genome engineering of *Pseudomonas* using this system are as follows: (1) upon induction of the system *via* addition of 3-methylbenzoate (3-mBz), Cas9 is produced and the protein associates with the sgRNA to introduce a double-stranded (ds) break in the DNA at the target locus that is lethal for cells harboring the wild-type sequence. This counter-selecting action of CRISPR/Cas9 can be combined with methods of recombineering (2). Thereby, a linear DNA fragment [e.g., dsDNA obtained *via* PCR (2.1) or a chemically synthesized single-stranded (ss) DNA fragment, both containing homologous arms (HA) to the target locus on each end] is used to introduce genetic modifications (indicated with a red star) at a target site. If dsDNA is used (2.2), one of the two DNA strands can be digested with an exonuclease (e.g., Beta and Exo, from the λ Red system commonly used in *E. coli*), leaving a ssDNA fragment that is protected from further degradation by a ssDNA-binding protein (e.g., Beta, SSR, or RecT). The two flanking HAs mediate the introduction of the mutagenic ssDNA fragment into the chromosome *via* homologous recombination or as an Okazaki fragment during the replication of the bacterial chromosome (2.3). Moreover, CRISPR/Cas9 counter-selection of the wild-type sequence allows for the allelic exchange of a homologous DNA fragment containing the desired modifications directly from a plasmid *via* a double-crossover mechanism (3). If the sgRNA target is changed during this mutagenesis procedure, only the cells having the allelic exchange can survive in the presence of both Cas9 and the expressed sgRNA

system from phage λ (Murphy 2016). These genes encode (1) a $5' \rightarrow 3'$ exonuclease (Exo) that digests one of the two strands on the DNA fragment, (2) a dsDNA-binding protein (Gam) that protects the introduced fragment from the attack of endogenous nucleases, and (3) a single-stranded (ss) DNA-binding protein (Beta) that shields the single-stranded product from degradation and mediates recombination (Fig. 3). It was also shown that recombineering approaches work even more efficiently when using an ssDNA fragment (e.g., a synthetized oligonucleotide)—thereby requiring only the function of protein Beta (Ellis et al. 2001). This method can generate up to 6% of recombinant cells within a single round of treatment without any means for selection. Traditionally, λ Red recombineering is often used in a two-step approach, similar to strategies based on suicide plasmids, where an antibiotic resistance gene is first introduced at the target site together with a counter-selectable marker. With the second round of recombineering, using a different DNA fragment, the inserted gene cassette is removed, leaving behind only the desired mutation (i.e., scarless modification). Moreover, λ Red recombineering can be combined with CRISPR/Cas9 counter-selection, which enabled the introduction of any genetic change within a single step supplying only linear DNA fragments. This technology is particularly powerful when performed multi-cycled and in an automated way, e.g., with the *CRMAGE* method (*C*RISPR/Cas9 and λ *R*ed recombineering based *M*ultiplex *A*utomated *G*enome *E*ngineering) (Ronda et al. 2016).

In spite of its many advantages, all approaches employing CRISPR/Cas9 technologies suffer from a lack of understanding about the characteristics that qualify an effective sgRNA sequence. Consequently, these methods can only have low efficiency and extensive screening efforts are often required to identify mutant cells without the introduction of counter-selectable markers. An additional bottleneck for recombineering approaches in *P. putida* has been the identification of recombinase enzymes that function efficiently (Ricaurte et al. 2018). In contrast to *E. coli*, *P. putida* is missing (or they are yet to be identified) active bacteriophages that could provide DNA-editing enzymes that are specifically tailored to work optimally within its host. Attempts to render the λ Red system functional in *Pseudomonas* species were reported (Lesic and Rahme 2008; Liang and Liu 2010), but the information available in the literature would indicate an inferior performance when compared to conventional methods. Recent progress was made by identifying the RecET system in *Pseudomonas syringae*, where the genes are part of a putative prophage (Swingle et al. 2010). The protein pair resembles the λ Red system in that they share homologies with the Exo and Beta components of the λ Red system, respectively, and thus provide the same functions. Due to the close relationship of its native host and *P. putida*, there are good chances that the identified system performs well in the latter as well. The first reported attempt to employ RecET-based genome engineering in *P. putida* was reported by Choi et al. (2018), who used the system to enable recombineering with linear donor dsDNA. However, the reported efficiency of the RecET system was relatively low and was by far outperformed by a suicide plasmid-based approach that the authors combined with the action of RecET as well. However, since the characteristics and reported efficacy of the combined method resembled that of the native *recA*-mediated homologous recombination, the role of RecET in boosting HR is yet to be analyzed in detail. In a similar approach, another promising candidate was proposed with the λ Red Beta-like recombinase SSR found in *P. putida* strain DOT-T1E (Aparicio et al. 2018). The *ssr* gene was transplanted into the *Pseudomonas* cell factory of choice, strain KT2440, and shown to promote recombination with linear ssDNA fragments, although the procedure had lower efficiencies than in *E. coli* with the λ Red system. In general, further optimization would be required for both the RecET-based and the SSR-based systems in order to become standard, established methods for genome engineering in the laboratories.

A reliable, recombineering-based system for site-directed mutagenesis in the chromosome (especially when working with large libraries of mutagenic oligonucleotides) is thus far one of the most desired tools within the *Pseudomonas* community. In particular, it remains to be shown how CRISPR/Cas9 counter-selection, as well as a repetitive cyclization by process automation, could help to overcome the present limitations in generating mutants by recombineering. Nevertheless, if combined with the conventional two-step mutagenesis approach using suicide plasmids, the sequence-specific counter-selection capability that CRISPR/Cas9 systems offer enables the deletion of genes that before could be achieved only with significant efforts—if at all. Specifically directing Cas9 to the wild-type sequence that is intended to be deleted or modified selects for those cells, which resolve the previously co-integrated plasmid in a way that yields the desired mutation (Wirth et al. 2019). One way or the other, the constellation of available tools for genome engineering of *Pseudomonas* species and, in particular, of *P. putida*, has significantly enabled our ability to interrogate core functions of the cell. The genome manipulation of these species has paved the way to expanding the fundamental knowledge on bacterial physiology—and also enabled the design and construction of reduced-genome variants of *P. putida* with enhanced properties for biotechnological applications.

3 Biotechnological Applications of Reduced-Genome Variants of *P. putida*

With a wide palette of efficient molecular tools that are compatible with *P. putida* at hand, very extensive chromosome editing projects are possible. With the genetic engineering process itself no longer constituting any major bottleneck, scientists can take up the search for functions within the genome that are of no use or even undesired for a robust microbial cell factory. The concept of a synthetic minimal cell has been proposed for different organisms, further examples of which can be found throughout this book. A wide variety of experimental approaches has been used within the last decade to identify and study essential genes, e.g., in *E. coli*, *B. subtilis* or *P. aeruginosa*. These methods comprise targeted gene deletions, the generation of conditional knockout mutants, genome-wide RNA interference screens, and libraries acquired through saturation transposon mutagenesis (Yu et al. 2002; Kobayashi et al. 2003; Baba et al. 2006; Kato and Hashimoto 2007; Liberati et al. 2006; Goodall et al. 2018; Juhas et al. 2014). In *E. coli*, the list of genes considered to be essential consists of 300 open reading frames (ORFs) out of a number of 4288 (Goodall et al. 2018). These ORFs encode functions required for protein synthesis and quality control, cell wall biosynthesis, cell division, DNA replication and chromosome maintenance, RNA synthesis and degradation, as well as core metabolic functions (Juhas et al. 2014). However, while we understand more and more about the genes that allow a biological cell to survive and sustain growth, we still lack knowledge about functions that become essential in conditions different from the ones that were defined to study gene essentiality. Moreover, within the category of genes considered non-essential there can be found some that ensure cell robustness and resistance to different types of stress, and many genes still encode for functions that thus far have not been identified (Acevedo-Rocha et al. 2013; Danchin and Fang 2016). Because of these deficits in our understanding about complex biological systems, the idea of designing a "minimal cell" from its roots is currently less auspicious than identifying a suitable biological *chassis* that already displays all desired characteristics and remove unnecessary and unwanted properties—while adding others.

As indicated in Sect. 1.2, and being a ubiquitous bacterium that can be found in a large variety of different habitats, *P. putida* is adapted to survive and thrive under very diverse conditions and with using a wealth of different substrates. It has therefore acquired many functions that are not essential for survival under the controlled conditions of a bio-fermentation setup. Although a set of "essential" genes depends on the given application, one can identify certain gene sets that encode functions generally classified as having negative effects. First, these functions can be related with metabolic regulations and biochemical processes that affect the consumption of a chosen substrate and its conversion into a desired product (Nikel and de Lorenzo 2018). When adopting synthetic devices or systems to produce heterologous enzymes in order to access new metabolic routes or to produce such proteins themselves, the accompanying biochemical functions should not interfere with the extant metabolism. Consequently, they should be largely metabolically inert, or completely decoupled from the host biochemistry (Durante-Rodríguez et al. 2018). Molecular devices should furthermore not interfere with the native regulation patterns—which in some cases are poorly understood. A suitable approach to design a microbial platform for the production of certain aromatic compounds could thus be to first remove the comprehensive collection of functions related to their metabolism and to subsequently implement completely orthogonal systems for their production (Yu et al. 2019). When adopting the widely used XylS/*Pm* regulator/promoter system for the inducible expression of genes (Gawin et al. 2017), for instance, the native regulation system in *P. putida* was shown to respond by activating a range of different metabolic and regulatory genes upon exposure to one of the system's diverse effectors (benzoic acid and derivatives thereof) (Pérez-Pantoja et al. 2015; Volke et al. 2019). In addition, a well-known nuisance when using *P. putida* for the production of biofuels (and structurally related molecules) is its remarkable capacity to degrade the final products and intermediate metabolites by the action of many different and promiscuous oxido-reductases and other catabolic activities (Vallon et al. 2015). Thus, getting rid of such metabolic activities while at the same time taking advantage of the other metabolic and stress-endurance traits that the *chassis* provides is a key step toward creating an ideal whole-cell biocatalyst.

There is a variety of cellular processes that affect the internal supply of *metabolic currencies* that are needed within a myriad of biochemical processes [e.g., NAD(P)H and ATP, reflected in the energy charge and redox ratios] without providing a benefit in a production setup. Besides being used to facilitate biochemical reactions that would otherwise be unfavorable, ATP is required in large quantities to ensure the proper folding of polypeptides, mediated by the essential chaperon complex GroEL/ES. In fact, GroEL/ES is one of the cellular processes with the highest demand for ATP, especially when the proteome is compromised by physical and chemical stresses (Billerbeck et al. 2013). It seems thus obvious that one limiting factor for the quantitative production of foreign proteins is the supply of ATP. One example of an energy-wasting process is the flagellar motion that many bacteria use to move from a location of nutrient scarcity to one that provides substrates that are needed for growth as well as oxygen—a process that appears redundant in a stirred fermentation tank. In 2014, Martínez-García et al. (2014b) reported the removal of a large stretch of DNA (~70 kb, corresponding to ~1.1% of the genome) from the chromosome of *P. putida* KT2440, including 69 relevant structural and regulatory genes for the assembly and export of flagella as well as several chemotaxis functions. This operation resulted in a diverse set of different physiological changes in the reduced-genome strain. Most obviously, a complete absence of cell motility was observed, resulting in higher sedimentation rates. The loss of the outer membrane-associated flagella furthermore decreased surface hydrophobicity, a property that is of advantage for mixed planktonic culture systems because it reduces the formation of biofilms that complicate purification processes and impairs fermentation equipment. Indeed, non-flagellated cells demonstrated substantial decreased biofilm formation in an early phase of the cultivation. However, after 24 h of prolonged cultivation, the formation of biofilms was increased compared to the wild-type strain due to an elevated production of exopolysaccharides, possibly triggered by incipient nutrient starvation and oxygen limitation. It will have to be determined which effect is dominant in a large-scale production setup or if the long-term response can be countered by further strain engineering. Most significant were however the changes that could be observed within the mutant metabolism. The lag phase after exposure to various carbon sources was reduced, and the maximum growth rate was significantly altered (i.e., decreased for growth in complex medium and in minimal medium with glucose and succinate as sole carbon sources, increased for growth on fructose). This was accompanied by a change in the ATP/ADP ratio (i.e., the energy charge of the cell) by a factor of ~1.3. Concurrently, the non-flagellated mutant had a 1.2-fold higher $NADPH/NADP^+$ ratio than the wild-type strain while the catabolic charge (i.e., the $NADH/NAD^+$ ratio) remained essentially constant. An increased availability of NADPH does not only enhance the anabolic capacity of the cell for biosynthesis but also increase its tolerance toward oxidative stress and UV exposure (Martínez-García et al. 2014c)—a trait that is useful for most industrially relevant fermentation and biotransformation applications.

An essential property of a cell factory in large-scale production scenarios is its genetic stability (de Lorenzo and Couto 2019). Unless countered via the implementation of some sort of product-addiction mechanism that ensures that cells maintain a producing phenotype, genetically modified cell factories tend to escape the metabolic burden that was imposed on them through evolutionary mechanisms. These processes result in mutations that lead to a loss of function within the overexpressed genes or pathways. Genetic diversity within a population of bacterial cells is favorable in a natural context because it enables the adaptation of the organism to changing environments, but it is detrimental for a robust cell factory (Fernández-Cabezón et al. 2019). In nature, mutations in the genome can be caused through errors made by DNA polymerases during replication, by error-prone DNA repair mechanisms, or because of external stress factors, e.g., UV irradiation or mutagenic chemicals. Yet another source of genetic instability lies dormant within the chromosome of many bacteria in the form of viral DNA and transposing elements, where these elements can constitute up to 20% of the chromosome (Casjens 2003). Lysogenic phages can be found within the genomic sequence of many bacteria, where they usually become inactive until cells encounter certain stresses such as DNA damage or nutrient starvation, or because of stochastic events that trigger their excision and the resumption of a lytic cycle (Casjens 2003; Ilves et al. 2001). Although subject to a continuous decay while resting within the bacterial chromosome, they often retain some of their gene functions if they provide a benefit for the host. For example, prophages can express immunity and exclusion systems that provide a barrier for the superinfection with a related phage, or the prophage could have introduced fitness-enhancing genes that it had acquired horizontally from different sources [e.g., antibiotic or stress resistance determinants (Winstanley et al. 2009; Wang et al. 2010)]. Another function that is regularly retained is that of transposition within the genome. While mobile genetic elements might contain beneficial functions under certain (selective) conditions, thus being selected for in a positive way, they also represent a disruptive mutagenic force by randomly inserting into gene clusters that are crucial for a producing phenotype.

In *P. putida* KT2440, 2.6% of the genome was found to encode phage-related functions, distributed about four prophage elements, each one containing up to 72 ORFs (Martínez-García et al. 2014d). Intensive studies on their functionality revealed that none of the prophages was able to re-initiate a lytic cycle despite having the ability to be excised under specific environmental conditions. However, deletion of all prophage regions by means of the methods discussed in Sect. 2.1 led to an enhanced tolerance to a diverse set of stress factors. Prophage-less *P. putida* showed increased survival in stationary phase and a lower sensitivity to UV-irradiation and to various types of chemical mutagens. Moreover, the removal of the proviral load led to an increase in fitness in a set of different culture media compositions (Martínez-García et al. 2014d). Extensive analysis of the annotated genome of *P. putida* KT2440 further revealed the presence of 54 transposable elements, one of which has been reported to become particularly active in vivo under carbon starvation (Ilves et al. 2001). In a study published in 2014, Martínez-García et al. (2014d) removed all of these mobile genetic elements together with the previously established targets for genome reduction in *P. putida* (i.e., the flagellar system and proviral elements) as well as dsDNA-degrading systems that are likely to interfere with the introduction of foreign DNA during genetic engineering efforts. The deleted regions comprised a total number of 299 genes corresponding to ~4.3% of the genome in *P. putida* KT2440 and resulted in the cell-factory strain *P. putida*

EM42. This strain was subjected to extensive physiological and genetic characterization. As already described for a flagella-less strain, lag phases were consistently decreased on various carbon sources, likely due to an increased NADPH/$NADP^+$ ratio that enabled the cells to overcome oxidative stress during starvation (Martínez-García et al. 2014c; Rolfe et al. 2012). The growth performance of such reduced-genome variant in complex medium was slightly decreased, while growth in defined media was not affected for any of the various carbon sources tested. However, the genome-reduced strain was able to form significantly more biomass from the same amount of substrate, suggesting that the multiple deletions reduced the maintenance requirements during the cultivation. This experimental observation is in line with an elevated energy charge in strain EM42 (likely due to the absence of flagella, as described above), as well to as an increased intracellular concentration of the central metabolite acetyl-CoA—thereby indicating a better availability of resources for the synthesis of biomass components. The altered maintenance requirements of genome-reduced *P. putida* strains were later confirmed quantitatively (Lieder et al. 2015).

Because of the ATP-dependency that is inherent to the expression of heterologous proteins, an increased energy availability resulted in a more efficient production of foreign polypeptides by *P. putida* EM42, as shown within the same study as well as in further experiments (Martínez-García et al. 2014c; Lieder et al. 2015). Finally, the surplus of ATP allows strain EM42 to survive and even grow at elevated temperatures (42 °C) that are usually lethal for wild-type strain *P. putida* KT2440 (Aparicio et al. 2019). Altogether, these effects distinguish the genome-reduced *P. putida* strain as a powerful cell factory, combining the native metabolic versatility and stress resistance of the parental strain and adding further performance-enhancing characteristics as well as an increased genetic stability.

4 Conclusion and the Way Ahead

P. putida represents a striking example for the journey of an environmental bacterium from its natural habitat into becoming a cell factory useful for contemporary biotechnology. After its discovery in 1960 in Japan, it was first studied as a model organism for the biodegradation for a range of natural and later also xenobiotic arenes (Nakazawa 2002). It was only later that researchers recognized that these outstanding metabolic capacities go hand in hand with physiological characteristics that give it the potential as a potent industrial cell factory. Examples of cellular and molecular features that distinguish *P. putida* from other microbial chassis are abundant, and some were characterized in further detail (a selection of traits was discussed in the previous chapters). However, since the publication of the full genome sequence of strain KT2440 in 2002, only 79% of all genes were associated with a (potential) function, many of which are only putative based on homologies with proteins whose function was identified in other organisms (Belda et al. 2016).

The full metabolic potential of this host has still to be fully exploited. Traditionally, the identification of specific physiological aspects of a microbe relied on methods that were established within the fields of classical microbiology and biochemistry (including laborious and time-consuming experiments, studying one function at a time). The current progress made within the data sciences (with all varieties within the "omics" field), accompanied by technologies for the automation of wet-laboratory tasks, are taking these endeavors to a new level. More and more effort is put into the development of high-throughput screening methods for the identification of new biochemical functions, and their combination with transcriptional and proteomic profiles allows for the identification of new genotype–phenotype relationships. With DNA synthesis becoming accessible to everyone at very low costs and even at large scale, any gene or variants thereof can be simply synthesized and cloned into a suitable expression system, ready to be delivered into the cell factory of choice in a very short time for further analysis of their function (s). Especially if manual tasks are performed by robots, the combination of DNA synthesis and high-throughput cloning, expression and screening allows for the collection of innumerable data that can be combined with the analytic power of newly developed machine-learning algorithms to systematically explore genotype–phenotype relations (Riordon et al. 2019; Presnell et al. 2019). These developments will be of particular value to harness the full metabolic potential *P. putida* provides. It is expected that more biochemical functions and pathways will be revealed with the collective volume of data gathered in "omics" experiments under various cultivation conditions and with a diversity of substrates. The combination of knowledge-based, machine-learning guided design of efficient catalysts will thus result in minimal-genome strains with enhanced performance to fulfill the requirements of contemporary biotechnology—marking the transition into a veritable bioeconomy (de Lorenzo et al. 2018; Martinelli and Nikel 2019).

Acknowledgments The financial support from The Novo Nordisk Foundation (grant NNF10CC1016517, and *LiFe*, NNF18OC0034818), the Danish Council for Independent Research (*SWEET*, DFF-Research Project 8021-00039B), and the European Union's *Horizon 2020* Research and Innovation Programme under grant agreement No. 814418 (*SinFonia*) to P.I.N. is gratefully acknowledged. The authors declare that there is no conflict of interest related to the contents of this chapter.

References

Abram KZ, Udaondo Z (2019) Towards a better metabolic engineering reference: the microbial *chassis*. Microb Biotechnol. https://doi.org/10.1111/1751-7915.13363

Acevedo-Rocha CG, Fang G, Schmidt M, Ussery DW, Danchin A (2013) From essential to persistent genes: a functional approach to constructing synthetic life. Trends Genet 29(5):273–279

Aparicio T, de Lorenzo V, Martínez-García E (2018) CRISPR/Cas9-based counterselection boosts recombineering efficiency in *Pseudomonas putida*. Biotechnol J 13(5):1700161

Aparicio T, de Lorenzo V, Martínez-García E (2019) Improved thermotolerance of genome-reduced *Pseudomonas putida* EM42 enables effective functioning of the P_L/cI*857* system. Biotechnol J 14(1):e1800483

Ayres EK, Thomson VJ, Merino G, Balderes D, Figurski DH (1993) Precise deletions in large bacterial genomes by vector-mediated excision (VEX)—the *trfA* gene of promiscuous plasmid *RK2* is essential for replication in several gram-negative hosts. J Mol Biol 230(1):174–185

Baba T, Ara T, Hasegawa M, Takai Y, Okumura Y, Baba M, Datsenko KA, Tomita M, Wanner BL, Mori H (2006) Construction of *Escherichia coli* K-12 in-frame, single-gene knockout mutants: the Keio collection. Mol Syst Biol 2:2006.0008

Baritugo KAG, Kim HT, David YC, Choi JH, Choi JI, Kim TW, Park C, Hong SH, Na JG, Jeong KJ, Joo JC, Park SJ (2018) Recent advances in metabolic engineering of *Corynebacterium glutamicum* as a potential platform microorganism for biorefinery. Biofuels Bioprod Biorefin 12(5):899–925

Beckham GT, Johnson CW, Karp EM, Salvachúa D, Vardon DR (2016) Opportunities and challenges in biological lignin valorization. Curr Opin Biotechnol 42:40–53

Beites T, Mendes MV (2015) *Chassis* optimization as a cornerstone for the application of synthetic biology based strategies in microbial secondary metabolism. Front Microbiol 6:906

Belda E, van Heck RGA, López-Sánchez MJ, Cruveiller S, Barbe V, Fraser C, Klenk HP, Petersen J, Morgat A, Nikel PI, Vallenet D, Rouy Z, Sekowska A, Martins dos Santos VAP, de Lorenzo V, Danchin A, Médigue C (2016) The revisited genome of *Pseudomonas putida* KT2440 enlightens its value as a robust metabolic *chassis*. Environ Microbiol 18(10):3403–3424

Benedetti I, de Lorenzo V, Nikel PI (2016) Genetic programming of catalytic *Pseudomonas putida* biofilms for boosting biodegradation of haloalkanes. Metab Eng 33:109–118

Billerbeck S, Calles B, Müller CL, de Lorenzo V, Panke S (2013) Towards functional orthogonalisation of protein complexes: individualisation of GroEL monomers leads to distinct quasihomogeneous single rings. Chembiochem 14(17):2310–2321

Blank LM, Ionidis G, Ebert BE, Bühler B, Schmid A (2008) Metabolic response of *Pseudomonas putida* during redox biocatalysis in the presence of a second octanol phase. FEBS J 275 (20):5173–5190

Calero P, Nikel PI (2019) Chasing bacterial *chassis* for metabolic engineering: a perspective review from classical to non-traditional microorganisms. Microb Biotechnol 12(1):98–124

Casjens S (2003) Prophages and bacterial genomics: what have we learned so far? Mol Microbiol 49(2):277–300

Chavarría M, Goñi-Moreno A, de Lorenzo V, Nikel PI (2016) A metabolic widget adjusts the phosphoenolpyruvate-dependent fructose influx in *Pseudomonas putida*. mSystems 1(6): e00154–16

Choi KR, Cho JS, Cho IJ, Park D, Lee SY (2018) Markerless gene knockout and integration to express heterologous biosynthetic gene clusters in *Pseudomonas putida*. Metab Eng 47:463–474

Chua JW, Hsieh JH (1990) Oxidative bioconversion of toluene to 1,3-butadiene-1,4-dicarboxylic acid (*cis,cis*-muconic acid). World J Microbiol Biotechnol 6(2):127–143

Conway T (1992) The Entner-Doudoroff pathway: history, physiology and molecular biology. FEMS Microbiol Rev 9(1):1–27

Cook TB, Rand JM, Nurani W, Courtney DK, Liu SA, Pfleger BF (2018) Genetic tools for reliable gene expression and recombineering in *Pseudomonas putida*. J Ind Microbiol Biotechnol 45(7):1–11

Danchin A (2012) Scaling up synthetic biology: do not forget the *chassis*. FEBS Lett 586 (15):2129–2137

Danchin A, Fang G (2016) Unknown unknowns: essential genes in quest for function. Microb Biotechnol 9(5):530–540

Davy AM, Kildegaard HF, Andersen MR (2017) Cell factory engineering. Cell Syst 4(3):262–275

de Lorenzo V, Couto J (2019) The important versus the exciting: reining contradictions in contemporary biotechnology. Microb Biotechnol 12(1):32–34

de Lorenzo V, Prather KL, Chen GQ, O'Day E, von Kameke C, Oyarzún DA, Hosta-Rigau L, Alsafar H, Cao C, Ji W, Okano H, Roberts RJ, Ronaghi M, Yeung K, Zhang F, Lee SY (2018) The power of synthetic biology for bioproduction, remediation and pollution control: the UN's sustainable development goals will inevitably require the application of molecular biology and biotechnology on a global scale. EMBO Rep 19(4):e45658

Durante-Rodríguez G, de Lorenzo V, Nikel PI (2018) A post-translational metabolic switch enables complete decoupling of bacterial growth from biopolymer production in engineered *Escherichia coli*. ACS Synth Biol 7:2686–2697

Dvořák P, de Lorenzo V (2018) Refactoring the upper sugar metabolism of *Pseudomonas putida* for co-utilization of cellobiose, xylose, and glucose. Metab Eng 48:94–108

Ellis HM, Yu D, DiTizio T, Court DL (2001) High efficiency mutagenesis, repair, and engineering of chromosomal DNA using single-stranded oligonucleotides. Proc Natl Acad Sci U S A 98 (12):6742–6746

Elmore JR, Furches A, Wolff GN, Gorday K, Guss AM (2017) Development of a high efficiency integration system and promoter library for rapid modification of *Pseudomonas putida* KT2440. Metab Eng Commun 5:1–8

Fernández-Cabezón L, Cros A, Nikel PI (2019) Evolutionary approaches for engineering industrially-relevant phenotypes in bacterial cell factories. Biotechnol J 14(9):1800439

Flamholz A, Noor E, Bar-Even A, Liebermeister W, Milo R (2013) Glycolytic strategy as a tradeoff between energy yield and protein cost. Proc Natl Acad Sci U S A 110(24):10039–10044

Galvão TC, de Lorenzo V (2005) Adaptation of the yeast *URA3* selection system to Gram-negative bacteria and generation of a Δ *betCDE Pseudomonas putida* strain. Appl Environ Microbiol 71 (2):883–892

Gasiunas G, Barrangou R, Horvath P, Siksnys V (2012) Cas9-crRNA ribonucleoprotein complex mediates specific DNA cleavage for adaptive immunity in bacteria. Proc Natl Acad Sci U S A 109(39):E2579–E2586

Gawin A, Valla S, Brautaset T (2017) The XylS/*Pm* regulator/promoter system and its use in fundamental studies of bacterial gene expression, recombinant protein production and metabolic engineering. Microb Biotechnol 10(4):702–718

Goodall ECA, Robinson A, Johnston IG, Jabbari S, Turner KA, Cunningham AF, Lund PA, Cole JA, Henderson IR (2018) The essential genome of *Escherichia coli* K-12. mBio 9(1):e02096–17

Graf N, Altenbuchner J (2011) Development of a method for markerless gene deletion in *Pseudomonas putida*. Appl Environ Microbiol 77(15):5549–5552

Gu Y, Xu X, Wu Y, Niu T, Liu Y, Li J, Du G, Liu L (2018) Advances and prospects of *Bacillus subtilis* cellular factories: from rational design to industrial applications. Metab Eng 50:109–121

Heipieper HJ, Neumann G, Cornelissen S, Meinhardt F (2007) Solvent-tolerant bacteria for biotransformations in two-phase fermentation systems. Appl Microbiol Biotechnol 74(5):961–973

Hoang TT, Karkhoff-Schweizer RR, Kutchma AJ, Schweizer HP (1998) A broad-host-range Flp-FRT recombination system for site-specific excision of chromosomally-located DNA sequences: application for isolation of unmarked *Pseudomonas aeruginosa* mutants. Gene 212(1):77–86

Ilves H, Hõrak R, Kivisaar M (2001) Involvement of σ^S in starvation-induced transposition of *Pseudomonas putida* transposon Tn*4652*. J Bacteriol 183(18):5445–5448

Jakočiūnas T, Jensen MK, Keasling JD (2017) System-level perturbations of cell metabolism using CRISPR/Cas9. Curr Opin Biotechnol 46:134–140

Jiménez JI, Miñambres B, García JL, Díaz E (2002) Genomic analysis of the aromatic catabolic pathways from *Pseudomonas putida* KT2440. Environ Microbiol 4(12):824–841

Jinek M, Chylinski K, Fonfara I, Hauer M, Doudna JA, Charpentier E (2012) A programmable dual-RNA-guided DNA endonuclease in adaptive bacterial immunity. Science 337 (6096):816–821

Johnson CW, Salvachúa D, Khanna P, Smith H, Peterson DJ, Beckham GT (2016) Enhancing muconic acid production from glucose and lignin-derived aromatic compounds *via* increased protocatechuate decarboxylase activity. Metab Eng Commun 3:111–119

Juhas M, Reuss DR, Zhu B, Commichau FM (2014) *Bacillus subtilis* and *Escherichia coli* essential genes and minimal cell factories after one decade of genome engineering. Microbiology 160 (11):2341–2351

Kato J, Hashimoto M (2007) Construction of consecutive deletions of the *Escherichia coli* chromosome. Mol Syst Biol 3:132

Kim J, Salvador M, Saunders E, González J, Avignone-Rossa C, Jiménez JI (2016) Properties of alternative microbial hosts used in synthetic biology: towards the design of a modular *chassis*. Essays Biochem 60(4):303–313

Kobayashi K, Ehrlich SD, Albertini A, Amati G, Andersen KK, Arnaud M, Asai K, Ashikaga S, Aymerich S, Bessieres P, Boland F, Brignell SC, Bron S, Bunai K, Chapuis J, Christiansen LC, Danchin A, Débarbouille M, Dervyn E, Deuerling E, Devine K, Devine SK, Dreesen O, Errington J, Fillinger S, Foster SJ, Fujita Y, Galizzi A, Gardan R, Eschevins C, Fukushima T, Haga K, Harwood CR, Hecker M, Hosoya D, Hullo MF, Kakeshita H, Karamata D, Kasahara Y, Kawamura F, Koga K, Koski P, Kuwana R, Imamura D, Ishimaru M, Ishikawa S, Ishio I, Le Coq D, Masson A, Mauël C, Meima R, Mellado RP, Moir A, Moriya S, Nagakawa E, Nanamiya H, Nakai S, Nygaard P, Ogura M, Ohanan T, O'Reilly M, O'Rourke M, Pragai Z, Pooley HM, Rapoport G, Rawlins JP, Rivas LA, Rivolta C, Sadaie A, Sadaie Y, Sarvas M, Sato T, Saxild HH, Scanlan E, Schumann W, Seegers JFML, Sekiguchi J, Sekowska A, Séror SJ, Simon M, Stragier P, Studer R, Takamatsu H, Tanaka T, Takeuchi M, Thomaides HB, Vagner V, van Dijl JM, Watabe K, Wipat A, Yamamoto H, Yamamoto M, Yamamoto Y, Yamane K, Yata K, Yoshida K, Yoshikawa H, Zuber U, Ogasawara N (2003) Essential *Bacillus subtilis* genes. Proc Natl Acad Sci U S A 100(8):4678–4683

Kohlstedt M, Starck S, Barton N, Stolzenberger J, Selzer M, Mehlmann K, Schneider R, Pleissner D, Rinkel J, Dickschat JS, Venus J, van Duuren JNJH, Wittmann C (2018) From lignin to nylon: cascaded chemical and biochemical conversion using metabolically engineered *Pseudomonas putida*. Metab Eng 47:279–293

Lesic B, Rahme LG (2008) Use of the λ red recombinase system to rapidly generate mutants in *Pseudomonas aeruginosa*. BMC Mol Biol 9:20

Li L, Liu X, Wei K, Lu Y, Jiang W (2019) Synthetic biology approaches for chromosomal integration of genes and pathways in industrial microbial systems. Biotechnol Adv 37 (5):730–745

Liang R, Liu J (2010) Scarless and sequential gene modification in *Pseudomonas* using PCR product flanked by short homology regions. BMC Microbiol 10:209

Liberati NT, Urbach JM, Miyata S, Lee DG, Drenkard E, Wu G, Villanueva J, Wei T, Ausubel FM (2006) An ordered, nonredundant library of *Pseudomonas aeruginosa* strain PA14 transposon insertion mutants. Proc Natl Acad Sci U S A 103(8):2833–2838

Lieder S, Nikel PI, de Lorenzo V, Takors R (2015) Genome reduction boosts heterologous gene expression in *Pseudomonas putida*. Microb Cell Fact 14(1):23

Linger JG, Vardon DR, Guarnieri MT, Karp EM, Hunsinger GB, Franden MA, Johnson CW, Chupka G, Strathmann TJ, Pienkos PT, Beckham GT (2014) Lignin valorization through integrated biological funneling and chemical catalysis. Proc Natl Acad Sci U S A 111(33):12013

Marqués S, Ramos JL (1993) Transcriptional control of the *Pseudomonas putida* TOL plasmid catabolic pathways. Mol Microbiol 9(5):923–929

Martinelli L, Nikel PI (2019) Breaking the state-of-the-art in the chemical industry with new-to-nature products *via* synthetic microbiology. Microb Biotechnol 12(2):187–190

Martínez-García E, de Lorenzo V (2011) Engineering multiple genomic deletions in gram-negative bacteria: analysis of the multi-resistant antibiotic profile of *Pseudomonas putida* KT2440. Environ Microbiol 13(10):2702–2716

Martínez-García E, de Lorenzo V (2017) Molecular tools and emerging strategies for deep genetic/genomic refactoring of *Pseudomonas*. Curr Opin Biotechnol 47:120–132

Martínez-García E, Aparicio T, Goñi-Moreno A, Fraile S, de Lorenzo V (2014a) SEVA 2.0: an update of the standard European vector architecture for de−/re-construction of bacterial functionalities. Nucleic Acids Res 43(D1):D1183–D1189

Martínez-García E, Nikel PI, Aparicio T, de Lorenzo V (2014b) *Pseudomonas* 2.0: genetic upgrading of *P. putida* KT2440 as an enhanced host for heterologous gene expression. Microb Cell Fact 13(1):159

Martínez-García E, Nikel PI, Chavarría M, de Lorenzo V (2014c) The metabolic cost of flagellar motion in *Pseudomonas putida* KT2440. Environ Microbiol 16(1):291–303

Martínez-García E, Jatsenko T, Kivisaar M, de Lorenzo V (2014d) Freeing *Pseudomonas putida* KT2440 of its proviral load strengthens endurance to environmental stresses. Environ Microbiol 17(1):76–90

Marx CJ, Lidstrom ME (2002) Broad-host-range Cre-*lox* system for antibiotic marker recycling in gram-negative bacteria. BioTechniques 33(5):1062–1067

Moore ERB, Tindall BJ, Martins dos Santos VAP, Pieper DH, Ramos JL, Palleroni NJ (2006) Nonmedical *Pseudomonas*. In: Dworkin M, Falkow S, Rosenberg E, Schleifer KH, Stackebrandt E (eds) The prokaryotes—volume 6: proteobacteria gamma subclass. Springer, New York, NY, pp 646–703

Murphy KC (2016) λ recombination and recombineering. In: Slauch JM (ed) *EcoSal Plus* domain 7: genetics and genetic tools, 2016/05/26 edn. ASM, Washington D.C. https://doi.org/10.1128/ecosalplus.ESP-0011-2015

Mutalik VK, Guimaraes JC, Cambray G, Lam C, Christoffersen MJ, Mai QA, Tran AB, Paull M, Keasling JD, Arkin AP, Endy D (2013) Precise and reliable gene expression *via* standard transcription and translation initiation elements. Nat Methods 10:354–360

Nakazawa T (2002) Travels of a *Pseudomonas*, from Japan around the world. Environ Microbiol 4 (12):782–786

Nelson KE, Weinel C, Paulsen IT, Dodson RJ, Hilbert H, dos Santos VAPM, Fouts DE, Gill SR, Pop M, Holmes M, Brinkac L, Beanan M, DeBoy RT, Daugherty S, Kolonay J, Madupu R, Nelson W, White O, Peterson J, Khouri H, Hance I, Lee PC, Holtzapple E, Scanlan D, Tran K, Moazzez A, Utterback T, Rizzo M, Lee K, Kosack D, Moestl D, Wedler H, Lauber J, Stjepandic D, Hoheisel J, Straetz M, Heim S, Kiewitz C, Eisen JA, Timmis KN, Düsterhöft A, Tümmler B, Fraser CM (2002) Complete genome sequence and comparative analysis of the metabolically versatile *Pseudomonas putida* KT2440. Environ Microbiol 4(12):799–808

Nikel PI, de Lorenzo V (2018) *Pseudomonas putida* as a functional *chassis* for industrial biocatalysis: from native biochemistry to *trans*-metabolism. Metab Eng 50:142–155

Nikel PI, Chavarría M, Fuhrer T, Sauer U, de Lorenzo V (2015) *Pseudomonas putida* KT2440 strain metabolizes glucose through a cycle formed by enzymes of the Entner-Doudoroff, Embden-Meyerhof-Parnas, and pentose phosphate pathways. J Biol Chem 290(43):25920–25932

Nogales J, Palsson BØ, Thiele I (2008) A genome-scale metabolic reconstruction of *Pseudomonas putida* KT2440: *i*JN746 as a cell factory. BMC Syst Biol 2:79

Noor E, Eden E, Milo R, Alon U (2010) Central carbon metabolism as a minimal biochemical walk between precursors for biomass and energy. Cell 39(5):809–820

Oberhardt MA, Puchałka J, Martins dos Santos VAP, Papin JA (2011) Reconciliation of genome-scale metabolic reconstructions for comparative systems analysis. PLoS Comput Biol 7(3):e1001116

Palazzotto E, Tong Y, Lee SY, Weber T (2019) Synthetic biology and metabolic engineering of actinomycetes for natural product discovery. Biotechnol Adv 37(6):107366

Palleroni NJ (2010) The *Pseudomonas* story. Environ Microbiol 12(6):1377–1383

Palleroni NJ, Kunisawa R, Contopoulou R, Doudoroff M (1973) Nucleic acid homologies in the genus *Pseudomonas*. Int J Syst Bacteriol 23:333–339

Peña DA, Gasser B, Zanghellini J, Steiger MG, Mattanovich D (2018) Metabolic engineering of *Pichia pastoris*. Metab Eng 50:2–15

Pérez-Pantoja D, Kim J, Silva-Rocha R, de Lorenzo V (2015) The differential response of the P_{ben} promoter of *Pseudomonas putida* mt-2 to BenR and XylS prevents metabolic conflicts in *m*-xylene biodegradation. Environ Microbiol 17(1):64–75

Poblete-Castro I, Becker J, Dohnt K, Martins dos Santos VAP, Wittmann C (2012) Industrial biotechnology of *Pseudomonas putida* and related species. Appl Microbiol Biotechnol 93 (6):2279–2290

Poblete-Castro I, Borrero de Acuña JM, Nikel PI, Kohlstedt M, Wittmann C (2017) Host organism: *Pseudomonas putida*. In: Wittmann C, Liao JC (eds) Industrial biotechnology: microorganisms. Wiley, Weinheim

Poblete-Castro I, Wittmann C, Nikel PI (2019) Biochemistry, genetics and biotechnology of glycerol utilization in *Pseudomonas* species. Microb Biotechnol. https://doi.org/10.1111/1751-7915.13400

Presnell KV, Alper HS (2019) Systems metabolic engineering meets machine learning: a new era for data-driven metabolic engineering. Biotechnol J 14(9):e1800416

Puchałka J, Oberhardt MA, Godinho M, Bielecka A, Regenhardt D, Timmis KN, Papin JA, Martins dos Santos VAP (2008) Genome-scale reconstruction and analysis of the *Pseudomonas putida* KT2440 metabolic network facilitates applications in biotechnology. PLoS Comput Biol 4(10): e1000210

Rakowski SA, Filutowicz M (2013) Plasmid R6K replication control. Plasmid 69(3):231–242

Ricaurte DE, Martínez-García E, Nyerges A, Pal C, de Lorenzo V, Aparicio T (2018) A standardized workflow for surveying recombinases expands bacterial genome-editing capabilities. Microb Biotechnol 11(1):176–188

Riordon J, Sovilj D, Sanner S, Sinton D, Young EWK (2019) Deep learning with microfluidics for biotechnology. Trends Biotechnol 37(3):310–324

Rojo F (2010) Carbon catabolite repression in *Pseudomonas*: optimizing metabolic versatility and interactions with the environment. FEMS Microbiol Rev 34(5):658–684

Rolfe MD, Rice CJ, Lucchini S, Pin C, Thompson A, Cameron AD, Alston M, Stringer MF, Betts RP, Baranyi J, Peck MW, Hinton JC (2012) Lag phase is a distinct growth phase that prepares bacteria for exponential growth and involves transient metal accumulation. J Bacteriol 194 (3):686–701

Ronda C, Pedersen LE, Sommer MOA, Nielsen AT (2016) *CRMAGE*: CRISPR optimized MAGE recombineering. Sci Rep 6:19452

Segura A, Molina L, Fillet S, Krell T, Bernal P, Muñoz-Rojas J, Ramos JL (2012) Solvent tolerance in gram-negative bacteria. Curr Opin Biotechnol 23(3):415–421

Silva-Rocha R, Martínez-García E, Calles B, Chavarría M, Arce-Rodríguez A, de las Heras A, Páez-Espino AD, Durante-Rodríguez G, Kim J, Nikel PI, Platero R, de Lorenzo V (2013) The standard European vector architecture (SEVA): a coherent platform for the analysis and deployment of complex prokaryotic phenotypes. Nucleic Acids Res 41(D1):D666–D675

Simon O, Klaiber I, Huber A, Pfannstiel J (2014) Comprehensive proteome analysis of the response of *Pseudomonas putida* KT2440 to the flavor compound vanillin. J Proteome 109:212–227

Sohn SB, Kim TY, Park JM, Lee SY (2010) *In silico* genome-scale metabolic analysis of *Pseudomonas putida* KT2440 for polyhydroxyalkanoate synthesis, degradation of aromatics and anaerobic survival. Biotechnol J 5(7):739–750

Steinmetz M, Le Coq D, Djemia HB, Gay P (1983) Genetic analysis of *sacB*, the structural gene of a secreted enzyme, levansucrase of *Bacillus subtilis* Marburg. Mol Gen Genet 191(1):138–144

Sudarsan S, Dethlefsen S, Blank LM, Siemann-Herzberg M, Schmid A (2014) The functional structure of central carbon metabolism in *Pseudomonas putida* KT2440. Appl Environ Microbiol 80(17):5292–5303

Swingle B, Bao Z, Markel E, Chambers A, Cartinhour S (2010) Recombineering using RecTE from *Pseudomonas syringae*. Appl Environ Microbiol 76(15):4960–4968

Timmis KN (2002) *Pseudomonas putida*: a cosmopolitan opportunist *par excellence*. Environ Microbiol 4(12):779–781

Vallon T, Simon O, Rendgen-Heugle B, Frana S, Mückschel B, Broicher A, Siemann-Herzberg M, Pfannenstiel J, Hauer B, Huber A, Breuer M, Takors R (2015) Applying systems biology tools to study *n*-butanol degradation in *Pseudomonas putida* KT2440. Eng Life Sci 15(8):760–771

Vardon DR, Franden MA, Johnson CW, Karp EM, Guarnieri MT, Linger JG, Salm MJ, Strathmann TJ, Beckham GT (2015) Adipic acid production from lignin. Energy Environ Sci 8(2):617–628

Volke DC, Nikel PI (2018) Getting bacteria in shape: synthetic morphology approaches for the design of efficient microbial cell factories. Adv Biosyst 2(11):1800111

Volke DC, Turlin J, Mol V, Nikel PI (2019) Physical decoupling of XylS/*Pm* regulatory elements and conditional proteolysis enable precise control of gene expression in *Pseudomonas putida*. Microb Biotechnol. https://doi.org/10.1111/1751-7915.13383

Wang X, Kim Y, Ma Q, Hong SH, Pokusaeva K, Sturino JM, Wood TK (2010) Cryptic prophages help bacteria cope with adverse environments. Nat Commun 1:147–147

Wang Y, Horlamus F, Henkel M, Kovacic F, Schläfle S, Hausmann R, Wittgens A, Rosenau F (2019) Growth of engineered *Pseudomonas putida* KT2440 on glucose, xylose, and arabinose: hemicellulose hydrolysates and their major sugars as sustainable carbon sources. GCB Bioenergy 11(1):249–259

Winstanley C, Langille MG, Fothergill JL, Kukavica-Ibrulj I, Paradis-Bleau C, Sanschagrin F, Thomson NR, Winsor GL, Quail MA, Lennard N, Bignell A, Clarke L, Seeger K, Saunders D, Harris D, Parkhill J, Hancock RE, Brinkman FS, Levesque RC (2009) Newly introduced genomic prophage islands are critical determinants of in vivo competitiveness in the Liverpool epidemic strain of *Pseudomonas aeruginosa*. Genome Res 19(1):12–23

Wirth NT, Kozaeva E, Nikel PI (2019) Accelerated genome engineering of *Pseudomonas putida* by I-*Sce*I—mediated recombination and CRISPR-Cas9 counterselection. Microb Biotechnol. https://doi.org/10.1111/1751-7915.13396

Yang Z, Zhang Z (2018) Engineering strategies for enhanced production of protein and bio-products in *Pichia pastoris:* a review. Biotechnol Adv 36(1):182–195

Yu BJ, Sung BH, Koob MD, Lee CH, Lee JH, Lee WS, Kim MS, Kim SC (2002) Minimization of the *Escherichia coli* genome using a Tn5-targeted Cre/*loxP* excision system. Nat Biotechnol 20 (10):1018–1023

Yu LP, Wu FQ, Chen GQ (2019) Next generation industrial biotechnology—transforming the current industrial biotechnology into competitive processes. Biotechnol J 14(9):e1800437

Zobel S, Benedetti I, Eisenbach L, de Lorenzo V, Wierckx NJP, Blank LM (2015) Tn7-based device for calibrated heterologous gene expression in *Pseudomonas putida*. ACS Synth Biol 4 (12):1341–1351

Genome-Reduced *Corynebacterium glutamicum* Fit for Biotechnological Applications

Volker F. Wendisch

Abstract Genome minimization ultimately leads to the smallest genome sustaining life of a given cell; however, growth of this cell may be very slow and may require multiple supplements, e.g., to overcome amino acid auxotrophies. By contrast, genome reduction of industrially relevant bacteria such as *Corynebacterium glutamicum* does not aim at generating minimal cells. Rather chassis cells are developed that are as fit as the wild type with respect to a target function, for example, growth of *C. glutamicum* in glucose minimal medium. Thus, a balance between reducing the burden of expressed genes and maintaining fast growth with glucose without the requirement for supplements such as amino acids is required. Here, the application of this concept to *C. glutamicum* is discussed. Moreover, an outlook on how the advent of genome editing by CRISPR-Cas9 or CRISPR-Cas12a(Cpf1) impacts genome reduction and how highly parallel genome editing must be met by highly parallel strain characterization is presented. Finally, metabolic engineering approaches for the overproduction of amino acids, organic acids, terpenoids, and diamines making use of genome-reduced *C. glutamicum* strains are detailed.

Keywords *Corynebacterium glutamicum* · Genome reduction · Amino acid production · Metabolic engineering · Fine chemicals · Two-step homologous recombination, CRISPR/Cas9

Google Scholar Profile: XBj3q30AAAAJ; Researcher ID: M-5940-2016; Scopus Author ID: 6701420809

V. F. Wendisch (✉)
Genetics of Prokaryotes, Faculty of Biology, Bielefeld University, Bielefeld, Germany

Center for Biotechnology (CeBiTec), Bielefeld University, Bielefeld, Germany
e-mail: volker.wendisch@uni-bielefeld.de

A. R. Lara, G. Gosset (eds.), *Minimal Cells: Design, Construction, Biotechnological Applications*, https://doi.org/10.1007/978-3-030-31897-0_4

1 *Corynebacterium glutamicum*: One of the Pillars of Biotechnology

1.1 Role of C. glutamicum *in the Bioeconomy*

It is believed that bioeconomy will play an important role in the world's future. White biotechnology, also known as industrial biotechnology, makes use of biotechnology for the sustainable processing and production of chemicals, materials, and fuel (Frazzetto 2003). *Corynebacterium glutamicum* is a central pillar of white biotechnology. *C. glutamicum* has a history of more than 50 years of safe production of food and feed amino acids, an industrial process which operates at the million-ton scale per annum (Lee and Wendisch 2017) and shows a compound annual growth rate of 5.6% over 2017–2022 reaching US$25.6 billion by 2022 (Wendisch 2019).

Strain development for *C. glutamicum* has embraced and driven technological development in the classical (Leuchtenberger et al. 2005; Ohnishi et al. 2008), genetic engineering (Kirchner and Tauch 2003; Ikeda 2003), systems biology (Ma et al. 2017), synthetic biology (Lee and Wendisch 2017; Wendisch 2014), and systems metabolic engineering eras (Lee and Wendisch 2017; Becker and Wittmann 2012; Contador et al. 2009). Currently, this is obvious by the application and further development of CRISPR interference (Cleto et al. 2016), CRISPR-Cas9 (Cho et al. 2017) and CRISPR-Cas12a (Jiang et al. 2017) genome editing and CRISPR multiplexing (Wang et al. 2018a), biosensor-driven strain selection (Zhao et al. 2016; Steffen et al. 2016; Mahr et al. 2015; Eggeling et al. 2015; Siedler et al. 2014; Mustafi et al. 2012, 2014) and flux control (Zhou and Zeng 2015a, b), and new process concepts such as coproduction (Henke et al. 2018a) and synthetic consortia (Sgobba et al. 2018) that have been applied to *C. glutamicum*.

1.2 C. glutamicum *as Host for a Multitude of Production Processes*

C. glutamicum has been engineered for the production of a broad spectrum of value-added compounds including specialty amino acids (Perez-Garcia et al. 2016, 2017) such as N-alkylated amino acids (Mindt et al. 2018a, b, 2019a, b) and omega-amino acids (Jorge et al. 2016, 2017a, b; Rohles et al. 2016); diamines such as putrescine and cadaverine (Imao et al. 2017; Schneider and Wendisch 2011); organic acids such as pyruvate (Wieschalka et al. 2013), succinate (Tsuge et al. 2013; Kim et al. 2015; Litsanov et al. 2012, 2013), glutarate (Perez-Garcia et al. 2018), and itaconate (Otten et al. 2015); alcohols such as isobutanol (Blombach et al. 2011; Smith et al. 2010) and n-propanol (Siebert and Wendisch 2015); aromatic compounds such as PHBA (Syukur Purwanto et al. 2018; Kallscheuer and Marienhagen 2018; Kitade et al. 2018), 7-bromo- or 7-chloro-L-tryptophan (Veldmann et al. 2019a, b), phenylpropanoids (Kallscheuer et al. 2016a), and anthocyanine (Zha et al. 2018);

vitamins such as pantothenate (Chassagnole et al. 2002) and riboflavin (Taniguchi and Wendisch 2015); terpenoids such as patchoulol (Henke et al. 2018b) and astaxanthin (Henke et al. 2016); and polymers such polyhydroxyalkanoates (Jo et al. 2007), hyaluronic acids (Hoffmann and Altenbuchner 2014), chondroitin (Cheng et al. 2019), and proteins (Freudl 2017, 2018). To facilitate biorefinery applications, a flexible carbon feedstock concept has been realized for production processes from various second-generation feedstocks without competing uses in human and animal nutrition (Zahoor et al. 2012; Wendisch et al. 2016).

1.3 **C. glutamicum** *Genome and Genome-Scale Tools*

C. glutamicum possesses a single circular chromosome with 3.3 Mb (Kalinowski et al. 2003; Ikeda and Nakagawa 2003) and more than 3000 protein encoding sequences (CDS). Genome-scale methods have been developed early (Wendisch 2003; Wendisch et al. 2006). Based on the complete genome sequence (Kalinowski et al. 2003), genome-scale metabolic models were reconstructed. The first genome-scale metabolic models followed the approach for the *E. coli* genome-scale metabolic model (Feist et al. 2007) and comprised 446 and 502 reactions, respectively, involving 441 and 423 metabolites, respectively (Shinfuku et al. 2009; Kjeldsen and Nielsen 2009). The genome-scale model *iEZ475* added balances for protons and water (https://www.13cflux.net/models/Corynebacterium_glutamicum/index.jsp) and contains 475 metabolic reactions involving 408 metabolites (340 intra- and 68 extracellular) that could be grouped to central carbon metabolism (about 42 reactions), amino acid synthesis (about 110 reactions) as well as to oxidative phosphorylation, membrane lipid metabolism, nucleotide salvage pathway, cofactor biosynthesis, biomass formation, alternate carbon metabolism, and about 90 transport reactions (https://www.13cflux.net/models/Corynebacterium_glutamicum/index.jsp). Biosynthesis reactions leading to protein, DNA, RNA, and cell wall components were accounted for based on their weight fraction of the biomass. The most advanced model (iCW773) has recently been described and reconstructs 773 genes, 950 metabolites, and 1207 reactions, of which 252 are transport reactions (Zhang et al. 2019). Although all these models are named genome-scale, only about 26% of all ORFs are covered by the most advanced model. These stoichiometric models were complemented by a regulatory model involving 97 transcriptional regulator proteins and 1432 regulatory interactions which later was extended to include other corynebacterial species and *E. coli* (Pauling et al. 2012).

Transcriptomics was developed for *C. glutamicum*, first based on DNA microarrays (Wendisch 2003) and later by RNAseq (Pfeifer-Sancar et al. 2013). A landscape RNAseq study helped to refine genome annotation with a re-annotation of 200 gene starts and the finding that among the 2000 transcriptional start sites identified, about 33% belonged to leaderless transcripts (Pfeifer-Sancar et al. 2013). Differential RNAseq is nowadays used to compare global gene expression patterns (Lee et al. 2013; Neshat et al. 2014; Freiherr von Boeselager et al. 2018;

Taniguchi et al. 2017). Proteomics for cytoplasmic proteins, membrane fraction proteins, cell wall-associated proteins, and secreted proteins are now available (Hermann et al. 2000; Schaffer et al. 2001; Schluesener et al. 2007; Fischer and Poetsch 2006; Hansmeier et al. 2006). This, for example, led to the discovery of pupylation as posttranslational modification that is relevant for iron release from the iron storage protein ferritin independent of degradation (Kuberl et al. 2014, 2016). Metabolomics has been developed for *C. glutamicum* (Klatt et al. 2018; Zhang et al. 2018) and, for example, helped to identify a new pathway involving γ-glutamyl transpeptidase with γ-glutamyl dipeptides (γ-Glu-Glu, γ-Glu-Gln, γ-Glu-Val, γ-Glu-Leu, γ-Glu-Met) having been detected by HPLC-MS in concentrations from 0.15 to 0.4 mg/g CDW (Walter et al. 2016).

2 Prophage-Cured Strains

2.1 MB001 Derived from Wild-Type ATCC 13032

The *C. glutamicum* genome contains three prophage DNA islands (CGP1, CGP2, and CGP3). CGP1 comprises genes cg1507 to cg1524 (13.5 kbp), CGP2 comprises genes cg1746 to cg1752 (3.9 kbp), and CGP3 is the largest prophage region with 187.3 kbp (comprising genes cg1890 to cg2071 (Kalinowski et al. 2003; Frunzke et al. 2008). The activity of bacteriophages and phage-related mobile elements is a major source for genome rearrangements and genetic instability of their bacterial hosts. Genome-wide expression analysis often revealed differential expression of phage genes (Sindelar and Wendisch 2007; Krings et al. 2006). Moreover, the large prophage CGP3 has recently been shown to be excised under SOS-response-inducing conditions (Frunzke et al. 2008). Single-cell analyses with transcriptional fusions of promoters of phage genes (Pint2 and Plysin) to fluorescent protein reporter genes revealed that 0.01–0.08% of the cells grown in standard minimal medium induced CGP3 spontaneously, which reduced their viability. Apparently, spontaneously occurring DNA damage induced the SOS response and as consequence prophage induction (Nanda et al. 2014). This process required actively proliferating cells, whereas sporadic SOS induction was still observed in resting cells (Helfrich et al. 2015). The prophage CGP3-encoded nucleoid-associated protein CgpS binds AT-rich DNA as prevails in the entire CGP3 prophage region, but is scarce throughout the rest of the genome. In its absence, a significantly increased induction frequency of the CGP3 prophage resulted, whereas a strain lacking the CGP3 prophage displayed stable growth (Pfeifer et al. 2016). Based on the properties of the prophages and the resulting genetic instability, the first target for genome reduction was the deletion of these prophage DNA islands (Baumgart et al. 2013).

Deletion of the three prophage DNA islands reduced the genome size of *C. glutamicum* ATCC 13032 by 6% and resulted in strain MB001. Its growth properties were unchanged under standard and stress conditions. Under SOS-response-inducing conditions that trigger CGP3 induction in the *C. glutamicum* wild type, strain MB001 fared better than the wild type showing improved growth and fitness. In addition, strain MB001 exhibited increased transformation efficiency. This was attributed to the loss of the restriction-modification system (cg1996–cg1998) located within CGP3. Furthermore, plasmid copy number appeared to be increased since production of a heterologous model protein (enhanced yellow fluorescent protein, eYFP) was 30% higher than in the wild type. Similarly, deletion of the genes for restriction-modification system (cg1996–cg1998) improved eYFP production (Baumgart et al. 2013).

These results characterized MB001 as an intermediate strain to be improved by further genome reduction (s. below), e.g., by targeting mobile IS elements, and as a suitable strain for metabolic engineering (stable, growing as fast as wild type on glucose minimal medium, higher plasmid copy number, and better transformation efficiency). *C. glutamicum* MB001 was used as host for the production of various value-added compounds: amino acids (Eberhardt et al. 2014; Lubitz et al. 2016; Jensen et al. 2015; Wu et al. 2019; Lubitz and Wendisch 2016), phenylpropanoids (Kallscheuer and Marienhagen 2018; Kallscheuer et al. 2016a, b), isoprenoids (Henke et al. 2016, 2018a; Heider et al. 2014a, b; Binder et al. 2016), alcohols (Huang et al. 2017), carboxylic acids (Lubitz and Wendisch 2016; Chen et al. 2017), and proteins (Kortmann et al. 2015; Hemmerich et al. 2016, 2018a, 2019). In addition, MB001 and derivatives have been used to study Mu-transposition (Gorshkova et al. 2018), assembly of the septal cell envelope (Zhou et al. 2019), infection with phages φ673 and φ674 phages (Yomantas et al. 2018), identification of an isoprenoid pyrophosphate-dependent transcriptional regulator (Henke et al. 2017), cAMP phosphodiesterase CpdA (Schulte et al. 2017), and cryptic prophages (Pfeifer et al. 2016), as basis for ALE toward higher growth rates on glucose minimal medium (Pfeifer et al. 2017) and to assemble bacterial microcompartments (Huber et al. 2017).

2.2 Prophage-Cured Lysine-Producing Model Strain GRLys1

The concept of prophage island DNA deletion was transferred from the wild type (see above) to the lysine-producing model strain DM1933 (Unthan et al. 2015a). The prophage DNA sequences of the three phages CGP1, CGP2, and CGP3 were deleted from the base strain DM1933 that contained the following genomic modifications promoting lysine overproduction: Δ*pck*, pyc^{P458S}, hom^{V59A}, *2 copies of* $lysC^{T311I}$, *asd, dapA, dapB, ddh, lysA,* and *lysE* (Unthan et al. 2015a). Derivatives of GRLys1 were used to overproduce L-pipecolic acid (L-PA) (Perez-Garcia et al. 2016, 2017, 2019), 5-aminovaleric acid (5AVA) (Jorge et al. 2017b), glutarate (Perez-Garcia et al. 2018), and for the coproduction of astaxanthin with lysine (Henke et al. 2018a).

3 IS Element-Free Strain

3.1 MB001-Derived IS Element-Free Strain CR099

All copies of IS elements ISCg1 and ISCg2 were deleted from the genome of strain MB001. In addition, it contains mutation A468T in Cg1720 which was inadvertently introduced. Cg1720 encodes the ATPase component of an uncharacterized ABC transporter. This strain was used to characterize synthetases and a hydrolase of the small alarmone (pp)pGpp (Ruwe et al. 2017, 2018). In a similar approach two IS element-free *C. glutamicum* strains were derived from ATCC 13032: one lacking IS elements IS*Cg1a*, IS*Cg1b*, IS*Cg1c*, IS*Cg1e* and another lacking ISCg2b, ISCg2c, ISCg2e, ISCg2f (Choi et al. 2015). Increased protein production was demonstrated in the IS element-free strains (Choi et al. 2015).

4 *C. glutamicum* Chassis Strain C1∗ Derived from ATCC 13032

A chassis strain based on *C. glutamicum* ATCC 13032 was constructed in a targeted top-down approach. As target function, uncompromised growth in glucose minimal medium was chosen. *C. glutamicum* MB001 was used as starting strain. Next, genes were classified either as (a) known to be nonessential from prior experiments, (b) likely nonessential based on transposon mutagenesis screens, (c) unclassifiable or (d) likely essential due to high conservation (Fig. 1). From these, genomic clusters with genes classified as (likely) nonessential were chosen for deletion from the genome of MB001. The generated deletion mutants were evaluated with respect to growth in glucose minimal medium. This phenotyping step proved crucial to identify nonessential gene clusters that are irrelevant for maintaining the biological fitness of the wild type (WT). A total of 26 gene clusters were found to be nonessential and their individual deletions shown not to compromise growth in glucose minimal medium.

Based on this mutant collection, combinatorial deletions of these gene clusters were performed resulting in a library of 28 strains. After statistical analysis of a thorough phenotypic screen and one genetic correction, the final chassis strain C1∗, exhibiting a genome reduction of 13.4% (412 deleted genes; Fig. 2) but showing wild-type-like growth behavior in glucose minimal medium, robustness against several stresses (including oxygen limitation), and long-term growth stability in defined and complex growth media, was selected (Baumgart et al. 2018).

Notably, genome sequencing of the penultimate strain, named C1, revealed a mutation in the promoter region of regulatory gene *ramA* (Auchter et al. 2011), i.e., a promoter down mutation (TGCACT instead of the conserved −10-region TACACT). Moreover, this mutation is located in the SugR binding sites overlapping the −10

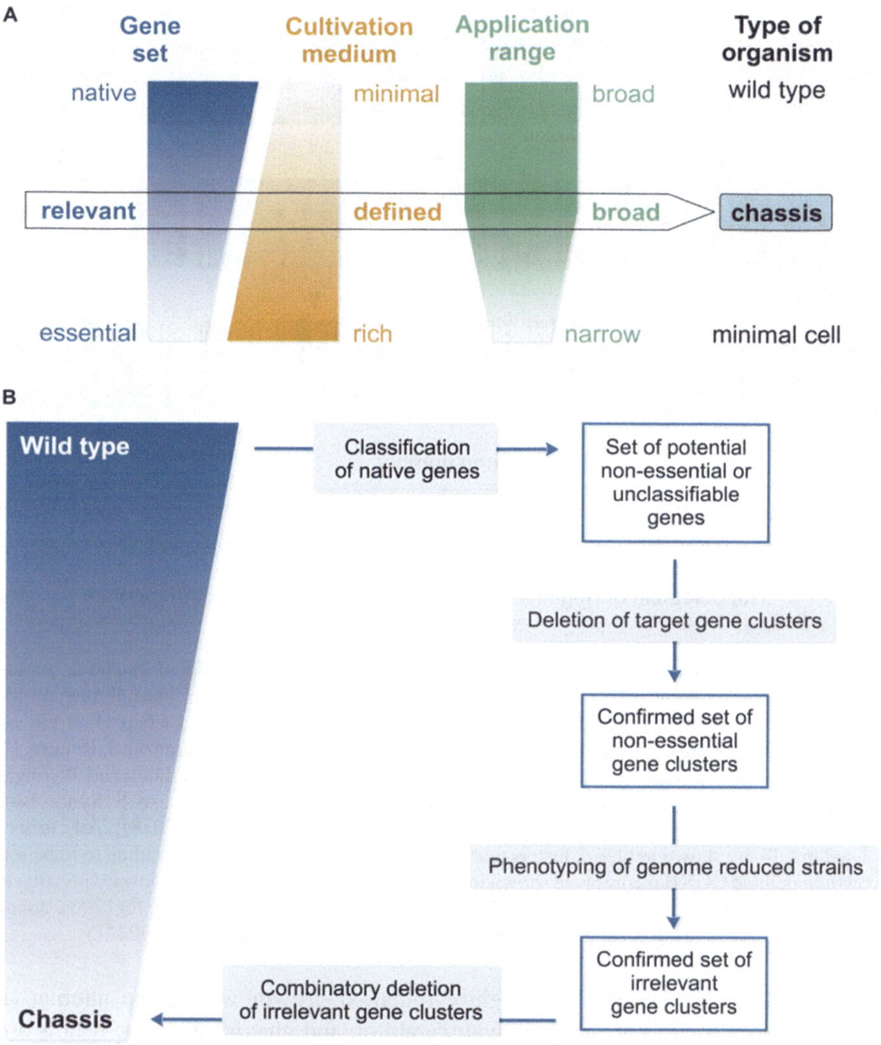

Fig. 1 Definitions and workflow for the construction of a chassis organism of *Corynebacterium glutamicum* [Copyright © 2015 Unthan, Baumgart, Radek, Herbst, Siebert, Brühl, Bartsch, Bott, Wiechert, Marin, Hans, Krämer, Seibold, Frunzke, Kalinowski, Rückert, Wendisch, Noack; reproduced from (Unthan et al. 2015a)]. (**a**) Definitions considering the interplay of gene set, cultivation medium, and application range for different types of organisms. (**b**) Scheme of our targeted top-down approach toward a chassis covering only genes that are relevant for growth on defined medium and maintaining the broad application range of the wild-type organism

region (Engels et al. 2008; Engels and Wendisch 2007). A transcriptome analysis revealed sixfold reduced *ramA* RNA levels and reduced RNA levels for several genes of the *ramA* regulon. Therefore, this point mutation in *C. glutamicum* C1 was reversed to yield the chassis strain *C. glutamicum* C1∗ (Baumgart et al. 2018).

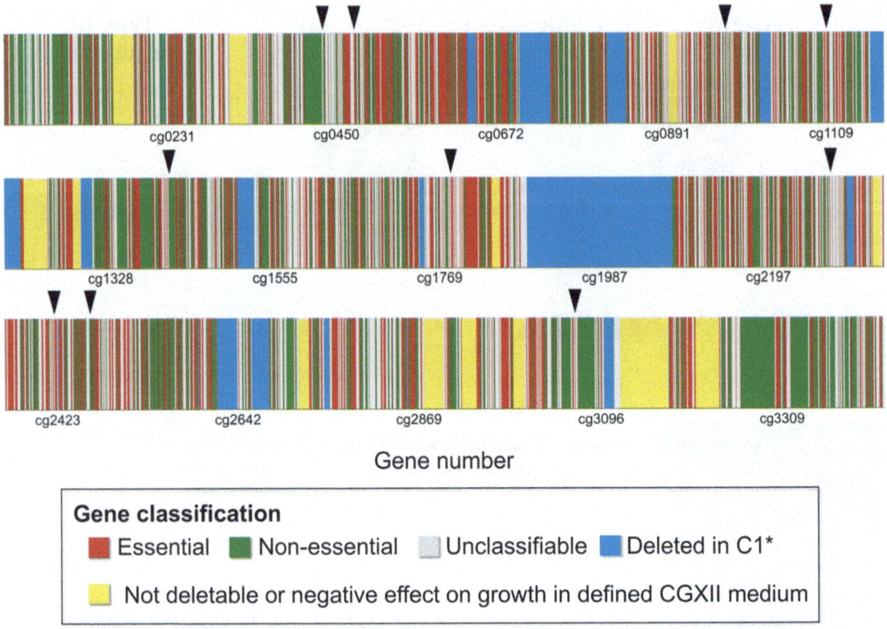

Fig. 2 *C. glutamicum* ATCC 13032 genome map with classification results of essential, nonessential, and unclassifiable genes. [Copyright © Reprinted with permission from Baumgart M, Unthan S, Kloss R, Radek A, Polen T, Tenhaef N, Muller MF, Kuberl A, Siebert D, Bruhl N, Marin K, Hans S, Kramer R, Bott M, Kalinowski J, Wiechert W, Seibold G, Frunzke J, Ruckert C, Wendisch VF, Noack S (2018) *Corynebacterium glutamicum* Chassis C1∗: Building and Testing a Novel Platform Host for Synthetic Biology and Industrial Biotechnology. ACS Synth Biol 7 (1):132–144. Copyright 2018 American Chemical Society (Baumgart et al. 2018)]. All clusters deleted in C1∗ are shown in blue. Clusters that could not be deleted or deletions leading to impaired growth in defined CGXII medium are shown in yellow. Black arrows are pointing toward glycolysis genes *pgi* (cg0973), *pfkA* (cg1409), *fda* (cg3068), *tpi* (cg1789), *gap* (cg1791), *pgk* (cg1790), *gpmA* (cg0482), *eno* (cg1111), *pyk* (cg2291), *aceE* (cg2466), *lpd* (cg0441), and *sucB* (cg2421)

C. glutamicum C1∗ showed slightly impaired growth with some alternative carbon sources such as acetate, pyruvate, arabitol, and gluconate. These results are possible since the target function chosen was uncompromised growth with glucose as sole carbon and energy source. However, these physiological peculiarities have to be remembered when constructing and evaluating C1∗-derived strains for production purposes. As in the case of reversion of the *ramA* promoter down mutation present in C1, other SNPs may have to be reverted to allow for fast growth with acetate, pyruvate, gluconate, or arabitol.

For all glucose-based production purposes, *C. glutamicum* C1∗ is an ideal starting point for metabolic engineering as a biotechnologically relevant chassis.

5 Applications of Genome-Reduced Strains

5.1 Applications of Prophage-Cured Strain MB001 and Derivatives

C. glutamicum MB001 found manifold biotechnological applications (Table 1). Derivatives of this prophage-cured strain were used for the production of proteins (Kortmann et al. 2015), citrulline (Eberhardt et al. 2014; Lubitz et al. 2016), proline (Jensen et al. 2015), lysine (Wu et al. 2019), decaprenoxanthin (Heider et al. 2014a, b), astaxanthin (Henke et al. 2016, 2018a), ciprofloxacin-triggered glutamate and oxoglutarate production (Lubitz and Wendisch 2016), valencene (Binder et al. 2016), 3-hydroxypropionic acid (Chen et al. 2017), coproduction of 1,3-propanediol and glutamate (Huang et al. 2017), and phenylpropanoids (Kallscheuer and Marienhagen 2018; Kallscheuer et al. 2016a, b).

As an example, the construction and use of strain MB001(DE3) for protein production based on an IPTG-inducible T7 expression system will be discussed. Part of the DE3 region from the protein production host *E. coli* BL21(DE3) including the T7 RNA polymerase *gene 1* driven by the *E. coli* lacUV5 promoter, which also is active in *C. glutamicum*, was integrated into the chromosome of *C. glutamicum* MB001 (Kortmann et al. 2015). The corresponding expression vector pMKEx2 was developed to express (a) the *lacI* gene encoding *E. coli lac* repressor and (b) genes of interest under the control of a T7 promoter followed by *lacO1* for induction by IPTG (Kortmann et al. 2015). The inducibility of the system was shown to be 450-fold when expression of the fluorescence protein reporter gene *eyfp* was analyzed. Fully IPTG-induced T7 RNA polymerase-dependent expression was about 3.5 times higher than the expression from the fully IPTG-induced *tac* promoter in a control strain with the endogenous RNA polymerase. Importantly, fully IPTG-induced T7 RNA polymerase-dependent expression led to a uniform population with 99% of all cells showing high fluorescence as shown by flow cytometry (Kortmann et al. 2015). As an impressive application example, overexpression of the endogenous pyruvate kinase gene *pyk* was demonstrated. The already very high *pyk* gene expression in the wild type (leading to a specific pyruvate kinase activity of 2.6 U/mg) was boosted about 50-fold (135 U/mg) (Kortmann et al. 2015).

5.2 CORYNEX

C. glutamicum strain ATCC13869 was commercialized as a protein expression system under the trademark CORYNEX® by the Japanese company Ajinomoto. When using the CORYNEX® strain YDK010, secretion of the Fab fragment of human anti-HER2 was low. Deletion of the genes encoding penicillin-binding protein (PBP1a), which is involved in cell wall peptidoglycan synthesis, and the surface (S)-layer protein CspB, showed a synergistic effect allowing efficient Fab production

Table 1 Biotechnological applications using genome-reduced *C. glutamicum* strains

Product	Base strain	Production parameter(s)	References
3-hydroxy-propionic acid			Chen et al. (2017)
Arginine	MB001	Y: 0.30 $g \cdot g^{-1}$	Jensen et al. (2015)
Astaxanthin	MB001	C: 1.7 mg g^{-1} DCW; V: 0.4 mg L^{-1} h^{-1};	Henke et al. (2016, 2018a)
Noreugenin	MB001	T: 53 mg/L	Milke et al. (2019)
Citrulline	MB001	T: 44.1 ± 0.5 mM; Y: 0.38 ± 0.01 $g \cdot g^{-1}$; P: 0.32 ± 0.01 $g \cdot l^{-1} \cdot h^{-1}$	Eberhardt et al. (2014), Lubitz et al. (2016)
Coproduction of 1,3-propanediol and glutamate			Huang et al. (2017)
Coproduction of astaxanthin with glutamate	MB001	Astaxanthin: T: 2.33 $mg \cdot L^{-1}$; Y = 2.22 $g \cdot g^{-1}$; P: 0.12 $mg \cdot L^{-1} \cdot h^{-1}$ Glutamate: T: 0.05 $g \cdot L^{-1}$; Y: 0.13 $g \cdot g^{-1}$; V: 005 $g \cdot L^{-1} \cdot h^{-1}$	Henke et al. (2018a)
Coproduction of astaxanthin with lysine	GRLys1	Astaxanthin: T: 10 $mg \cdot L^{-1}$; C: 0.4 $mg \cdot g^{-1}$; Y = 0.07 $g \cdot g^{-1}$ Lysine: T: 48 $g \cdot L^{-1}$; Y: 0.35 $g \cdot g^{-1}$	Henke et al. (2018a)
Coproduction of decaprenoxanthin with glutamate	MB001	Decaprenoxanthin: T: 8.66 $mg \cdot L^{-1}$; Y = 0.97 $g \cdot g^{-1}$; P: 0.05 $mg \cdot L^{-1} \cdot h^{-1}$ Glutamate: T: 0.02 $g \cdot L^{-1}$; Y: 0.48 $g \cdot g^{-1}$; V: 0.18 $g \cdot L^{-1} \cdot h^{-1}$	Henke et al. (2018a)
Coproduction of decaprenoxanthin with lysine	GRLys1	Decaprenoxanthin: T: 6.10 $mg \cdot L^{-1}$; Y = 0.34 $g \cdot g^{-1}$; P: 0.19 $mg \cdot L^{-1} \cdot h^{-1}$ Lysine: T: 2.79 $g \cdot L^{-1}$; Y: 0.15 $g \cdot g^{-1}$; V: 0.09 $g \cdot L^{-1} \cdot h^{-1}$	Henke et al. (2018a)
Decaprenoxanthin	MB001	C: 0.4 mg g^{-1} DCW	Heider et al. (2014a, b)
Glutamate (triggered by ciprofloxycin)	MB001	T: 37 mM; Y: 0.13 g g^{-1}	Lubitz and Wendisch (2016)
Lycopene	MB001	C: 0.43 mg g^{-1} DCW	Henke et al. (2016, 2018a)
Lysine			Wu et al. (2019)
Ornithine	MB001	Y: 0.52 $g \cdot g^{-1}$	Jensen et al. (2015)
Oxoglutarate (triggered by ciprofloxycin)	MB001	T: 18 mM	Lubitz and Wendisch (2016)
4-hydroxy-butyrate	MB001	T: 3.3 g g^{-1}	Kallscheuer and Marienhagen (2018)
Resveratrol	MB001	T: 158 mg L^{-1}	Kallscheuer et al. (2016a, b)
Proline	MB001	Y: 0.29 $g \cdot g^{-1}$	Jensen et al. (2015)

(continued)

Table 1 (continued)

Product	Base strain	Production parameter(s)	References
Proteins	MB001	Pyruvate kinase: Sp.act. 135 U/mg	Kortmann et al. (2015)
Protocatechuate	MB001	T: 2 g g^{-1}	Kallscheuer and Marienhagen (2018)
Putrescine	MB001	Y: 0.17 g·g^{-1}	Jensen et al. (2015)
Zeaxanthin	MB001	C: 1.2 mg g^{-1} DCW	Heider et al. (2014a, b)
β-Carotene	MB001	C: 12 mg g^{-1} DCW; V: 3.4 mg L^{-1} h^{-1}	Henke et al. (2016, 2018a)

Abbreviations: *T* titer or concentration in culture broth, *Y* product yield on substrate (unless otherwise indicated glucose was used as substrate), *V* volumetric productivity, *C* cellular content, *CDW* cell dry weight

using the CORYNEX® system. This indicated at least two major permeability barriers to Fab secretion, i.e. peptidoglycan and the S-layer (Matsuda et al. 2014).

5.3 Applications of Prophage-Cured Strain GRLys1 and Derivatives

Derivatives of GRLys1 were used to overproduce 5AVA (Jorge et al. 2017b), L-PA (Perez-Garcia et al. 2016, 2017, 2019), glutarate (Perez-Garcia et al. 2018), and for the coproduction of astaxanthin with lysine (Henke et al. 2018a).

As example, glutarate production based on the prophage-cured, lysine-producing model strain GRLys1 will be discussed (Fig. 3). Systems metabolic engineering included flux enforcement, which refers to coupling a biosynthetic production pathway to a metabolite pathway required for growth. This strategy has previously been applied to amino acid production by *E. coli* and *C. glutamicum*. Coupling of a production pathway involving a 2-oxoglutarate-dependent hydroxylase to growth by deletion of 2-oxoglutarate dehydrogenase subunit gene *sucA* has first been shown for 4-hydroxy-L-isoleucine production by *E. coli* (Smirnov et al. 2010) and later for 4-hydroxy-L-proline production (Theodosiou et al. 2017). Thus, these production pathways became part of an artificial TCA cycle. This concept was extended in succinyl-CoA synthetase-negative (Δ*sucCD*), lysine-producing *C. glutamicum* strains. In this case, the succinylase branch of L-lysine production metabolically complemented the TCA cycle disrupted due to the *sucCD* deletion (Kind et al. 2013). Also coupling of the major ammonium assimilating enzyme glutamate dehydrogenase to transamination reactions was used for flux enforcement when cadaverine/putrescine transaminase PutA and GABA/5AVA aminotransferase GabT introduced for glutarate production metabolically complemented for the

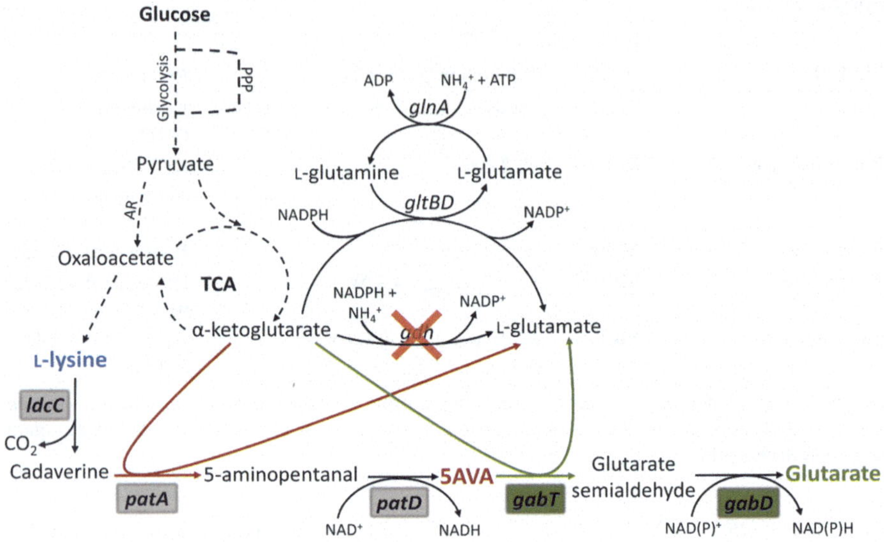

Fig. 3 Schematic representation of the metabolic engineering strategy for glutarate production by recombinant *C. glutamicum* [Copyright © 2018 Pérez-García, Jorge, Dreyszas, Risse and Wendisch; reproduced from (Perez-Garcia et al. 2018)]. The biosynthetic pathway for glutarate production was implemented by heterologous expression in an L-lysine producer and coupled with endogenous L-glutamate synthesis. *PPP* pentose phosphate pathway, *TCA* tricarboxylic acid cycle, *AR* anaplerotic reactions, *glnA* glutamine synthase gene, *gltBD* glutamine aminotransferase complex genes, *gdh* glutamate dehydrogenase, *ldcC* L-lysine decarboxylase, *patA* putrescine transaminase, *patD*γ-aminobutyraldehyde dehydrogenase, *gabT* GABA/5AVA aminotransferase gene, *gabD* succinate/glutarate-semialdehyde dehydrogenase gene. Magenta arrows depict transamination reaction in the 5AVA pathway. Green arrows depict transamination reaction in the glutarate pathway. Gray-shadowed genes are originally from *E. coli* and were added by heterologous overexpression. Green-shadowed genes are originally from *C. glutamicum*, *P. putida*, *P. syringae*, or *P. stutzeri* and were added by heterologous overexpression

absence of glutamate dehydrogenase (Perez-Garcia et al. 2018). This prophage-cured, flux-enforced strain in addition required expression of a heterologous gene for lysine decarboxylase for glutarate production. In this five-step synthetic pathway, lysine was decarboxylated to cadaverine by lysine decarboxylase, and cadaverine converted to glutarate by two transamination (catalyzed PutA, GabT) and two oxidation steps (catalyzed by PutD and GabD) to the targeted product glutarate (Perez-Garcia et al. 2018).

6 Outlook on Construction and Testing of New Genome-Reduced Strains

Targets for gene deletions relevant for genome reduction can be scored by CRISPR interference (Wiedenheft et al. 2012) as applied first to *C. glutamicum* with respect to lysine production (Cleto et al. 2016). Evaluation of groups of genes for combined deletion can be done by multiplex CRISPRi (Park et al. 2018). Sequential or parallel targeted genome deletions and replacements in *C. glutamicum* by CRISPR genome editing are facile since this bacterium lacks efficient nonhomologous end-joining. Although genome reduction in *C. glutamicum* has until now relied on genome editing by two-step homologous recombination using the conditionally lethal levansucrase (*sacB*) for positive selection (Jäger et al. 1992), genome editing by CRISPR/Cas9 or CRISPR/Cas12a as developed for *C. glutamicum* (Cho et al. 2017; Jiang et al. 2017; Wang et al. 2018a, b; Cameron Coates et al. 2019; Liu et al. 2017) will find application in further genome streamlining.

Highly parallel strain characterization relies on microbioreactor systems that are based either on shaken microtiter plate cultivation devices or on downscaled stirred tank reactors (Hemmerich et al. 2018b). These systems allow for optical, noninvasive, online monitoring of important process parameters such as biomass concentration, dissolved oxygen, pH, or reporter protein fluorescence. Their use is potentiated by combination with liquid handling robots for automatization of operation procedures. On-line and off-line strain phenotyping under industrially relevant conditions enables identification of the optimal combination of producer strain and bioprocess control strategy. Of course, the strain collections generated in genome reduction projects can be scored very well using microbioreactor systems as has been shown for characterizing growth (Hemmerich et al. 2017), protein secretion (Hemmerich et al. 2016, 2019), or amino acid production (Steffen et al. 2016; Baumgart et al. 2013, 2018; Unthan et al. 2015a, b).

Acknowledgment VFW gratefully acknowledges support by ERACoBiotech grant INDIE (BMEL 22023517), by the Indo-German project BIOCON (BMBF 01DQ17009) and by funding from the state of North Rhine-Westphalia (NRW) and the "European Regional Development Fund (EFRE)", Project "ClusterIndustrial Biotechnology (CLIB) Kompetenzzentrum Biotechnologie (CKB)" (34.EFRE-0300095/1703FI04).

References

Auchter M, Cramer A, Huser A, Ruckert C, Emer D, Schwarz P, Arndt A, Lange C, Kalinowski J, Wendisch VF, Eikmanns BJ (2011) RamA and RamB are global transcriptional regulators in *Corynebacterium glutamicum* and control genes for enzymes of the central metabolism. J Biotechnol 154(2–3):126–139. https://doi.org/10.1016/j.jbiotec.2010.07.001

Baumgart M, Unthan S, Ruckert C, Sivalingam J, Grunberger A, Kalinowski J, Bott M, Noack S, Frunzke J (2013) Construction of a prophage-free variant of *Corynebacterium glutamicum*

ATCC 13032 for use as a platform strain for basic research and industrial biotechnology. Appl Environ Microbiol 79(19):6006–6015. https://doi.org/10.1128/AEM.01634-13

Baumgart M, Unthan S, Kloss R, Radek A, Polen T, Tenhaef N, Muller MF, Kuberl A, Siebert D, Bruhl N, Marin K, Hans S, Kramer R, Bott M, Kalinowski J, Wiechert W, Seibold G, Frunzke J, Ruckert C, Wendisch VF, Noack S (2018) *Corynebacterium glutamicum*chassis C1∗: building and testing a novel platform host for synthetic biology and industrial biotechnology. ACS Synth Biol 7(1):132–144. https://doi.org/10.1021/acssynbio.7b00261

Becker J, Wittmann C (2012) Bio-based production of chemicals, materials and fuels -*Corynebacterium glutamicum* as versatile cell factory. Curr Opin Biotechnol 23(4):631–640. https://doi.org/10.1016/j.copbio.2011.11.012

Binder D, Frohwitter J, Mahr R, Bier C, Grunberger A, Loeschcke A, Peters-Wendisch P, Kohlheyer D, Pietruszka J, Frunzke J, Jaeger KE, Wendisch VF, Drepper T (2016) Light-controlled cell factories: employing photocaged isopropyl-beta-D-thiogalactopyranoside for light-mediated optimization of *lac* promoter-based gene expression and (+)-Valencene biosynthesis in *Corynebacterium glutamicum*. Appl Environ Microbiol 82(20):6141–6149. https://doi.org/10.1128/AEM.01457-16

Blombach B, Riester T, Wieschalka S, Ziert C, Youn JW, Wendisch VF, Eikmanns BJ (2011) *Corynebacterium glutamicum* tailored for efficient isobutanol production. Appl Environ Microbiol 77(10):3300–3310. https://doi.org/10.1128/AEM.02972-10

Cameron Coates R, Blaskowski S, Szyjka S, van Rossum HM, Vallandingham J, Patel K, Serber Z, Dean J (2019) Systematic investigation of CRISPR-Cas9 configurations for flexible and efficient genome editing in *Corynebacterium glutamicum* NRRL-B11474. J Ind Microbiol Biotechnol 46(2):187–201. https://doi.org/10.1007/s10295-018-2112-7

Chassagnole C, Letisse F, Diano A, Lindley ND (2002) Carbon flux analysis in a pantothenate overproducing *Corynebacterium glutamicum* strain. Mol Biol Rep 29(1–2):129–134

Chen Z, Huang J, Wu Y, Wu W, Zhang Y, Liu D (2017) Metabolic engineering of *Corynebacterium glutamicum* for the production of 3-hydroxypropionic acid from glucose and xylose. Metab Eng 39:151–158. https://doi.org/10.1016/j.ymben.2016.11.009

Cheng F, Luozhong S, Yu H, Guo Z (2019) Biosynthesis of chondroitin in engineered *Corynebacterium glutamicum*. J Microbiol Biotechnol 29:392. https://doi.org/10.4014/jmb.1810.10062

Cho JS, Choi KR, Prabowo CPS, Shin JH, Yang D, Jang J, Lee SY (2017) CRISPR/Cas9-coupled recombineering for metabolic engineering of *Corynebacterium glutamicum*. Metab Eng 42:157–167. https://doi.org/10.1016/j.ymben.2017.06.010

Choi JW, Yim SS, Kim MJ, Jeong KJ (2015) Enhanced production of recombinant proteins with *Corynebacterium glutamicum* by deletion of insertion sequences (IS elements). Microb Cell Factories 14:207. https://doi.org/10.1186/s12934-015-0401-7

Cleto S, Jensen JV, Wendisch VF, Lu TK (2016) *Corynebacterium glutamicum*metabolic engineering with CRISPR interference (CRISPRi). ACS Synth Biol 5(5):375–385. https://doi.org/10.1021/acssynbio.5b00216

Contador CA, Rizk ML, Asenjo JA, Liao JC (2009) Ensemble modeling for strain development of L-lysine-producing*Escherichia coli*. Metab Eng 11(4–5):221–233. https://doi.org/10.1016/j.ymben.2009.04.002

Eberhardt D, Jensen JV, Wendisch VF (2014) L-citrulline production by metabolically engineered *Corynebacterium glutamicum* from glucose and alternative carbon sources. AMB Express 4:85

Eggeling L, Bott M, Marienhagen J (2015) Novel screening methods--biosensors. Curr Opin Biotechnol 35:30–36. https://doi.org/10.1016/j.copbio.2014.12.021

Engels V, Wendisch VF (2007) The DeoR-type regulator SugR represses expression of *ptsG* in *Corynebacterium glutamicum*. J Bacteriol 189(8):2955–2966

Engels V, Lindner SN, Wendisch VF (2008) The global repressor SugR controls expression of genes of glycolysis and of the L-lactate dehydrogenase LdhA in *Corynebacterium glutamicum*. J Bacteriol 190(24):8033–8044. https://doi.org/10.1128/JB.00705-08

Feist AM, Henry CS, Reed JL, Krummenacker M, Joyce AR, Karp PD, Broadbelt LJ, Hatzimanikatis V, Palsson BO (2007) A genome-scale metabolic reconstruction for *Escherichia coli* K-12 MG1655 that accounts for 1260 ORFs and thermodynamic information. Mol Syst Biol 3:121. https://doi.org/10.1038/msb4100155

Fischer F, Poetsch A (2006) Protein cleavage strategies for an improved analysis of the membrane proteome. Proteome Sci 4:2

Frazzetto G (2003) White biotechnology. EMBO Rep 4(9):835–837. https://doi.org/10.1038/sj.embor.embor928

Freiherr von Boeselager R, Pfeifer E, Frunzke J (2018) Cytometry meets next-generation sequencing - RNA-Seq of sorted subpopulations reveals regional replication and iron-triggered prophage induction in *Corynebacterium glutamicum*. Sci Rep 8(1):14856. https://doi.org/10.1038/s41598-018-32997-9

Freudl R (2017) Beyond amino acids: use of the *Corynebacterium* glutamicum cell factory for the secretion of heterologous proteins. J Biotechnol 258:101–109. https://doi.org/10.1016/j.jbiotec.2017.02.023

Freudl R (2018) Signal peptides for recombinant protein secretion in bacterial expression systems. Microb Cell Factories 17:52. https://doi.org/10.1186/s12934-018-0901-3

Frunzke J, Bramkamp M, Schweitzer JE, Bott M (2008) Population heterogeneity in *Corynebacterium glutamicum* ATCC 13032 caused by prophage CGP3. J Bacteriol 190(14):5111–5119

Gorshkova NV, Lobanova JS, Tokmakova IL, Smirnov SV, Akhverdyan VZ, Krylov AA, Mashko SV (2018) Mu-driven transposition of recombinant mini-Mu unit DNA in the *Corynebacterium glutamicum* chromosome. Appl Microbiol Biotechnol 102(6):2867–2884. https://doi.org/10.1007/s00253-018-8767-1

Hansmeier N, Chao TC, Puhler A, Tauch A, Kalinowski J (2006) The cytosolic, cell surface and extracellular proteomes of the biotechnologically important soil bacterium *Corynebacterium efficiens* YS-314 in comparison to those of *Corynebacterium glutamicum* ATCC 13032. Proteomics 6(1):233–250

Heider SA, Peters-Wendisch P, Netzer R, Stafnes M, Brautaset T, Wendisch VF (2014a) Production and glucosylation of C50 and C 40 carotenoids by metabolically engineered *Corynebacterium glutamicum*. Appl Microbiol Biotechnol 98(3):1223–1235. https://doi.org/10.1007/s00253-013-5359-y

Heider SA, Wolf N, Hofemeier A, Peters-Wendisch P, Wendisch VF (2014b) Optimization of the IPP precursor supply for the production of lycopene, decaprenoxanthin and astaxanthin by *Corynebacterium glutamicum*. Front Bioeng Biotechnol 2:28. https://doi.org/10.3389/fbioe.2014.00028

Helfrich S, Pfeifer E, Kramer C, Sachs CC, Wiechert W, Kohlheyer D, Noh K, Frunzke J (2015) Live cell imaging of SOS and prophage dynamics in isogenic bacterial populations. Mol Microbiol 98(4):636–650. https://doi.org/10.1111/mmi.13147

Hemmerich J, Rohe P, Kleine B, Jurischka S, Wiechert W, Freudl R, Oldiges M (2016) Use of a Sec signal peptide library from Bacillus subtilis for the optimization of cutinase secretion in *Corynebacterium glutamicum*. Microb Cell Factories 15(1):208. https://doi.org/10.1186/s12934-016-0604-6

Hemmerich J, Wiechert W, Oldiges M (2017) Automated growth rate determination in high-throughput microbioreactor systems. BMC Res Notes 10(1):617. https://doi.org/10.1186/s13104-017-2945-6

Hemmerich J, Tenhaef N, Steffens C, Kappelmann J, Weiske M, Reich SJ, Wiechert W, Oldiges M, Noack S (2018a) Less sacrifice, more insight: repeated low-volume sampling of microbioreactor cultivations enables accelerated deep phenotyping of microbial strain libraries. Biotechnol J. https://doi.org/10.1002/biot.201800428

Hemmerich J, Noack S, Wiechert W, Oldiges M (2018b) Microbioreactor systems for accelerated bioprocess development. Biotechnol J 13(4):e1700141. https://doi.org/10.1002/biot.201700141

Hemmerich J, Moch M, Jurischka S, Wiechert W, Freudl R, Oldiges M (2019) Combinatorial impact of sec signal peptides from *Bacillus subtilis* and bioprocess conditions on heterologous cutinase secretion by *Corynebacterium glutamicum*. Biotechnol Bioeng 116(3):644–655. https://doi.org/10.1002/bit.26873

Henke NA, Heider SA, Peters-Wendisch P, Wendisch VF (2016) Production of the marine carotenoid astaxanthin by metabolically engineered *Corynebacterium glutamicum*. Mar Drugs 14(7):124. https://doi.org/10.3390/md14070124

Henke NA, Heider SAE, Hannibal S, Wendisch VF, Peters-Wendisch P (2017) Isoprenoid pyrophosphate-dependent transcriptional regulation of Carotenogenesis in *Corynebacterium glutamicum*. Front Microbiol 8:633. https://doi.org/10.3389/fmicb.2017.00633

Henke NA, Wiebe D, Perez-Garcia F, Peters-Wendisch P, Wendisch VF (2018a) Coproduction of cell-bound and secreted value-added compounds: simultaneous production of carotenoids and amino acids by *Corynebacterium glutamicum*. Bioresour Technol 247:744–752. https://doi.org/10.1016/j.biortech.2017.09.167

Henke NA, Wichmann J, Baier T, Frohwitter J, Lauersen KJ, Risse JM, Peters-Wendisch P, Kruse O, Wendisch VF (2018b) Patchoulol production with metabolically engineered *Corynebacterium glutamicum*. Genes (Basel) 9(4):219. https://doi.org/10.3390/genes9040219

Hermann T, Finkemeier M, Pfefferle W, Wersch G, Kramer R, Burkovski A (2000) Two-dimensional electrophoretic analysis of *Corynebacterium glutamicum* membrane fraction and surface proteins. Electrophoresis 21(3):654–659

Hoffmann J, Altenbuchner J (2014) Hyaluronic acid production with *Corynebacterium glutamicum*: effect of media composition on yield and molecular weight. J Appl Microbiol 117(3):663–678. https://doi.org/10.1111/jam.12553

Huang J, Wu Y, Wu W, Zhang Y, Liu D, Chen Z (2017) Cofactor recycling for co-production of 1,3-propanediol and glutamate by metabolically engineered *Corynebacterium glutamicum*. Sci Rep 7:42246. https://doi.org/10.1038/srep42246

Huber I, Palmer DJ, Ludwig KN, Brown IR, Warren MJ, Frunzke J (2017) Construction of recombinant Pdu metabolosome shells for small molecule production in *Corynebacterium glutamicum*. ACS Synth Biol 6(11):2145–2156. https://doi.org/10.1021/acssynbio.7b00167

Ikeda M (2003) Amino acid production processes. Adv Biochem Eng Biotechnol 79:1–35

Ikeda M, Nakagawa S (2003) The *Corynebacterium glutamicum* genome: features and impacts on biotechnological processes. Appl Microbiol Biotechnol 62(2–3):99–109

Imao K, Konishi R, Kishida M, Hirata Y, Segawa S, Adachi N, Matsuura R, Tsuge Y, Matsumoto T, Tanaka T, Kondo A (2017) 1,5-Diaminopentane production from xylooligosaccharides using metabolically engineered *Corynebacterium glutamicum* displaying beta-xylosidase on the cell surface. Bioresour Technol 245(Pt B):1684–1691. https://doi.org/10.1016/j.biortech.2017.05.135

Jäger W, Schäfer A, Pühler A, Labes G, Wohlleben W (1992) Expression of the *Bacillus subtilis sacB* gene leads to sucrose sensitivity in the gram-positive bacterium *Corynebacterium glutamicum* but not in *Streptomyces lividans*. J Bacteriol 174(16):5462–5465

Jensen JV, Eberhardt D, Wendisch VF (2015) Modular pathway engineering of *Corynebacterium glutamicum* for production of the glutamate-derived compounds ornithine, proline, putrescine, citrulline, and arginine. J Biotechnol 214:85–94. https://doi.org/10.1016/j.jbiotec.2015.09.017

Jiang Y, Qian F, Yang J, Liu Y, Dong F, Xu C, Sun B, Chen B, Xu X, Li Y, Wang R, Yang S (2017) CRISPR-Cpf1 assisted genome editing of *Corynebacterium glutamicum*. Nat Commun 8:15179. https://doi.org/10.1038/ncomms15179

Jo SJ, Matsumoto K, Leong CR, Ooi T, Taguchi S (2007) Improvement of poly(3-hydroxybutyrate) [P(3HB)] production in *Corynebacterium glutamicum* by codon optimization, point mutation and gene dosage of P(3HB) biosynthetic genes. J Biosci Bioeng 104(6):457–463

Jorge JM, Leggewie C, Wendisch VF (2016) A new metabolic route for the production of gamma-aminobutyric acid by *Corynebacterium glutamicum* from glucose. Amino Acids 48:2519. https://doi.org/10.1007/s00726-016-2272-6

Jorge JM, Nguyen AQ, Perez-Garcia F, Kind S, Wendisch VF (2017a) Improved fermentative production of gamma-aminobutyric acid via the putrescine route: systems metabolic engineering for production from glucose, amino sugars, and xylose. Biotechnol Bioeng 114(4):862–873. https://doi.org/10.1002/bit.26211

Jorge JMP, Perez-Garcia F, Wendisch VF (2017b) A new metabolic route for the fermentative production of 5-aminovalerate from glucose and alternative carbon sources. Bioresour Technol 245(Pt B):1701–1709. https://doi.org/10.1016/j.biortech.2017.04.108

Kalinowski J, Bathe B, Bartels D, Bischoff N, Bott M, Burkovski A, Dusch N, Eggeling L, Eikmanns BJ, Gaigalat L, Goesmann A, Hartmann M, Huthmacher K, Kramer R, Linke B, McHardy AC, Meyer F, Mockel B, Pfefferle W, Puhler A, Rey DA, Ruckert C, Rupp O, Sahm H, Wendisch VF, Wiegrabe I, Tauch A (2003) The complete *Corynebacterium glutamicum* ATCC 13032 genome sequence and its impact on the production of L-aspartate-derived amino acids and vitamins. J Biotechnol 104(1–3):5–25

Kallscheuer N, Marienhagen J (2018) *Corynebacterium glutamicum* as platform for the production of hydroxybenzoic acids. Microb Cell Factories 17(1):70. https://doi.org/10.1186/s12934-018-0923-x

Kallscheuer N, Vogt M, Stenzel A, Gatgens J, Bott M, Marienhagen J (2016a) Construction of a *Corynebacterium glutamicum* platform strain for the production of stilbenes and (2S)-flavanones. Metab Eng 38:47–55. https://doi.org/10.1016/j.ymben.2016.06.003

Kallscheuer N, Vogt M, Kappelmann J, Krumbach K, Noack S, Bott M, Marienhagen J (2016b) Identification of the *phd* gene cluster responsible for phenylpropanoid utilization in *Corynebacterium glutamicum*. Appl Microbiol Biotechnol 100(4):1871–1881. https://doi.org/10.1007/s00253-015-7165-1

Kim EM, Um Y, Bott M, Woo HM (2015) Engineering of *Corynebacterium glutamicum* for growth and succinate production from levoglucosan, a pyrolytic sugar substrate. FEMS Microbiol Lett 362(19):fnv161. https://doi.org/10.1093/femsle/fnv161

Kind S, Becker J, Wittmann C (2013) Increased lysine production by flux coupling of the tricarboxylic acid cycle and the lysine biosynthetic pathway--metabolic engineering of the availability of succinyl-CoA in *Corynebacterium glutamicum*. Metab Eng 15:184–195. https://doi.org/10.1016/j.ymben.2012.07.005

Kirchner O, Tauch A (2003) Tools for genetic engineering in the amino acid-producing bacterium *Corynebacterium glutamicum*. J Biotechnol 104(1–3):287–299

Kitade Y, Hashimoto R, Suda M, Hiraga K, Inui M (2018) Production of 4-Hydroxybenzoic acid by an aerobic growth-arrested bioprocess using metabolically engineered *Corynebacterium glutamicum*. Appl Environ Microbiol 84(6):e02587–e02517. https://doi.org/10.1128/AEM.02587-17

Kjeldsen KR, Nielsen J (2009) In silico genome-scale reconstruction and validation of the *Corynebacterium glutamicum* metabolic network. Biotechnol Bioeng 102(2):583–597. https://doi.org/10.1002/bit.22067

Klatt S, Brammananth R, O'Callaghan S, Kouremenos KA, Tull D, Crellin PK, Coppel RL, McConville MJ (2018) Identification of novel lipid modifications and intermembrane dynamics in *Corynebacterium glutamicum* using high-resolution mass spectrometry. J Lipid Res 59 (7):1190–1204. https://doi.org/10.1194/jlr.M082784

Kortmann M, Kuhl V, Klaffl S, Bott M (2015) A chromosomally encoded T7 RNA polymerase-dependent gene expression system for *Corynebacterium glutamicum*: construction and comparative evaluation at the single-cell level. Microb Biotechnol 8(2):253–265. https://doi.org/10.1111/1751-7915.12236

Krings E, Krumbach K, Bathe B, Kelle R, Wendisch VF, Sahm H, Eggeling L (2006) Characterization of myo-inositol utilization by *Corynebacterium glutamicum*: the stimulon, identification of transporters, and influence on L-lysine formation. J Bacteriol 188(23):8054–8061

Kuberl A, Franzel B, Eggeling L, Polen T, Wolters DA, Bott M (2014) Pupylated proteins in *Corynebacterium glutamicum* revealed by MudPIT analysis. Proteomics 14(12):1531–1542. https://doi.org/10.1002/pmic.201300531

Kuberl A, Polen T, Bott M (2016) The pupylation machinery is involved in iron homeostasis by targeting the iron storage protein ferritin. Proc Natl Acad Sci U S A 113(17):4806–4811. https://doi.org/10.1073/pnas.1514529113

Lee JH, Wendisch VF (2017) Production of amino acids –genetic and metabolic engineering approaches. Bioresour Technol 245(Pt B):1575–1587. https://doi.org/10.1016/j.biortech.2017.05.065

Lee JY, Seo J, Kim ES, Lee HS, Kim P (2013) Adaptive evolution of *Corynebacterium glutamicum* resistant to oxidative stress and its global gene expression profiling. Biotechnol Lett 35 (5):709–717. https://doi.org/10.1007/s10529-012-1135-9

Leuchtenberger W, Huthmacher K, Drauz K (2005) Biotechnological production of amino acids and derivatives: current status and prospects. Appl Microbiol Biotechnol 69(1):1–8

Litsanov B, Brocker M, Bott M (2012) Toward homosuccinate fermentation: metabolic engineering of *Corynebacterium glutamicum* for anaerobic production of succinate from glucose and formate. Appl Environ Microbiol 78(9):3325–3337. https://doi.org/10.1128/AEM.07790-11

Litsanov B, Brocker M, Bott M (2013) Glycerol as a substrate for aerobic succinate production in minimal medium with *Corynebacterium glutamicum*. Microb Biotechnol 6(2):189–195. https://doi.org/10.1111/j.1751-7915.2012.00347.x

Liu J, Wang Y, Lu Y, Zheng P, Sun J, Ma Y (2017) Development of a CRISPR/Cas9 genome editing toolbox for *Corynebacterium glutamicum*. Microb Cell Factories 16(1):205. https://doi.org/10.1186/s12934-017-0815-5

Lubitz D, Wendisch VF (2016) Ciprofloxacin triggered glutamate production by *Corynebacterium glutamicum*. BMC Microbiol 16(1):235. https://doi.org/10.1186/s12866-016-0857-6

Lubitz D, Jorge JM, Perez-Garcia F, Taniguchi H, Wendisch VF (2016) Roles of export genes *cgmA* and *lysE* for the production of L-arginine and L-citrulline by *Corynebacterium glutamicum*. Appl Microbiol Biotechnol 100(19):8465–8474. https://doi.org/10.1007/s00253-016-7695-1

Ma Q, Zhang Q, Xu Q, Zhang C, Li Y, Fan X, Xie X, Chen N (2017) Systems metabolic engineering strategies for the production of amino acids. Synth Syst Biotechnol 2(2):87–96. https://doi.org/10.1016/j.synbio.2017.07.003

Mahr R, Gätgens C, Gätgens J, Polen T, Kalinowski J, Frunzke J (2015) Biosensor-driven adaptive laboratory evolution of L-valine production in *Corynebacterium glutamicum*. Metab Eng 32:184–194. https://doi.org/10.1016/j.ymben.2015.09.017

Matsuda Y, Itaya H, Kitahara Y, Theresia NM, Kutukova EA, Yomantas YA, Date M, Kikuchi Y, Wachi M (2014) Double mutation of cell wall proteins CspB and PBP1a increases secretion of the antibody fab fragment from *Corynebacterium glutamicum*. Microb Cell Factories 13(1):56. https://doi.org/10.1186/1475-2859-13-56

Milke L, Kallscheuer N, Kappelmann J, Marienhagen J (2019) Tailoring *Corynebacterium glutamicum* towards increased malonyl-CoA availability for efficient synthesis of the plant pentaketide noreugenin. Microb Cell Factories 18(1):71. https://doi.org/10.1186/s12934-019-1117-x

Mindt M, Risse JM, Gruss H, Sewald N, Eikmanns BJ, Wendisch VF (2018a) One-step process for production of N-methylated amino acids from sugars and methylamine using recombinant *Corynebacterium glutamicum* as biocatalyst. Sci Rep 8(1):12895. https://doi.org/10.1038/s41598-018-31309-5

Mindt M, Walter T, Risse JM, Wendisch VF (2018b) Fermentative production of N-Methylglutamate from glycerol by recombinant *Pseudomonas putida*. Front Bioeng Biotechnol 6:159. https://doi.org/10.3389/fbioe.2018.00159

Mindt M, Heuser M, Wendisch VF (2019a) Xylose as preferred substrate for sarcosine production by recombinant *Corynebacterium glutamicum*. Bioresour Technol 281:135–142. https://doi.org/10.1016/j.biortech.2019.02.084

Mindt M, Hannibal S, Heuser M, Risse JM, Keerthi S, Madhavan Nampoothiri K, Wendisch VF (2019b) Fermentative production of N-alkylated glycine derivatives by recombinant Corynebacterium glutamicum using a mutant of imine reductase DpkA from Pseudomonas putida. Front Bioeng Biotechnol 7

Mustafi N, Grünberger A, Kohlheyer D, Bott M, Frunzke J (2012) The development and application of a single-cell biosensor for the detection of L-methionine and branched-chain amino acids. Metab Eng 14(4):449–457. https://doi.org/10.1016/j.ymben.2012.02.002

Mustafi N, Grünberger A, Mahr R, Helfrich S, Noh K, Blombach B, Kohlheyer D, Frunzke J (2014) Application of a genetically encoded biosensor for live cell imaging of L-valine production in pyruvate dehydrogenase complex-deficient *Corynebacterium glutamicum* strains. PLoS One 9 (1):e85731. https://doi.org/10.1371/journal.pone.0085731

Nanda AM, Heyer A, Kramer C, Grunberger A, Kohlheyer D, Frunzke J (2014) Analysis of SOS-induced spontaneous prophage induction in *Corynebacterium glutamicum* at the single-cell level. J Bacteriol 196(1):180–188. https://doi.org/10.1128/JB.01018-13

Neshat A, Mentz A, Ruckert C, Kalinowski J (2014) Transcriptome sequencing revealed the transcriptional organization at ribosome-mediated attenuation sites in *Corynebacterium glutamicum* and identified a novel attenuator involved in aromatic amino acid biosynthesis. J Biotechnol 190:55–63. https://doi.org/10.1016/j.jbiotec.2014.05.033

Ohnishi J, Mizoguchi H, Takeno S, Ikeda M (2008) Characterization of mutations induced by *N*-methyl-*N′*-nitro-*N*-nitrosoguanidine in an industrial *Corynebacterium glutamicum* strain. Mutat Res 649(1–2):239–244

Otten A, Brocker M, Bott M (2015) Metabolic engineering of *Corynebacterium glutamicum* for the production of itaconate. Metab Eng 30:156–165. https://doi.org/10.1016/j.ymben.2015.06.003

Park J, Shin H, Lee SM, Um Y, Woo HM (2018) RNA-guided single/double gene repressions in *Corynebacterium glutamicum* using an efficient CRISPR interference and its application to industrial strain. Microb Cell Factories 17(1):4. https://doi.org/10.1186/s12934-017-0843-1

Pauling J, Rottger R, Tauch A, Azevedo V, Baumbach J (2012) CoryneRegNet 6.0--updated database content, new analysis methods and novel features focusing on community demands. Nucleic Acids Res 40(Database issue):D610–D614. https://doi.org/10.1093/nar/gkr883

Perez-Garcia F, Peters-Wendisch P, Wendisch VF (2016) Engineering *Corynebacterium glutamicum* for fast production of L-lysine and L-pipecolic acid. Appl Microbiol Biotechnol 100(18):8075–8090. https://doi.org/10.1007/s00253-016-7682-6

Perez-Garcia F, Max Risse J, Friehs K, Wendisch VF (2017) Fermentative production of L-pipecolic acid from glucose and alternative carbon sources. Biotechnol J 12(7). https://doi.org/10.1002/biot.201600646

Perez-Garcia F, Jorge JMP, Dreyszas A, Risse JM, Wendisch VF (2018) Efficient production of the dicarboxylic acid Glutarate by *Corynebacterium glutamicum* via a novel synthetic pathway. Front Microbiol 9:2589. https://doi.org/10.3389/fmicb.2018.02589

Perez-Garcia F, Brito LF, Wendisch VF (2019) Function of L-pipecolic acid as compatible solute in *Corynebacterium glutamicum* as basis for its production under hyperosmolar conditions. Front Microbiol 10:340. https://doi.org/10.3389/fmicb.2019.00340

Pfeifer E, Hunnefeld M, Popa O, Polen T, Kohlheyer D, Baumgart M, Frunzke J (2016) Silencing of cryptic prophages in Corynebacterium glutamicum. Nucleic Acids Res 44(21):10117–10131. https://doi.org/10.1093/nar/gkw692

Pfeifer E, Gatgens C, Polen T, Frunzke J (2017) Adaptive laboratory evolution of *Corynebacterium glutamicum* towards higher growth rates on glucose minimal medium. Sci Rep 7(1):16780. https://doi.org/10.1038/s41598-017-17014-9

Pfeifer-Sancar K, Mentz A, Ruckert C, Kalinowski J (2013) Comprehensive analysis of the *Corynebacterium glutamicum* transcriptome using an improved RNAseq technique. BMC Genomics 14(1):888. https://doi.org/10.1186/1471-2164-14-888

Rohles CM, Giesselmann G, Kohlstedt M, Wittmann C, Becker J (2016) Systems metabolic engineering of *Corynebacterium glutamicum* for the production of the carbon-5 platform chemicals 5-aminovalerate and glutarate. Microb Cell Factories 15(1):154. https://doi.org/10.1186/s12934-016-0553-0

Ruwe M, Kalinowski J, Persicke M (2017) Identification and functional characterization of small alarmone synthetases in *Corynebacterium glutamicum*. Front Microbiol 8:1601. https://doi.org/10.3389/fmicb.2017.01601

Ruwe M, Ruckert C, Kalinowski J, Persicke M (2018) Functional characterization of a small alarmone hydrolase in *Corynebacterium glutamicum*. Front Microbiol 9:916. https://doi.org/10.3389/fmicb.2018.00916

Schaffer S, Weil B, Nguyen VD, Dongmann G, Gunther K, Nickolaus M, Hermann T, Bott M (2001) A high-resolution reference map for cytoplasmic and membrane-associated proteins of *Corynebacterium glutamicum*. Electrophoresis 22(20):4404–4422

Schluesener D, Rogner M, Poetsch A (2007) Evaluation of two proteomics technologies used to screen the membrane proteomes of wild-type *Corynebacterium glutamicum* and an L-lysine-producing strain. Anal Bioanal Chem 389(4):1055–1064

Schneider J, Wendisch VF (2011) Biotechnological production of polyamines by bacteria: recent achievements and future perspectives. Appl Microbiol Biotechnol 91(1):17–30. https://doi.org/10.1007/s00253-011-3252-0

Schulte J, Baumgart M, Bott M (2017) Identification of the cAMP phosphodiesterase CpdA as novel key player in cAMP-dependent regulation in *Corynebacterium glutamicum*. Mol Microbiol 103(3):534–552. https://doi.org/10.1111/mmi.13574

Sgobba E, Stumpf AK, Vortmann M, Jagmann N, Krehenbrink M, Dirks-Hofmeister ME, Moerschbacher B, Philipp B, Wendisch VF (2018) Synthetic *Escherichia coli-Corynebacterium glutamicum* consortia for L-lysine production from starch and sucrose. Bioresour Technol 260:302–310. https://doi.org/10.1016/j.biortech.2018.03.113

Shinfuku Y, Sorpitiporn N, Sono M, Furusawa C, Hirasawa T, Shimizu H (2009) Development and experimental verification of a genome-scale metabolic model for *Corynebacterium glutamicum*. Microb Cell Factories 8:43. https://doi.org/10.1186/1475-2859-8-43

Siebert D, Wendisch VF (2015) Metabolic pathway engineering for production of 1,2-propanediol and 1-propanol by *Corynebacterium glutamicum*. Biotechnol Biofuels 8:91. https://doi.org/10.1186/s13068-015-0269-0

Siedler S, Schendzielorz G, Binder S, Eggeling L, Bringer S, Bott M (2014) SoxR as a single-cell biosensor for NADPH-consuming enzymes in *Escherichia coli*. ACS Synth Biol 3(1):41–47. https://doi.org/10.1021/sb400110j

Sindelar G, Wendisch VF (2007) Improving lysine production by *Corynebacterium glutamicum* through DNA microarray-based identification of novel target genes. Appl Microbiol Biotechnol 76(3):677–689. https://doi.org/10.1007/s00253-007-0916-x

Smirnov SV, Kodera T, Samsonova NN, Kotlyarova VA, Rushkevich NY, Kivero AD, Sokolov PM, Hibi M, Ogawa J, Shimizu S (2010) Metabolic engineering of *Escherichia coli* to produce (2S, 3R, 4S)-4-hydroxyisoleucine. Appl Microbiol Biotechnol 88(3):719–726. https://doi.org/10.1007/s00253-010-2772-3

Smith KM, Cho KM, Liao JC (2010) Engineering *Corynebacterium glutamicum* for isobutanol production. Appl Microbiol Biotechnol 87(3):1045–1055. https://doi.org/10.1007/s00253-010-2522-6

Steffen V, Otten J, Engelmann S, Radek A, Limberg M, Koenig BW, Noack S, Wiechert W, Pohl M (2016) A toolbox of genetically encoded FRET-based biosensors for rapid L-lysine analysis. Sensors (Basel) 16(10):1604. https://doi.org/10.3390/s16101604

Syukur Purwanto H, Kang MS, Ferrer L, Han SS, Lee JY, Kim HS, Lee JH (2018) Rational engineering of the shikimate and related pathways in *Corynebacterium glutamicum* for 4-hydroxybenzoate production. J Biotechnol 282:92–100. https://doi.org/10.1016/j.jbiotec.2018.07.016

Taniguchi H, Wendisch VF (2015) Exploring the role of sigma factor gene expression on production by *Corynebacterium glutamicum*: sigma factor H and FMN as example. Front Microbiol 6:740. https://doi.org/10.3389/fmicb.2015.00740

Taniguchi H, Henke NA, Heider SAE, Wendisch VF (2017) Overexpression of the primary sigma factor gene *sigA* improved carotenoid production by *Corynebacterium glutamicum*: application to production of β-carotene and the non-native linear C50 carotenoid bisanhydrobacterioruberin. Metab Eng Comm 4:1–11. https://doi.org/10.1016/j.meteno.2017.01.001

Theodosiou E, Breisch M, Julsing MK, Falcioni F, Bühler B, Schmid A (2017) An artificial TCA cycle selects for efficient alpha-ketoglutarate dependent hydroxylase catalysis in engineered *Escherichia coli*. Biotechnol Bioeng 114(7):1511–1520. https://doi.org/10.1002/bit.26281

Tsuge Y, Tateno T, Sasaki K, Hasunuma T, Tanaka T, Kondo A (2013) Direct production of organic acids from starch by cell surface-engineered *Corynebacterium glutamicum* in anaerobic conditions. AMB Express 3(1):72. https://doi.org/10.1186/2191-0855-3-72

Unthan S, Baumgart M, Radek A, Herbst M, Siebert D, Brühl N, Bartsch A, Bott M, Wiechert W, Marin K, Hans S, Krämer R, Seibold G, Frunzke J, Kalinowski J, Rückert C, Wendisch VF, Noack S (2015a) Chassis organism from *Corynebacterium glutamicum*—a top-down approach to identify and delete irrelevant gene clusters. Biotechnol J 10(2):290–301. https://doi.org/10.1002/biot.201400041

Unthan S, Radek A, Wiechert W, Oldiges M, Noack S (2015b) Bioprocess automation on a mini pilot plant enables fast quantitative microbial phenotyping. Microb Cell Factories 14:32. https://doi.org/10.1186/s12934-015-0216-6

Veldmann KH, Minges H, Sewald N, Lee JH, Wendisch VF (2019a) Metabolic engineering of *Corynebacterium glutamicum* for the fermentative production of halogenated tryptophan. J Biotechnol 291:7–16. https://doi.org/10.1016/j.jbiotec.2018.12.008

Veldmann KH, Dachwitz S, Risse JM, Lee J-H, Sewald N, Wendisch VF (2019b) Bromination of L-tryptophan in a fermentative process with Corynebacterium glutamicum. Front Bioeng Biotechnol 7

Walter F, Grenz S, Ortseifen V, Persicke M, Kalinowski J (2016) *Corynebacterium glutamicum ggtB* encodes a functional gamma-glutamyl transpeptidase with gamma-glutamyl dipeptide synthetic and hydrolytic activity. J Biotechnol 232:99–109. https://doi.org/10.1016/j.jbiotec.2015.10.019

Wang Y, Liu Y, Liu J, Guo Y, Fan L, Ni X, Zheng X, Wang M, Zheng P, Sun J, Ma Y (2018a) MACBETH: multiplex automated *Corynebacterium glutamicum* base editing method. Metab Eng 47:200–210. https://doi.org/10.1016/j.ymben.2018.02.016

Wang B, Hu Q, Zhang Y, Shi R, Chai X, Liu Z, Shang X, Wen T (2018b) A RecET-assisted CRISPR-Cas9 genome editing in *Corynebacterium glutamicum*. Microb Cell Factories 17(1):63. https://doi.org/10.1186/s12934-018-0910-2

Wendisch VF (2003) Genome-wide expression analysis in *Corynebacterium glutamicum* using DNA microarrays. J Biotechnol 104(1–3):273–285

Wendisch VF (2014) Microbial production of amino acids and derived chemicals: synthetic biology approaches to strain development. Curr Opin Biotechnol 30C:51–58. https://doi.org/10.1016/j.copbio.2014.05.004

Wendisch VF (2019) Metabolic engineering advances and prospects for amino acid production. Metab Eng. https://doi.org/10.1016/j.ymben.2019.03.008

Wendisch VF, Bott M, Kalinowski J, Oldiges M, Wiechert W (2006) Emerging *Corynebacterium glutamicum* systems biology. J Biotechnol 124(1):74–92. https://doi.org/10.1016/j.jbiotec.2005.12.002

Wendisch VF, Brito LF, Gil Lopez M, Hennig G, Pfeifenschneider J, Sgobba E, Veldmann KH (2016) The flexible feedstock concept in industrial biotechnology: metabolic engineering of *Escherichia coli, Corynebacterium glutamicum, Pseudomonas, Bacillus* and yeast strains for access to alternative carbon sources. J Biotechnol 234:139–157. https://doi.org/10.1016/j.jbiotec.2016.07.022

Wiedenheft B, Sternberg SH, Doudna JA (2012) RNA-guided genetic silencing systems in bacteria and archaea. Nature 482(7385):331–338. https://doi.org/10.1038/nature10886

Wieschalka S, Blombach B, Bott M, Eikmanns BJ (2013) Bio-based production of organic acids with Corynebacterium glutamicum. Microb Biotechnol 6(2):87–102. https://doi.org/10.1111/1751-7915.12013

Wu W, Zhang Y, Liu D, Chen Z (2019) Efficient mining of natural NADH-utilizing dehydrogenases enables systematic cofactor engineering of lysine synthesis pathway of *Corynebacterium glutamicum*. Metab Eng 52:77–86. https://doi.org/10.1016/j.ymben.2018.11.006

Yomantas YAV, Abalakina EG, Lobanova JS, Mamontov VA, Stoynova NV, Mashko SV (2018) Complete nucleotide sequences and annotations of phi673 and phi674, two newly characterised lytic phages of *Corynebacterium glutamicum* ATCC 13032. Arch Virol 163(9):2565–2568. https://doi.org/10.1007/s00705-018-3867-x

Zahoor A, Lindner SN, Wendisch VF (2012) Metabolic engineering of *Corynebacterium glutamicum* aimed at alternative carbon sources and new products. Comput Struct Biotechnol J 3:e201210004. https://doi.org/10.5936/csbj.201210004

Zha J, Zang Y, Mattozzi M, Plassmeier J, Gupta M, Wu X, Clarkson S, Koffas MAG (2018) Metabolic engineering of *Corynebacterium glutamicum* for anthocyanin production. Microb Cell Factories 17(1):143. https://doi.org/10.1186/s12934-018-0990-z

Zhang Q, Zheng X, Wang Y, Yu J, Zhang Z, Dele-Osibanjo T, Zheng P, Sun J, Jia S, Ma Y (2018) Comprehensive optimization of the metabolomic methodology for metabolite profiling of *Corynebacterium glutamicum*. Appl Microbiol Biotechnol 102(16):7113–7121. https://doi.org/10.1007/s00253-018-9095-1

Zhang Y, Shang X, Wang B, Hu Q, Liu S, Wen T (2019) Reconstruction of tricarboxylic acid cycle in *Corynebacterium glutamicum* with a genome-scale metabolic network model for trans-4-hydroxyproline production. Biotechnol Bioeng 116(1):99–109. https://doi.org/10.1002/bit.26818

Zhao FL, Zhang C, Tang Y, Ye BC (2016) A genetically encoded biosensor for in vitro and in vivo detection of NADP. Biosens Bioelectron 77:901–906. https://doi.org/10.1016/j.bios.2015.10.063

Zhou LB, Zeng AP (2015a) Exploring lysine riboswitch for metabolic flux control and improvement of L-lysine synthesis in *Corynebacterium glutamicum*. ACS Synth Biol 4(6):729–734. https://doi.org/10.1021/sb500332c

Zhou LB, Zeng AP (2015b) Engineering a lysine-ON riboswitch for metabolic control of lysine production in *Corynebacterium glutamicum*. ACS Synth Biol 4(12):1335–1340. https://doi.org/10.1021/acssynbio.5b00075

Zhou X, Rodriguez-Rivera FP, Lim HC, Bell JC, Bernhardt TG, Bertozzi CR, Theriot JA (2019) Sequential assembly of the septal cell envelope prior to V snapping in *Corynebacterium glutamicum*. Nat Chem Biol 15(3):221–231. https://doi.org/10.1038/s41589-018-0206-1

Reduction of the *Saccharomyces cerevisiae* Genome: Challenges and Perspectives

Luis Caspeta and Prisciluis Caheri Salas Navarrete

Abstract The fact that genomes contain many nonessential genes of limited or occasional importance for a cell has become the statement of the minimal genome idea, which postulates that cell genomes can be reduced to an unembellished minimum. Following François Jacob's observations, the minimal genome of *Saccharomyces cerevisiae* could be drawn up as a sort of ancestral genome that acquired complexity through the acquisition of new properties to sustain the nucleus, organelles, cytoplasmic structures, as well as cell cycle. This imposes new restrictions with several consequences. What is the nature of these restrictions and limitations that may make yeast to have a minimal genome 15 times larger than bacterial? Some answers are sketched out in this perspective.

Keywords *Saccharomyces cerevisiae* · Minimal genome · Genome reduction · Synthetic biology · Gene essentiality

1 Synthetic Biology and Minimal Genomes

1.1 Synthetic Biology, Minimal Cell, and Cell Factory

Molecular biology has given the means for studying the molecular basis of synthesis and interactions among molecules, specifically DNA, RNA, and proteins, with the "leading idea of explaining large-scale manifestations of classical biology" (Astbury 1961). Whole genome sequencing and genomic tools have been providing with a catalog of genes as well as their functions and interactions that make life possible and

L. Caspeta (✉)
Departamento de Ingeniería Celular y Biocatálisis, Instituto de Biotecnología, Universidad Nacional Autónoma de México, Cuernavaca, Morelos, Mexico
e-mail: lcaspeta@ibt.unam.mx

P. C. S. Navarrete
Centro de Investigación en Biotecnología, Universidad Autónoma del Estado de Morelos, Cuernavaca, Morelos, Mexico

A. R. Lara, G. Gosset (eds.), *Minimal Cells: Design, Construction, Biotechnological Applications*, https://doi.org/10.1007/978-3-030-31897-0_5

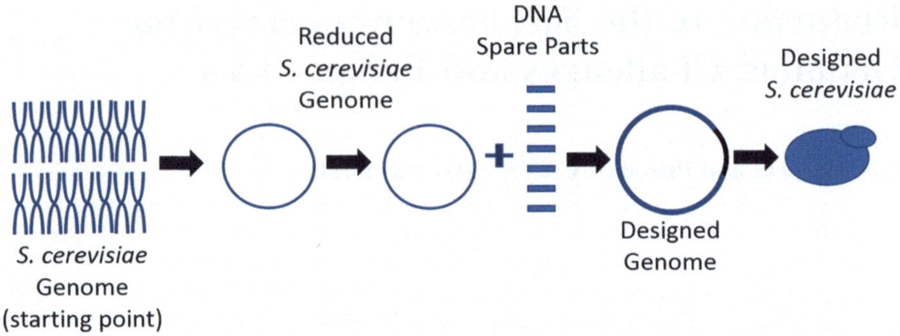

Fig. 1 In the future, cell factories will be constructed on the base of a minimal cell as chassis and by adding DNA modules with the required functions

diverse. The next challenge in biology is "the synthetic phase, [where] we will devise new control elements and add these . . . To the existing genomes or built up wholly new synthetic genomes and finally, other synthetic organisms"—a Wacław Szybalski thought quoted in Vickers (2016). Accordingly, synthetic biology is a discipline aiming to design and synthesize predictable, controllable, and transformable biological elements to construct new biological systems (Lee et al. 2013; Lachance et al. 2019). As a conspicuous way to ensure maximal control over these tasks, the synthetic biology will use cells with minimal genomes, i.e., minimal cells, as base frameworks.

Minimal cells contain a bare minimum of genetic information to sustain free-living, self-replicating life, and hence will serve as "programmable biological chassis" (Maniloff 1996; Lee et al. 2013; Vickers 2016; Lachance et al. 2019). Minimal cells have been suggested to be advantageous for having the potential of faster-replicating genomes to produce larger progeny cell yields, reduce the energy burden and loss of genome sequences with deleterious mutations (Maniloff 1996). Therefore, the future of biotechnological applications is the design of minimal genome species, e.g., a well-known model organism with this chassis generates a cell factory by adding genetic modules that include product-specific pathways, regulatory networks, and biosensors, among others (Fig. 1). In line with these ideas, cell factories using minimal genomes will have the potential for ruling the development of commercially relevant products like biofuels, chemicals, bioplastics, and medicines, as well as novel medical treatments.

1.2 Genome Reduction to Generate a Minimal Cell

Using comparative/functional genomics for analyzing families of gene orthologs in *M. genitalium*, *Escherichia coli*, and *Haemophilus influenzae*, the first set of 256 essential genes to sustain life was proposed (Mushegian and Koonin 1996). This number increased to ~360 genes obtained from gene lethality assays (Hutchison et al. 1999;

Glass et al. 2006) and then to 473 genes to construct a fully functional cell (Hutchison et al. 2016). The latter minimal-functional genome has unknown biological functions in 30% of its genes, which might be shortly revealed, and hence provide knowledge of the fundamental principles of life with a very reduced genome (0.53 Megabase pairs, Mbp)—almost one-tenth of *Escherichia coli* genome (4.64 Mbp) and ~23 times smaller than the *S. cerevisiae* genome (Blattner et al. 1997).

Following the former reasoning for constructing a minimal prokaryotic cell, in the construction of a reduced eukaryote cell a good model could be *Encephalitozoon cuniculi*, which possess the smallest known eukaryote genome so far (~2.9 Mbp) (Keeling et al. 2010). Comparative and functional genomics using bacteria, archaea, and *E. cuniculi*, among other eukaryotes, can provide information about the genes necessary for sustaining life in a minimal prokaryotic cell and the genetic modules necessary to construct a eukaryotic cell. *S. cerevisiae* with an accumulated functional gene annotation of 85% could serve as a useful model for minimal cell designing purposes. In this chapter, we discuss the minimal parts required to construct a minimal cell of *S. cerevisiae*.

2 Eukaryotes with Reduced Genomes

2.1 Small Genomes in Microsporidia

Genome compaction and reduction have been shaping very small genomes among eukaryotes (È Katinka et al. 2001; Worden et al. 2009; Nakjang et al. 2013). Microsporidia is a division of unicellular parasites from the fungi kingdom whose representatives possess one of the smallest genomes known for the Eukarya domain (Peyretaillade et al. 2011). For example, *E. intestinalisis* and *E. cuniculi* have chromosome sizes of 2.3 Mbp and 2.9 Mbp, respectively (È Katinka et al. 2001; Corradi et al. 2010). Compared with the genome sizes of *S. cerevisiae* (12 Mbp) and *Amoeba dubia* (the largest known genome 670,000 Mbp), *Encephalitozoon* genomes are ~4 to ~300,000 times smaller. However, these small Eukaryotic genomes are larger than those found in the small genomes of parasitic prokaryotes ranging from 0.16 Mbp to 0.6 Mbp (Tamas et al. 2002). Curiously, mitochondrial genomes from yeast and plants have ~0.1 Mpb and 0.6 Mbp of DNA, respectively (Foury et al. 1998; Morley and Nielsen 2017). Parasitic life then seems to have been shaping small genomes in both domains and driven permanent endosymbiotic relationships, such as that between the specialized bacteria that originated mitochondria. Curiously, mitochondrial absence and an atypical Golgi apparatus are relevant features of the *Encephalitozoon* sp. (Peyretaillade et al. 2011).

The extreme reduction of DNA content in Microsporidia is associated with a large reduction in eukaryotic-like biological functions; a comparison of genome characteristics and associated gene functions in *E. cuniculi* and *S. cerevisiae* is shown in Table 1. Particularly, *Encephalitozoon* species do not have mitochondria and associated functions, namely the electron transport chain and the tricarboxylic-acid cycle

Table 1 Comparison of some relevant features between *E. cuniculi* and *S. cerevisiae*

	E. cuniculi	*S. cerevisiae*
Genome-relevant features		
Genome size (Mbp)	2.9	12.1
Total ORFs	1999	5807
Average gene density (gene/kb)	0.84	0.5
Number of chromosomes	11	16
Gene content in illustrative biological processes[a]		
Glycolytic process	12	25
Pentose-phosphate pathway	5	10
Trehalose metabolism	4	7
Fatty acid biosynthesis	20	27
Transcriptional control	44	226
60s and 40s ribosomal proteins	77	130
tRNA synthetases	21	39
rRNA processing	20	72
tRNA modification	7	19
tRNAs	46	299
Subtotal	256	829

[a]Data corresponding to the number of genes for *E. cuniculi* biological processes were taken from supplementary table NC from (Keeling et al. 2010)

(Van de Peer et al. 2000; È Katinka et al. 2001). They also lack peroxisomes, and to contend against oxidative stress, they have glutathione, thioredoxin-based systems, and a manganese superoxide dismutase as antioxidant in a pseudo-organelle called mitosome (È Katinka et al. 2001). Also, and due to a decrease in gene paralogs (Nakjang et al. 2013), they have a much lower number of tRNA coding genes, compared with *S. cerevisiae*, a lower number of protein-coding genes for metabolism, and a very reduced number of gene-coding functions of the transcription and translation apparatuses (Van de Peer et al. 2000; È Katinka et al. 2001). The loss of these and other biological functions typically found in a eukaryote-like *S. cerevisiae* has resulted in a reduction of *Encephalitozoon* genome to ~2.6 Mbp on average. Moreover, genome evolution in *E. cuniculi* includes its compaction to 0.84 genes/kbp, compared with 0.5 genes/kbp of yeast and 0.013 genes/kbp of human. Therefore, this microsporidium genome encodes one-third of yeast functions but with one fifth of the size of the yeast genome. Notwithstanding the reduction in cellular functions, Microsporidia is a very robust microorganism capable of adapting to multiple environmental conditions and is parasitizing many animal species (Van de Peer et al. 2000; Texier et al. 2010; Peyretaillade et al. 2011). This specialization, however, makes these parasites highly dependent on their hosts.

During the evolution of microsporidia genomes, it seems there was a massive loss of gene families when the Microsporidia phyla separated from the Ascomycota (Nakjang et al. 2013). However, Microsporidia retain ancestral genes that are highly connected and expressed and therefore appear to be essential. Among the conserved gene features, the prokaryote like ribosomal RNAs can be highlighted, i.e., the small subunit of rRNA; hence they were considered as primitive protozoa which emerged before mitochondrial endosymbiosis (Van de Peer et al. 2000). The invalidity of this hypothesis was suggested from genome sequencing that revealed the presence of gene-coding proteins for assembling the Fe-S cluster, necessary for cytoplasmic ribosome biogenesis (È Katinka et al. 2001). Therefore, these parasites are cataloged as fungi with the highly reduced genome (Van de Peer et al. 2000; Peyretaillade et al. 2011).

2.2 Some Yeasts Have Reduced Genomes

It is important to understand the universality of life through gene families analysis between distant species, but it is also relevant to pay attention of the natural diversity and evolutionary origin of yeasts, which can provide clues on yeast gene variability, chromosomal evolution, and conservation of the basic core of genes (Dujon and Louis 2017; Legras et al. 2018). The Saccharomycotina subphylum contains *S. cerevisiae* and many more yeast genera with more than a couple of hundred million years of history (Prieto and Wedin 2013). Through this history, yeasts have experienced massive genome duplications, tandem gene repeat formations, segmental duplications, and extensive gene loss and gain, which have been critical for functional differentiation among yeast species, mainly modeled by wild and anthropogenic niches (Wolfe and Shields 1997; Dujon and Louis 2017). Despite these events, yeast genome sizes do not vary meaningfully (Table 2). A paradigmatic event around some yeasts is the whole genome duplication (WGD), which occurred ~100 million years ago, leading to the apparition of the *Saccharomyces* (i.e., *S. cerevisiae* (S)) and *Candida* (*C. glabrata* (C)) genera, which separated from the *Kluyveromyces* genus (i.e., *K. lactis* (K) and *K. waltii*) (Table 2) (Fischer et al. 2000; Kellis et al. 2003; Dujon et al. 2004). This genomic event defines one subgroup of the four major subgroups of Saccharomycotina defined from genome architectures (Dujon and Louis 2017); the others are the CTG—(CTG) (i.e., *D. hansenii* (D)), methylotrophs (MT) (i.e., *komagataella phaffii* (*Pichia pastoris*)), and the basal lineages (BL) (i.e., *Yarrowia lipolytica* (Y)) subgroups.

Functional genomics analysis of homologous gene families (HGFs) in five yeasts, i.e., S, C, K, D, and Y (see prefixes in the previous paragraph), representing the four subgroups, i.e. Y, WGD, CTG and BL was performed (Dujon et al. 2004). This study showed that 40% of HGFs are common in SCKDY, i.e., 805 (from 2014 families and 17,153 genes). These genes seem to be universally or largely conserved during evolution in yeasts and thus will be part of a reduced genome. The most reduced genome among these five yeasts is *C. glabrata* (Table 2), which has also experienced dynamic compaction, as it can be seen by the smaller number of introns; probably caused by its type of life as a human pathogen. Interspecific gene conservations allow distinguishing genes conserved by the yeast lifestyle (Gabaldón et al. 2013). In fact, most abundant patterns were among SCK (WGD subgroup) and DY (CTG and BL subgroups) followed by SCKD (WGD and CTG subgroups),

Table 2 Gene compaction and reductions among yeasts from the Saccharomycotina phylum

Species	Chromosomes	Genome size (Mbp)	ORFs	Gene density (Genes/Mbp)	Introns[a]	tRNAs
S. cerevisiae[b]	16	12.1	5807	455	247	274
C. glabrata[c]	13	10.6	5283	417	76	207
K. waltii[d]	8	10.7	5230	488	222	240
K. lactis[e]	6	10.6	5329	488	125	162
A. gossypii[f]	7	9.2	4917	538	211	199
D. hansenii[e]	7	11.2	6906	548	334	205
Y. lipolytica[e]	6	20.5	6703	291	632	510
S. pombe[g]	3	12.5	4824	392	2238	174
E. cuniculi[h]	11	2.9	1999	799	14	46

The features of *E. cuniculi* genome are included for comparisons
[a]Ivashchenko et al. (2009)
[b]Goffeau et al. (1996)
[c]Koszul et al. (2003)
[d]Kellis et al. (2004)
[e]Dujon et al. (2004)
[f]Dietrich et al. (2004)
[g]Wood et al. (2002)
[h]È Katinka et al. (2001)

suggesting that genome evolution accompanied specific lifestyle among yeasts, shaping different reproduction and physiological properties (Dujon et al. 2004).

Learning on genome yeast evolution is crucial to understand the role of gene variation, especially gene conservation, loss and gain events. Analysis of horizontal gene exchanges and interspecies hybridization on yeast genomes sequenced until 2017 revealed that there are genes of bacterial origin in almost all yeast genomes, many of them are associated with basic metabolic functions (Koonin et al. 2004; Dujon and Louis 2017). In Saccharomycotina, evolution led to the loss of essential gene functions associated with RNA processing and chromatin modification which are critical in other eukaryotes (Dujon and Louis 2017). Gene paralog analysis of protoploid yeasts, i.e., pre-WGD, revealed that these conserve around 100 pairs of gene paralogs (Dujon and Louis 2017); in comparison *S. cerevisiae* contains around 500. Gene loss and gain can result in novel adaptive functions after alternating regulatory networks (Gabaldón et al. 2013). For example, the acquisition of the bacterial gene *URA1*, which compared to the yeast *URA9*, this is strictly anaerobic, and suggests that its incorporation allowed yeast evolution under anaerobiosis (Gojković et al. 2004). The Saccharomycetaceae family lost the genes for complex I of the mitochondrial DNA (Marcet-Houben et al. 2009; Dujon and Louis 2017). Loss of complex I can be associated with the duplication of nuclear genes for alternative dehydrogenases (Marcet-Houben et al. 2009; Legras et al. 2018), which are basic in respire-fermentative metabolism (Larsson et al. 1998). It is interesting that massive gene loss is accompanied with a small number of essential genes, but functionally coordinated genes tend to be lost in parallel (Lafontaine and Dujon 2010; Gabaldón et al. 2013). The evidence overall suggests that, despite this observation, a massive gene loss occurred among yeast species, and there is not a sort of minimal yeast genome; it seems that this event was compensated with gene duplications.

3 Deciphering Essential Parts of the *S. cerevisiae* Genome

The knowledge on yeast *S. cerevisiae*'s physiology and molecular biology was determinant for a group of scientists to take the decision of completing its genome sequence in 1996 (Goffeau et al. 1996). The sequence consisted of 12,068 kilobases including ~5885 potential protein-coding genes, i.e., open reading frames (ORFs), distributed in 16 chromosomes. Comparative genomics through *Escherichia coli* proteome and expression analysis using Northern blot let researchers recognize that around 50% of ORFs were unidentified (Goffeau et al. 1996; Velculescu et al. 1997). Analysis of gene expression patterns of the great majority of genes under classical fermentation conditions allowed recognizing temporal expression and expression patterns for unknown genes, which provide clues to their functions (DeRisi et al. 1997; Velculescu et al. 1997; Eisen et al. 1998; Gasch et al. 2000). Large-scale gene deletions linked to growth defects have been another approach to identify gene function (Winzeler et al. 1999; Giaever et al. 2002; St Onge et al. 2007). Moreover, the increasing number of bioinformatics tools for analyzing cell phenotypes in silico, i.e., genome-scale metabolic models, has accelerated functional recognition of unknown yeast genes (Förster et al. 2003a, b; Duarte et al. 2004; Bordel et al. 2010). Therefore, by employing these methods, the total of known ORF functions rapidly raised from ~50 to ~85% (Botstein and Fink 2011). Nowadays, this figure is around 90%. Hence, the yeast *S. cerevisiae* is the best-known eukaryote. With these antecedents on how the analysis of gene functions has evolved, the following sections in this chapter highlight on gene essentiality as a means to decipher indispensable parts of an *S. cerevisiae* minimal genome.

3.1 Clusters of Orthologous Groups (COG) Analysis

Orthologous groups of genes are collections of homologous genes that arise from a common DNA ancestral sequence without further specification of the evolutionary scenario and, therefore, the history of orthologous groups reflects the history of species (Fitch 1970). As each group is assumed to have evolved from an individual ancestral gene, its conservation presupposes a relevant function among major clades (Doolittle et al. 1996; Mushegian and Koonin 1996; Tatusov et al. 1997; Koonin 2005). In line with this idea, the 323 COGs found between species characteristic of main clades, i.e., bacteria, eukarya, and archaea (Tatusov et al. 1997), represent a set of gene functions that must be included in a minimal *S. cerevisiae* genome (Table 3).

Table 3 Phylogenetic patterns in COGs obtained from the analysis of 66 genomes of species from tree may or domains (data obtained from https://www.ncbi.nlm.nih.gov/COG/)

Biological processes/lineages[a]	sce	euk	euk (w/o ecu)	arch euk	bact euk
RNA processing and modification	21	20	1		
Chromatin structure and dynamics	8	8			
Energy production and conversion	73	20	44	5	
Cell cycle control, cell division, chromosome partitioning	14	10	4	2	1
Amino acid transport and metabolism	102	10	86	2	
Nucleotide transport and metabolism	48	11	35	5	1
Carbohydrate transport and metabolism	58	24	24	8	
Coenzyme transport and metabolism	67	5	57	2	
Lipid transport and metabolism	42	18	20	5	
Translation, ribosomal structure, and biogenesis	175	131	42	96	52
Transcription	62	57	4	11	4
Replication, recombination, and repair	61	43	13	14	4
Cell wall/membrane/envelope biogenesis	15	4	10	2	
Cell motility	3	1	1		
Posttranslational modification, protein turnover, chaperones	85	59	24	6	2
Inorganic ion transport and metabolism	45	11	28		
Secondary metabolites biosynthesis, transport, and catabolism	12		8		
General function prediction only	133	65	50	15	1
Function unknown	91	44	34	2	
Signal transduction mechanisms	27	16	9	1	
Intracellular trafficking, secretion, and vesicular transport	39	38	1	5	3
Defense mechanisms	4	3	1		
Nuclear structure	1	1			
Cytoskeleton	7	7			

[a]*sce S. cerevisiae*, *euk* eukaryotes, *ecu E. cuniculi*, *arch* archaea, *bact* bacteria

A deeper COG analysis with data from https://www.ncbi.nlm.nih.gov/COG/, which includes 66 genomes and 4873 COGs, makes it possible to conclude that there are 687 COGs found in eukaryotes, bacteria, and archaea (102); eukaryotes exclusively (*E. cuniculi*, *Schizosaccharomyces pombe*, and *S. cerevisiae*) (478), eukaryotes and archaea (187), and in bacteria and eukaryotes (63). As expected, COGs' relationship between eukaryotes and archaea includes 187 elements mainly associated with the translational machinery, archaeal/vacuolar-type H^+-ATPase. Many COGs of translational machinery, ribosomal proteins, and tRNA synthetases are shared among archaea, bacteria, and eukarya. Processes for RNA processing and modification, regulation of transcription, posttranslational modification, signal transduction mechanism, and intracellular trafficking are similar among eukaryotes.

196 out of the 262 universal genes required for prokaryotic cell life are present in *S. cerevisiae* (Velculescu et al. 1997).

An extensive analysis of COGs between three eukaryotes *E. cuniculi*, *S. pombe*, and *S. cerevisiae* leads to the recognition of 676 gene functions shared between them. This amount is close to the number of gene families shared among the Saccharomycotina phylum (Dujon et al. 2004)—804 (Sect. 2.2). Around 100 COGs are virtually exclusive of *S. cerevisiae*. Several COGs involved in DNA replication, recombination, and repair, transcription, lipid metabolism, and posttranslational modification and protein turnover are exclusive of eukaryotes. As expected, functions for RNA processing and modification and chromatin structure are also only present in eukaryotes. Translation, ribosomal structure, and biogenesis are highly conserved among three kingdoms. However, these processes are more closely related between archaea and eukarya.

3.2 Gene Ontology, Lethal Gene Deletions, and Gene Paralogs Analyses

3.2.1 Gene Ontology and Lethal Gene Deletions

An analysis of gene ontology (GO), i.e., genes and gene product attributes (Ashburner et al. 2000; www.yeastgenome.org), across *S. cerevisiae* can let us know that 21% of genome ORFs is dedicated to protein synthesis, processing/ modification, and protein synthesis regulation. Transcriptional processes with 17.4% of genome dedication are in second place, followed by DNA replication, maintenance, segregation, and repair with 14.5%; metabolic functions 7.1% and cell internal/external transport use 14.6%; cell budding, morphogenesis, cell division, and cell cycle 14.6%; organization of organelles 6.1%; and stress responses 6.7%. In terms of total GOs per cell structure and organelle, the cytoplasm contains 29% followed by nucleus with 27% and mitochondria with 19%; Golgi, vacuole, and peroxisome accumulate 7% (Table 4).

Table 4 GO, gene interactions, and inviable gene deletion analyses among cell structure and organelle (data obtained from www.yeastgenome.org)

Cell structure and organelle (major GOs)	Gene functions (% total)	Gene interactions (average)	Inviable gene deletions
Cytoplasm	1729 (29)	88	296
Nucleus	1612 (27)	104	391
Mitochondrion	1130 (19)	65	151
Endoplasmic reticulum	555 (9)	77	96
Plasma membrane	415 (7)	82	40
Fungal-type vacuole	287 (4)	42	4
Golgi apparatus	126 (2)	92	19
Peroxisome	60 (1)	50	8

Gene interaction measures the functional association between genes, i.e., epistasis. This parameter has been also associated with the essentiality of a gene. Most interactions occur between protein–protein, protein–RNA, protein–DNA, RNA–RNA, and RNA–DNA. Therefore, there are a large number of interactions in the nucleolus and, consequently, a higher number of essential genes (39% of total). An intriguing result of this analysis is the fact that the mitochondrion has many associated gene functions with a low gene interaction average; however, this organelle contains 151 essential genes, of which just 40 genes pertain to metabolism. With low inviable gene functions, the vacuole and peroxisome, associating 347 genes, are possibly indispensable cell parts.

3.2.2 Lethal Gene Deletions and Gene Ontology

Evaluation of the growth phenotypes in single gene-disruption mutants was done for 96% of annotated ORFs (Giaever et al. 2002). This study showed that ~83% of genes are nonessential for growth in rich medium at 30 °C, while the remaining ~1000 genes are essential. From the 83% nonessential genes, 15% of viable homozygous deletion strains grew at a slow rate (12–90% of wild-type growth). Interestingly, many of the genes included in this percentage encode proteins for ribosomal functions and for mitochondrial functions related to respiration, which means that these processes are required in high demand for *S. cerevisiae*. In fact, many of these genes are highly expressed during vigorous growth (Velculescu et al. 1997).

In the analysis performed for this chapter, it was calculated that 43% of cell functions dedicated to transcriptional processes are essential (~780) (Fig. 2). This percentage is the highest of function essentiality in yeast. For example, DNA processes and translational functions corresponded to 29.3 and 23.8% of essential functions (450 and 546, respectively)—notice that some functions overlap with transcriptional processes. It is interesting to notice that cellular transport and metabolism, which together have 2295 functions, included only ~370 gene essentials. In the case of metabolism, most essential functions concentrate on lipid metabolism (68 of 122). Transport essential genes accumulate 171 essential functions of a total of 250. These include transport of nucleobases, nucleosides, nucleotides, and nucleic acids, into and out of the cell, as well as Golgi vesicle and transmembrane transport. No lethal phenotypes in the transport of carbohydrates have been detected.

S. cerevisiae has evolved tolerance for certain stressful conditions through activating general and specific cell stress responses. The first response is devoted to preadapting the cell to further stress and the latter one is used for a specific type of stress (Mitchell et al. 2009). The robustness of these responses appears to be related with the number of components rather than their essentiality, since essential functions represent 5% of the genome. As the mitochondria have coevolved with the yeast genome and it is an essential part of yeast thermotolerance and resistance to oxidative stress, it is not surprising to find that, out of ~650 gene functions devoted to organelles organization, 46 of the 98 essential gene functions are found in mitochondria.

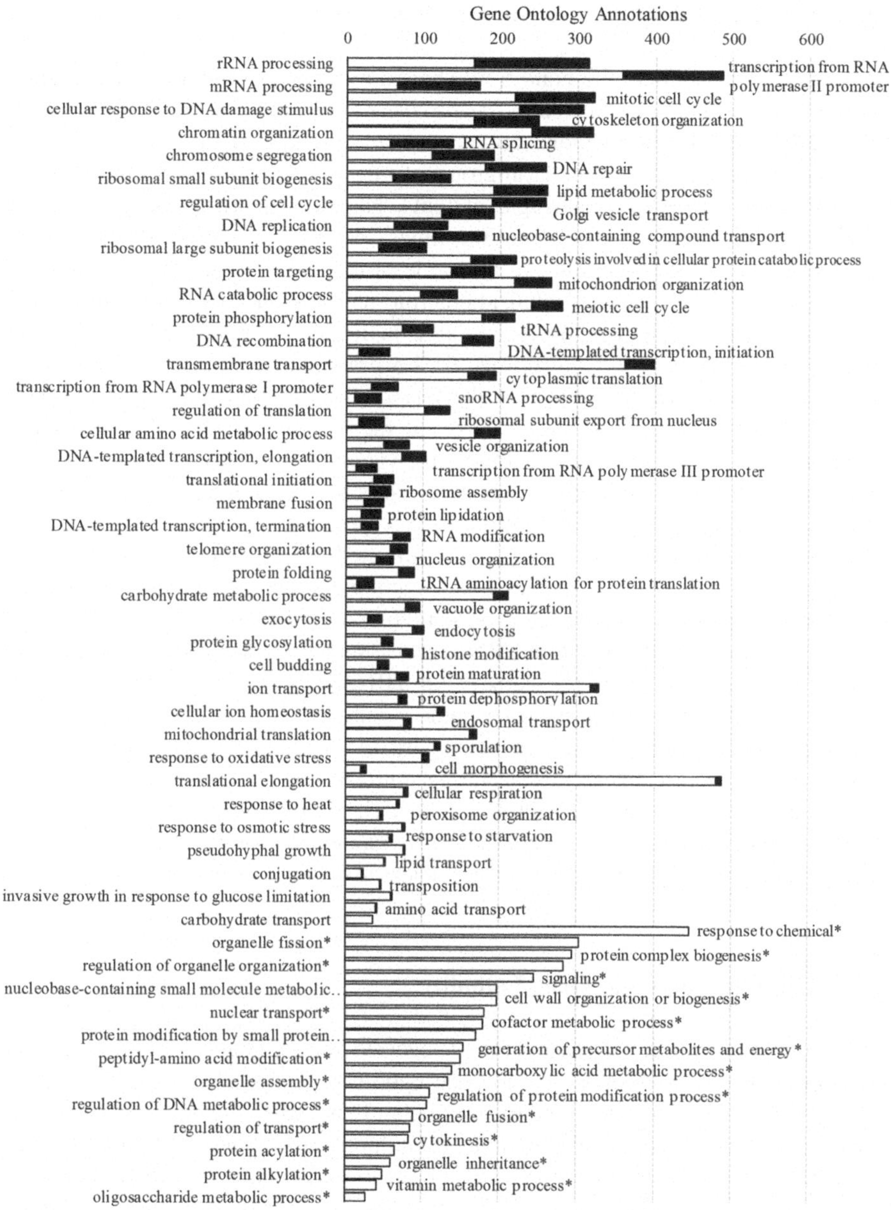

Fig. 2 Gene ontology and essentiality analyses in the yeast *S. cerevisiae*. These analyses were performed for this chapter, with information obtained from www.yeastgenome.org

3.2.3 Functional Relationship Among Genes

Synthetic lethalit, namely the condition when simultaneous mutation of two genes provoke cell death, was evaluated in *S. cerevisiae* (Tong et al. 2004; Costanzo et al. 2016). The results of these studies are summarized here. Simultaneous mutation of two genes from the nonessential subgroup of genes often showed a fitness defect resembling the corresponding single mutants. Negative interactions, i.e., when a double mutant displays a more extreme fitness defect than expected, were more enriched among genes belonging to the same biological process. This occurred in both a nonessential and an essential interaction network and between nonessential genes in different biological processes, like in linear metabolic pathways. From nearly one million genetic connections, 61% were negative interactions. Essential genes were more than 25-fold more densely connected than nonessential genes. Supervised and unsupervised clustering of interaction network identified that interactions among nonessential genes involving vacuolar transport, peroxisome, and mitochondria were less densely connected, whereas nonessential genes involving cell polarity, chromosome segregation, rRNA processing, mRNA splicing, and proteolysis functions were more densely connected—similar results are shown in Fig. 2. Interaction network identifies multifunctional genes which deletion is lethal for the cell. Pleiotropic genes, i.e., those involved in diverse functions, included genes with functions in translation, RNA processing, vesicle trafficking, and lipid and acetyl Co-A metabolism (Costanzo et al. 2016).

Nonessential gene group includes genes with the lowest interaction degree (Costanzo et al. 2016). These are often targeted with more deleterious single-nucleotide polymorphisms (SNPs), a higher ratio of nonsynonymous and synonymous nucleotide substitutions (dN/dS), and higher variations in gene expression under different genetic backgrounds and environmental changes. These genes are under reduced evolutionary and condition-specific constraints and regulation.

3.2.4 Gene Paralogs and Gene Lethality

Duplication of the ancestral yeast genome occurred approximately 100 million years ago. This event was followed by genome evolution leading *S. cerevisiae* to lose around 90% of the duplicated genome (Wolfe and Shields 1997; Seoighe and Wolfe 1998; Kellis et al. 2004). The remaining gene duplications generated paralogs, i.e., homologous genes as the result of gene duplication, which sum ~1509 (Gu et al. 2003). Analyzing large surveys of lethal phenotypes in single gene deletion assays (Giaever et al. 2002; Gu et al. 2003; Cherry et al. 2012), one can conclude that from the ~1100 unviable deletions, 5% involve duplicated genes and 95% occur in nonduplicated genes. These percentages suggest there is a significantly higher probability of functional compensation for a duplicate gene than for a singleton. Hence, the question arising here is whether one can delete copies of duplicated genes, which sum ~500 genes, namely, ~1.05 Mb (2.1 Kb/gene × 500 genes), whereas no significant negative impact would occur in yeast performance.

Ribosomal biogenesis, translation, regulation of translation, posttranslational modifications, and regulatory modifications are associated with ~2300 gene functions. Among these functions, 546 are essential. Protein synthesis is considered the most important cellular process to support vigorous growth and stress resistance. Therefore, it is not surprising that most of the gene paralogs are mainly found in this biological process. For instance, cytosolic translation possesses 113 gene paralog functions of 194 gene functions.

During long-term evolution of the yeast genome, the accumulation of mutations in both duplicated genes could result in differences of gene expression, function loss, and functional divergence between two copies (Hughes 1994; Koonin 2005; DeLuna et al. 2008). In distant paralogs, deletion of a functional copy can result in a lethal phenotype (Wagner 2000; Gu et al. 2003). There is also the possibility that one copy expressed at a very low rate when compared to the other paralog can also cause a strong negative or lethal phenotype, i.e., when duplicates are distributed in different compartments (Kuepfer et al. 2005). Furthermore, high-level expression of the two paralog members could suggest dosage amplification as the factor determining their retention in the genome (DeLuna et al. 2008). Another fact is that gene pairs in metabolism or translation increase average fitness, which will be advantageous during competition (DeLuna et al. 2008).

The association between redundancy and robustness or back-up functions does not seem to be the driving force to retain a pair of paralogs. Evidence suggest that gene retention occurred when one gene gains a new function, expression pattern, and/or localization (Kellis et al. 2004) and when gene dosage is needed to boost the activity of key functions (Seoighe and Wolfe 1998; DeLuna et al. 2008). Although it has been suggested that many paralogs play a relevant role in *S. cerevisiae* adaptation to fermentation during domestication (Wolfe 2004), deletion of 12 genes from the 27 related to fermentation caused no differences in fermentation capacity (Solis-Escalante et al. 2015). Conserved genes were, however, the highly expressed copy of a paralogous family. Null mutations of all genes in chromosome V showed that around 40% of them have little or no effect on growth rate in five different conditions (Smith et al. 1996).

3.3 Essentiality of Noncoding DNA

3.3.1 Introns

S. cerevisiae genome contains 295 introns located in 280 genes—0.147 Mbp in total, which is a very small intron density (0.04 introns/gene) compared with other fungi (0.96–2.42 intron per gene) (Kupfer et al. 2004). The systematic deletion of yeast introns showed that they are nonessential when yeast is growing in rich medium but promote stress resistance during starvation (Morgan et al. 2019; Parenteau et al.

2019). Although the presence of intron is not necessary for growth in rich media, genes lethally in the COG mRNA splicing via spliceosome are very high (52 out of the 88 genes associated are indispensable). However, it would be possible that combining intron deletion and gene deletion should avoid the accumulation of pre-mRNA, but some excised spliceosomal introns can have biological functions (Morgan et al. 2019). Thus, excised introns may have a regulatory role in cell adaptation, especially to starvation (Parenteau et al. 2019).

3.3.2 Long Terminal Repeats and Transposable Elements

Long terminal repeat (LTR) retrotransposons are widespread transposable elements in eukaryotes. Transposable elements in many systems appear to cause chromosomal rearrangements, such as deletions, inversions, and translocations. Since the replicative mode of transposition of these LTRs is by means of an RNA intermediate, the number of copies from these elements can rapidly increase and thereby the genome size of the host. Therefore, these elements show a correlation between their abundance and genome size and the coding information content (Boeke and Devine 1998). In yeast, there are 388 LTRs, which represent around ~1.5% of the genome—0.2 Mbp (Bleykasten-Grosshans et al. 2013). In humans, this percentage is ~45% (Consortium 2001). The deletion of LTRs from various yeast chromosomes did not show any negative impact in yeast growth (Dymond et al. 2011; Richardson et al. 2017).

Transposable elements (TE) are ubiquitous DNA elements in eukaryotes that can move to new genome locations and generate chromosomal deletions and rearrangements, sometimes affecting gene expression (Williamson 1983; Bleykasten-Grosshans and Neuvéglise 2011). In yeast these elements are called Ty (yeast TE) and are present in a relatively small number (~91 elements of ~5.7 kb each, 0.6 Mbp), generating disperse repeated sequences which can rearrange the yeast genome and thereby induce plasticity. For example, Ty-related chromosomal rearrangements, especially segmental duplications and hence gene amplifications, increase the fitness of strains under stressful selective pressure (Dunham et al. 2002; Caspeta et al. 2014). Interestingly, these chromosomal rearrangements are reversible and therefore are temporal solutions to stress (Yona et al. 2012). TE also generate genome insertions which can cause pleiotropic effects by interrupting gene expression of a, e.g. transcription factors (Kvitek et al. 2008). However, deletion in the region between Ty2 and RAHS causes respiratory deficiency (Oliver et al. 1992).

3.3.3 Autonomous Replicative Sequences

An autonomous replicative sequence (ARS) contains the origin of replication in yeast chromosomes. Spacing between ARS was estimated to be from 32 to 40 kb (Chan and Tye 1980), which is four to ten times shorter than the necessary to ensure chromosomal DNA replication (Williamson 1965; Newlon 1988). Based on reported fork rates (2.4–6.3 kb/min) and S phase length (25–40 min) at 30 °C, 120–500 kb of

DNA could be replicated from a single origin during the S phase (Newlon 1988; Dershowitz and Newlon 1993). Therefore, ~30 ARS of the 352–429 found in yeast chromosome could be required for chromosomal replication (0.23 Mbp) (Wyrick et al. 2001). In fact, chromosome III has nine replication origins. Remotion of some origins has little effect since replication continues over other origins. As more replication origins are deleted, the chromosome is gradually lost as cells divide, presumably because replication is too slow (Koshland et al. 1985; Newlon 1988). However, some ARS elements may be inactive during the replication process (Reynolds et al. 1989).

The ARSs have high A+T content (73–82%) when compared to chromosomal DNA (60%) and have 10–11 copies of consensus 5′-(A/T)TTTAT(A/G)TTT(A/T)-3′ (Broach et al. 1983), namely, 110–122 bp. In four ARSs of yeast, a small region of 25–65 bp seems to be essential for function. Deletion of 100–300 bp of the flanking region reduces function. These observations changed with ARN orientation: flanking sequences from the 3′ have a more profound effect than those on the 5′ (Newlon 1988). There is a possibility that ARS binding proteins are not required for ARS function and that their binding to DNA is fortuitous; instead, they may function in transcription termination (Buchman et al. 1988).

3.3.4 Telomeres and Chromosome Size

Telomeres are DNA fragments of repetitive sequences located at the end of each chromosome, and hence *S. cerevisiae* contains 32. These contain two kinds of subtelomeric elements (STEs) called X and Y′. X elements are heterogeneous and have a size ranging from 300 to 3 kb, whereas Y′ elements are only found in a half or two-thirds of yeast chromosomes (Newlon 1988). Although these sequences protect chromosomal ends from degradation and fusion to other chromosomes, they also function as fillers to increase chromosome size to some minimum required for its stability, and as barriers against transcriptional silencing, among others (Bussey et al. 1995; Louis 1995). For example, two DNA fragments of 245 bp and 252 bp found on the right arm and left end of chromosome III have high homology with an X element contained in yeast telomeres. This suggests that this chromosome was once shortened and that there is pressure against short chromosomes (Newlon 1988; Oliver et al. 1992). Artificial chromosomes of around 150 kb in size are mitotically unstable (Newlon 1988).

Deletions of chromosome III fragments resulting in lengths ranging from 50 to 300 kb were stable as the length increased over 120 kb and dramatically unstable when were smaller than 100 kb (Surosky et al. 1986). On the other hand, mutant yeast strains with telomeric paths shorter than normal showed a very long lag growth phase (Lustig and Petes 1986).

3.3.5 tRNAs

S. cerevisiae contains 299 tRNA coding genes. The tRNA pool is a key component of translation, since during translation elongation the selection of a tRNA for various codons is the rate-limiting step (Varenne et al. 1984). Among yeasts, the number of tRNA varies from 126 to 510 with *S. cerevisiae* in the middle of the distribution. Compared with *E. cuniculi*, this organism has one tent of tRNAs (46). Interestingly *K. lactis* only has 125 tRNAs and can grow at a rate of 0.34 1/h in defined medium, which is very similar to what is observed with *S. cerevisiae*. Hence, a relation between duplication time and a number of tRNAs is not proportional. Deletion of most of the tRNA genes showed no appreciable phenotype in rich medium (Bloom-Ackermann et al. 2014). However, the authors observed that stressful environments showed a set of condition-specific tRNA phenotypic defects. They also observed that the codeletion of tRNAs can be compensated by associates of the equivalent or different anticodon families.

4 Perspectives About Genome Reduction of *S. cerevisiae*

4.1 Synthetic Genomes of S. cerevisiae

As can be perceived after reading the previous paragraphs, whole genome analyses have provided a lot of insights about systems-level behavior of the yeast *S. cerevisiae*. However, a basic understanding of genome structure is not yet complete. Therefore, in the first attempt to generate a synthetic yeast genome (the Sc2.0 project), the researchers opted for a bottom-up approach. Hence, in that project, the replacement of each part of the genome by a synthetic one is followed by learning the relevance of every piece in the context of the phenotype (Murakami et al. 2007; Dymond et al. 2011; Annaluru et al. 2014; Richardson et al. 2017; Zhang et al. 2017; Luo et al. 2018). The core sequence specifications were predicted to generate a wild-type phenotype, with a highly stable genome and genetic flexibility. This approach also includes a Synthetic Chromosome Rearrangement and Modification by LoxP-mediated Evolution (SCRaMblE) process to generate a large level of genetic diversity, including deletions of chromosomes, chromosome arms, large and small portions, and genes (Dymond and Boeke 2012; Jia et al. 2018; Shen et al. 2018). Therefore, repeated rounds of SCRaMbLE could be effective to generate minimal genomes (Dymond and Boeke 2012).

Another attempt to generate an artificial yeast genome was the design and construction of an *S. cerevisiae* single circular genome. The authors created a circular chromosome by the successive end-to-end fusion of the 16 yeast chromosomes (Shao et al. 2018). Their design contemplated the deletion of 15 centromeres, 30 telomeres, and 19 long repeats and hence reduced the genome size to 11.8 Mb. Although significant changes were generated in the structure of the circular genome, the pattern of gene expression was similar to the wild-type strains containing the 16 chromosomes. These results support the idea that interchromosomal interactions have little effect on gene transcription.

4.2 Reduction of the S. cerevisiae *Genome*

Today, *S. cerevisiae* genome reductions account for 5%, 8%, and 9% of the wild-type genome (Murakami et al. 2007; Richardson et al. 2017; Shao et al. 2018). These reductions are mainly associated with deletion of Ty elements and LTR repeats to generate more stable genomes (Dymond et al. 2011; Richardson et al. 2017); removal of telomeres, long repeats, and centromeres (Shao et al. 2018); and deletion of 247 genes, which were predicted to improve ethanol production (Murakami et al. 2007).

Derived from data analyses completed for this synopsis, one can see that reducing yeast genome further than previous efforts is not apparently manageable. Many aspects related to, for example, complex gene interactions, genome structure, cellular stress responses, regulation of stress response and respiro-fermentative metabolism, organelle essentiality, and plasticity are not well understood. Therefore, one can conclude that the bottom-up strategy followed by the Sc2 consortium is the most likely path to asses a minimal genome of *S. cerevisiae*.

Summarizing the potentially nonessential genetic parts considered in Sect. 3, our calculations result in a genome reduction of ~3 Mbp to give a genome of 9 Mbp—close the chromosome of *A. gossypii*. Remarkably, for 95% of *A. gossypii* protein-coding genes, there are both homology and pattern of synteny with *S. cerevisiae* (Dietrich et al. 2004). Furthermore, the number of protein-coding genes from *A. gossypii* and *S. pombe*are is ~4800 which is close to the number of protein-coding genes remaining after the proposed gene deletion (~4500) calculated for this chapter. *S. pombe* and *A. gossypii* are both yeasts that speciated before the WGD event. After all, ~4500 genes and a genome of ~9 Mb could be potentially close to the minimum number of genes and minimal size of a free-living *S. cerevisiae*.

References

Annaluru N, Muller H, Mitchell LA et al (2014) Total synthesis of a functional designer eukaryotic chromosome. Science 344:55–58. https://doi.org/10.1126/science.1249252

Ashburner M, Ball CA, Blake JA et al (2000) Gene ontology: tool for the unification of biology. Nat Genet 25:25–29. https://doi.org/10.1038/75556

Astbury WT (1961) Molecular biology or ultrastructural biology? Nature 190:1124–1124. https://doi.org/10.1038/1901124a0

Blattner FR, Plunkett G, Bloch CA et al (1997) The complete genome sequence of *Escherichia coli* K-12. Science 277:1453–1462. https://doi.org/10.1126/SCIENCE.277.5331.1453

Bleykasten-Grosshans C, Neuvéglise C (2011) Transposable elements in yeasts. C R Biol 334:679–686. https://doi.org/10.1016/J.CRVI.2011.05.017

Bleykasten-Grosshans C, Friedrich A, Schacherer J (2013) Genome-wide analysis of intraspecific transposon diversity in yeast. BMC Genomics 14:399. https://doi.org/10.1186/1471-2164-14-399

Bloom-Ackermann Z, Navon S, Gingold H et al (2014) A comprehensive tRNA deletion library unravels the genetic architecture of the tRNA pool. PLoS Genet 10:e1004084. https://doi.org/10.1371/journal.pgen.1004084

Boeke JD, Devine SE (1998) Yeast retrotransposons: finding a nice quiet neighborhood. Cell 93:1087–1089. https://doi.org/10.1016/S0092-8674(00)81450-6

Bordel S, Agren R, Nielsen J (2010) Sampling the solution space in genome-scale metabolic networks reveals transcriptional regulation in key enzymes. PLoS Comput Biol 6:e1000859. https://doi.org/10.1371/journal.pcbi.1000859

Botstein D, Fink GR (2011) Yeast: an experimental organism for 21st century biology. Genetics 189:695–704. https://doi.org/10.1534/genetics.111.130765

Broach JR, Li Y-Y, Feldman J et al (1983) Localization and sequence analysis of yeast origins of DNA replication. Cold Spring Harb Symp Quant Biol 47:1165–1173. https://doi.org/10.1101/SQB.1983.047.01.132

Buchman AR, Kimmerly WJ, Rine J, Kornberg RD (1988) Two DNA-binding factors recognize specific sequences at silencers, upstream activating sequences, autonomously replicating sequences, and telomeres in *Saccharomyces cerevisiae*. Mol Cell Biol 8:210–225. https://doi.org/10.1128/MCB.8.1.210

Bussey H, Kaback DB, Zhong W et al (1995) The nucleotide sequence of chromosome I from *Saccharomyces cerevisiae*. Proc Natl Acad Sci USA 92:3809–3813

Caspeta L, Chen Y, Ghiaci P et al (2014) Altered sterol composition renders yeast thermotolerant. Science 346:75–78. https://doi.org/10.1126/science.1258137

Chan CS, Tye BK (1980) Autonomously replicating sequences in *Saccharomyces cerevisiae*. Proc Natl Acad Sci USA 77:6329–6333

Cherry JM, Hong EL, Amundsen C et al (2012) Saccharomyces genome database: the genomics resource of budding yeast. Nucleic Acids Res 40:D700–D705. https://doi.org/10.1093/nar/gkr1029

Consortium IHGS (2001) Initial sequencing and analysis of the human genome. Nature 409:860–921. https://doi.org/10.1038/35057062

Corradi N, Pombert J-F, Farinelli L et al (2010) The complete sequence of the smallest known nuclear genome from the microsporidian *Encephalitozoon intestinalis*. Nat Commun 1:77. https://doi.org/10.1038/ncomms1082

Costanzo M, VanderSluis B, Koch EN et al (2016) A global genetic interaction network maps a wiring diagram of cellular function. Science 353:aaf1420. https://doi.org/10.1126/science.aaf1420

DeLuna A, Vetsigian K, Shoresh N et al (2008) Exposing the fitness contribution of duplicated genes. Nat Genet 40:676–681. https://doi.org/10.1038/ng.123

DeRisi JL, Iyer VW, Brown PO (1997) Exploring the metabolic and genetic control of gene expression on a genomic scale. Science (80-) 278:680–686. https://doi.org/10.1126/science.278.5338.680

Dershowitz A, Newlon CS (1993) The effect on chromosome stability of deleting replication origins. Mol Cell Biol 13:391–398. https://doi.org/10.1128/mcb.13.1.391

Dietrich FS, Voegeli S, Brachat S et al (2004) The *Ashbya gossypii* genome as a tool for mapping the ancient *Saccharomyces cerevisiae* genome. Science 304:304–307. https://doi.org/10.1126/science.1095781

Doolittle RF, Feng DF, Tsang S et al (1996) Determining divergence times of the major kingdoms of living organisms with a protein clock. Science 271:470–477. https://doi.org/10.1126/SCIENCE.271.5248.470

Duarte NC, Herrgård MJ, Palsson BØ (2004) Reconstruction and validation of *Saccharomyces cerevisiae* iND750, a fully compartmentalized genome-scale metabolic model. Genome Res 14:1298–1309. https://doi.org/10.1101/gr.2250904

Dujon BA, Louis EJ (2017) Genome diversity and evolution in the budding yeasts (Saccharomycotina). Genetics 206:717–750. https://doi.org/10.1534/genetics.116.199216

Dujon B, Sherman D, Fischer G et al (2004) Genome evolution in yeasts. Nature 430:35–44. https://doi.org/10.1038/nature02579

Dunham MJ, Badrane H, Ferea T et al (2002) Characteristic genome rearrangements in experimental evolution of *Saccharomyces cerevisiae.* Proc Natl Acad Sci USA 99:16144–16149. https://doi.org/10.1073/pnas.242624799

Dymond J, Boeke J (2012) The *Saccharomyces cerevisiae* SCRaMbLE system and genome minimization. Bioeng Bugs 3(3):168–171. https://doi.org/10.4161/bbug.19543

Dymond JS, Richardson SM, Coombes CE et al (2011) Synthetic chromosome arms function in yeast and generate phenotypic diversity by design. Nature 477:471–476. https://doi.org/10.1038/nature10403

È Katinka MD, Duprat S, Cornillot E et al (2001) Genome sequence and gene compaction of the eukaryote parasite *Encephalitozoon cuniculi.* Nature 414:450–453. https://doi.org/10.1038/35106579

Eisen MB, Spellman PT, Brown PO, Botstein D (1998) Cluster analysis and display of genome-wide expression patterns. Proc Natl Acad Sci USA 95:14863–14868. https://doi.org/10.1073/pnas.95.25.14863

Fischer G, James SA, Roberts IN et al (2000) Chromosomal evolution in *Saccharomyces.* Nature 405:451–454. https://doi.org/10.1038/35013058

Fitch WM (1970) Distinguishing homologous from analogous proteins. Syst Zool 19:99. https://doi.org/10.2307/2412448

Förster J, Famili I, Fu P et al (2003a) Genome-scale reconstruction of the *Saccharomyces cerevisiae* metabolic network. Genome Res 13:244–253. https://doi.org/10.1101/gr.234503

Förster J, Famili I, Palsson BØ, Nielsen J (2003b) Large-scale evaluation of in silico gene deletions in *Saccharomyces cerevisiae.* Omi A J Integr Biol 7:193–202. https://doi.org/10.1089/153623103322246584

Foury F, Roganti T, Lecrenier N, Purnelle B (1998) The complete sequence of the mitochondrial genome of *Saccharomyces cerevisiae.* FEBS Lett 440:325–331. https://doi.org/10.1016/S0014-5793(98)01467-7

Gabaldón T, Martin T, Marcet-Houben M et al (2013) Comparative genomics of emerging pathogens in the *Candida glabrata* clade. BMC Genomics 14:623. https://doi.org/10.1186/1471-2164-14-623

Gasch AP, Spellman PT, Kao CM et al (2000) Genomic expression programs in the response of yeast cells to environmental changes. Mol Biol Cell 11:4241–4257. https://doi.org/10.1091/mbc.11.12.4241

Giaever G, Chu AM, Ni L et al (2002) Functional profiling of the *Saccharomyces cerevisiae* genome. Nature 418:387–391. https://doi.org/10.1038/nature00935

Glass JI, Assad-Garcia N, Alperovich N et al (2006) Essential genes of a minimal bacterium. Proc Natl Acad Sci USA 103:425–430. https://doi.org/10.1073/pnas.0510013103

Goffeau A, Barrell BG, Bussey H et al (1996) Life with 6000 genes. Science (80-) 274:546–567. https://doi.org/10.1126/science.274.5287.546

Gojković Z, Knecht W, Zameitat E et al (2004) Horizontal gene transfer promoted evolution of the ability to propagate under anaerobic conditions in yeasts. Mol Gen Genomics 271:387–393. https://doi.org/10.1007/s00438-004-0995-7

Gu Z, Steinmetz LM, Gu X et al (2003) Role of duplicate genes in genetic robustness against null mutations. Nature 421:63–66. https://doi.org/10.1038/nature01198

Hughes AL (1994) The evolution of functionally novel proteins after gene duplication. Proc R Soc Lond Ser B Biol Sci 256:119–124. https://doi.org/10.1098/rspb.1994.0058

Hutchison CA, Peterson SN, Gill SR et al (1999) Global transposon mutagenesis and a minimal mycoplasma genome. Science 286:2165–2169. https://doi.org/10.1126/SCIENCE.286.5447.2165

Hutchison CA, Chuang R-Y, Noskov VN et al (2016) Design and synthesis of a minimal bacterial genome. Science (80-) 351:6253–6253. https://doi.org/10.1126/science.aad6253

Ivashchenko AT, Tauasarova MI, Atambayeva SA (2009) Exon-intron structure of genes in complete fungal genomes. Mol Biol 43:24–31. https://doi.org/10.1134/S002689330901004X

Jia B, Wu Y, Li B-Z et al (2018) Precise control of SCRaMbLE in synthetic haploid and diploid yeast. Nat Commun 9:1933. https://doi.org/10.1038/s41467-018-03084-4

Keeling PJ, Corradi N, Morrison HG et al (2010) The reduced genome of the parasitic microsporidian *Enterocytozoon bieneusi* lacks genes for core carbon metabolism. Genome Biol Evol 2:304–309. https://doi.org/10.1093/gbe/evq022

Kellis M, Patterson N, Endrizzi M et al (2003) Sequencing and comparison of yeast species to identify genes and regulatory elements. Nature 423:241–254. https://doi.org/10.1038/nature01644

Kellis M, Birren BW, Lander ES (2004) Proof and evolutionary analysis of ancient genome duplication in the yeast *Saccharomyces cerevisiae*. Nature 428:617–624. https://doi.org/10.1038/nature02424

Koonin EV (2005) Orthologs, paralogs, and evolutionary genomics. Annu Rev Genet 39:309–338. https://doi.org/10.1146/annurev.genet.39.073003.114725

Koonin EV, Fedorova ND, Jackson JD et al (2004) A comprehensive evolutionary classification of proteins encoded in complete eukaryotic genomes. Genome Biol 5:R7. https://doi.org/10.1186/gb-2004-5-2-r7

Koshland D, Kent JC, Hartwell LH (1985) Genetic analysis of the mitotic transmission of minichromosomes. Cell 40:393–403. https://doi.org/10.1016/0092-8674(85)90153-9

Koszul R, Malpertuy A, Frangeul L, Bouchier C, Wincker P, Thierry A, Duthoy S, Ferris S, Hennequin C, Dujon B (2003) The complete mitochondrial genome sequence of the pathogenic yeast *Candida* (Torulopsis) *glabrata*. FEBS Lett 534(1–3):39–48. https://doi.org/10.1016/S0014-5793(02)03749-3

Kuepfer L, Sauer U, Blank LM (2005) Metabolic functions of duplicate genes in *Saccharomyces cerevisiae*. Genome Res 15:1421–1430. https://doi.org/10.1101/gr.3992505

Kupfer DM, Drabenstot SD, Buchanan KL et al (2004) Introns and splicing elements of five diverse fungi. Eukaryot Cell 3:1088–1100. https://doi.org/10.1128/EC.3.5.1088-1100.2004

Kvitek DJ, Will JL, Gasch AP (2008) Variations in stress sensitivity and genomic expression in diverse *S. cerevisiae* isolates. PLoS Genet 4:e1000223. https://doi.org/10.1371/journal.pgen.1000223

Lachance J-C, Rodrigue S, Palsson BO (2019) Minimal cells, maximal knowledge. elife 8. https://doi.org/10.7554/eLife.45379

Lafontaine I, Dujon B (2010) Origin and fate of pseudogenes in Hemiascomycetes: a comparative analysis. BMC Genomics 11:260. https://doi.org/10.1186/1471-2164-11-260

Larsson C, Påhlman IL, Ansell R et al (1998) The importance of the glycerol 3-phosphate shuttle during aerobic growth of *Saccharomyces cerevisiae*. Yeast 14:347–357. https://doi.org/10.1002/(SICI)1097-0061(19980315)14:4<347::AID-YEA226>3.0.CO;2-9

Lee B-R, Cho S, Song Y et al (2013) Emerging tools for synthetic genome design. Mol Cells 35:359–370. https://doi.org/10.1007/s10059-013-0127-5

Legras J-L, Galeote V, Bigey F et al (2018) Adaptation of *S. cerevisiae* to fermented food environments reveals remarkable genome plasticity and the footprints of domestication. Mol Biol Evol 35:1712–1727. https://doi.org/10.1093/molbev/msy066

Louis EJ (1995) The chromosome ends of *Saccharomyces cerevisiae*. Yeast 11:1553–1573. https://doi.org/10.1002/yea.320111604

Luo J, Sun X, Cormack BP, Boeke JD (2018) Karyotype engineering by chromosome fusion leads to reproductive isolation in yeast. Nature 560:392–396. https://doi.org/10.1038/s41586-018-0374-x

Lustig AJ, Petes TD (1986) Identification of yeast mutants with altered telomere structure. Proc Natl Acad Sci USA 83:1398–1402. https://doi.org/10.1073/pnas.83.5.1398

Maniloff J (1996) The minimal cell genome. Proc Natl Acad Sci USA 93:10004–10006. https://doi.org/10.1073/PNAS.93.19.10004

Marcet-Houben M, Marceddu G, Gabaldón T (2009) Phylogenomics of the oxidative phosphorylation in fungi reveals extensive gene duplication followed by functional divergence. BMC Evol Biol 9:295. https://doi.org/10.1186/1471-2148-9-295

Mitchell A, Romano GH, Groisman B et al (2009) Adaptive prediction of environmental changes by microorganisms. Nature 460:220–224. https://doi.org/10.1038/nature08112

Morgan JT, Fink GR, Bartel DP (2019) Excised linear introns regulate growth in yeast. Nature 565:606–611. https://doi.org/10.1038/s41586-018-0828-1

Morley SA, Nielsen BL (2017) Plant mitochondrial DNA. Front Biosci (Landmark Ed) 22:1023–1032

Murakami K, Tao E, Ito Y et al (2007) Large scale deletions in the *Saccharomyces cerevisiae* genome create strains with altered regulation of carbon metabolism. Appl Microbiol Biotechnol 75:589–597. https://doi.org/10.1007/s00253-007-0859-2

Mushegian AR, Koonin EV (1996) A minimal gene set for cellular life derived by comparison of complete bacterial genomes. Proc Natl Acad Sci USA 93:10268–10273. https://doi.org/10.1073/PNAS.93.19.10268

Nakjang S, Williams TA, Heinz E et al (2013) Reduction and expansion in microsporidian genome evolution: new insights from comparative genomics. Genome Biol Evol 5:2285–2303. https://doi.org/10.1093/gbe/evt184

Newlon CS (1988) Yeast chromosome replication and segregation. Microbiol Rev 52:568–601

Oliver SG, van der Aart QJM, Agostoni-Carbone ML et al (1992) The complete DNA sequence of yeast chromosome III. Nature 357:38–46. https://doi.org/10.1038/357038a0

Parenteau J, Maignon L, Berthoumieux M et al (2019) Introns are mediators of cell response to starvation. Nature 565:612–617. https://doi.org/10.1038/s41586-018-0859-7

Peyretaillade E, El Alaoui H, Diogon M et al (2011) Extreme reduction and compaction of microsporidian genomes. Res Microbiol 162:598–606. https://doi.org/10.1016/J.RESMIC.2011.03.004

Prieto M, Wedin M (2013) Dating the diversification of the major lineages of Ascomycota (Fungi). PLoS One 8:e65576. https://doi.org/10.1371/journal.pone.0065576

Reynolds AE, McCarroll RM, Newlon CS, Fangman WL (1989) Time of replication of ARS elements along yeast chromosome III. Mol Cell Biol 9:4488–4494

Richardson SM, Mitchell LA, Stracquadanio G et al (2017) Design of a synthetic yeast genome. Science 355:1040–1044. https://doi.org/10.1126/science.aaf4557

Seoighe C, Wolfe KH (1998) Extent of genomic rearrangement after genome duplication in yeast. Proc Natl Acad Sci USA 95:4447–4452. https://doi.org/10.1073/PNAS.95.8.4447

Shao Y, Lu N, Wu Z et al (2018) Creating a functional single-chromosome yeast. Nature 560:331–335. https://doi.org/10.1038/s41586-018-0382-x

Shen MJ, Wu Y, Yang K et al (2018) Heterozygous diploid and interspecies SCRaMbLEing. Nat Commun 9:1934. https://doi.org/10.1038/s41467-018-04157-0

Smith V, Chou KN, Lashkari D et al (1996) Functional analysis of the genes of yeast chromosome V by genetic footprinting. Science 274:2069–2074. https://doi.org/10.1126/SCIENCE.274.5295.2069

Solis-Escalante D, Kuijpers NGA, Barrajon-Simancas N et al (2015) A minimal set of glycolytic genes reveals strong redundancies in *Saccharomyces cerevisiae* central metabolism. Eukaryot Cell 14:804–816. https://doi.org/10.1128/EC.00064-15

St Onge RP, Mani R, Oh J et al (2007) Systematic pathway analysis using high-resolution fitness profiling of combinatorial gene deletions. Nat Genet 39:199–206. https://doi.org/10.1038/ng1948

Surosky RT, Newlon CS, Tye BK (1986) The mitotic stability of deletion derivatives of chromosome III in yeast. Proc Natl Acad Sci USA 83:414–418. https://doi.org/10.1073/pnas.83.2.414

Tamas I, Klasson L, Canbäck B et al (2002) 50 Million years of genomic stasis in endosymbiotic bacteria. Science (80-) 296:2376–2379. https://doi.org/10.1126/science.1071278

Tatusov RL, Koonin EV, Lipman DJ (1997) A genomic perspective on protein families. Science 278:631–637. https://doi.org/10.1126/SCIENCE.278.5338.631

Texier C, Vidau C, Viguès B et al (2010) Microsporidia: a model for minimal parasite–host interactions. Curr Opin Microbiol 13:443–449. https://doi.org/10.1016/J.MIB.2010.05.005

Tong AHY, Lesage G, Bader GD et al (2004) Global mapping of the yeast genetic interaction network. Science 303:808–813. https://doi.org/10.1126/science.1091317

Van de Peer Y, Ben Ali A, Meyer A (2000) Microsporidia: accumulating molecular evidence that a group of amitochondriate and suspectedly primitive eukaryotes are just curious fungi. Gene 246:1–8. https://doi.org/10.1016/S0378-1119(00)00063-9

Varenne S, Buc J, Lloubes R, Lazdunski C (1984) Translation is a non-uniform process: effect of tRNA availability on the rate of elongation of nascent polypeptide chains. J Mol Biol 180:549–576. https://doi.org/10.1016/0022-2836(84)90027-5

Velculescu VE, Zhang L, Zhou W et al (1997) Characterization of the yeast transcriptome. Cell 88:243–251. https://doi.org/10.1016/S0092-8674(00)81845-0

Vickers CE (2016) The minimal genome comes of age. Nat Biotechnol 34:623–624. https://doi.org/10.1038/nbt.3593

Wagner A (2000) Robustness against mutations in genetic networks of yeast. Nat Genet 24:355–361. https://doi.org/10.1038/74174

Williamson DH (1965) The timing of deoxyribonucleic acid synthesis in the cell cycle of *Saccharomyces cerevisiae*. J Cell Biol 25:517–528. https://doi.org/10.1083/jcb.25.3.517

Williamson VM (1983) Transposable elements in yeast. Int Rev Cytol 83:1–25. https://doi.org/10.1016/S0074-7696(08)61684-8

Winzeler EA, Shoemaker DD, Astromoff A et al (1999) Functional characterization of the *S. cerevisiae* genome by gene deletion and parallel analysis. Science 285:901–906

Wolfe K (2004) Evolutionary genomics: yeasts accelerate beyond BLAST. Curr Biol 14:R392–R394. https://doi.org/10.1016/J.CUB.2004.05.015

Wolfe KH, Shields DC (1997) Molecular evidence for an ancient duplication of the entire yeast genome. Nature 387:708–713. https://doi.org/10.1038/42711

Wood V, Gwilliam R, Rajandream M-A, Lyne M, Lyne R, Stewart A, Sgouros J, Peat N, Hayles J, Baker S, Basham D, Bowman S, Brooks K, Brown D, Brown S, Chillingworth T, Churcher C, Collins M, Connor R, Cronin A, Davis P, Feltwell T, Fraser A, Gentles S, Goble A, Hamlin N, Harris D, Hidalgo J, Hodgson G, Holroyd S, Hornsby T, Howarth S, Huckle EJ, Hunt S, Jagels K, James K, Jones L, Jones M, Leather S, McDonald S, McLean J, Mooney P, Moule S, Mungall K, Murphy L, Niblett D, Odell C, Oliver K, O'Neil S, Pearson D, Quail MA, Rabbinowitsch E, Rutherford K, Rutter S, Saunders D, Seeger K, Sharp S, Skelton J, Simmonds M, Squares R, Squares S, Stevens K, Taylor K, Taylor RG, Tivey A, Walsh S, Warren T, Whitehead S, Woodward J, Volckaert G, Aert R, Robben J, Grymonprez B, Weltjens I, Vanstreels E, Rieger M, Schäfer M, Müller-Auer S, Gabel C, Fuchs M, Fritzc C, Holzer E, Moestl D, Hilbert H, Borzym K, Langer I, Beck A, Lehrach H, Reinhardt R, Pohl TM, Eger P, Zimmermann W, Wedler H, Wambutt R, Purnelle B, Goffeau A, Cadieu E, Dréano S, Gloux S, Lelaure V, Mottier S, Galibert F, Aves SJ, Xiang Z, Hunt C, Moore K, Hurst SM, Lucas M, Rochet M, Gaillardin C, Tallada VA, Garzon A, Thode G, Daga RR, Cruzado L, Jimenez J, Sánchez M, del Rey F, Benito J, Domínguez A, Revuelta JL, Moreno S, Armstrong J, Forsburg SL, Cerrutti L, Lowe T, McCombie WR, Paulsen I, Potashkin J, Shpakovski GV, Ussery D, Barrell BG, Nurse P (2002) The genome sequence of *Schizosaccharomyces pombe*. Nature 415 (6874):871–880. https://doi.org/10.1038/nature724

Worden AZ, Lee J-H, Mock T et al (2009) Green evolution and dynamic adaptations revealed by genomes of the marine picoeukaryotes Micromonas. Science 324:268–272. https://doi.org/10.1126/science.1167222

Wyrick JJ, Aparicio JG, Chen T et al (2001) Genome-wide distribution of ORC and MCM proteins in *S. cerevisiae*: high-resolution mapping of replication origins. Science (80-) 294:2357–2360. https://doi.org/10.1126/science.1066101

Yona AH, Manor YS, Herbst RH et al (2012) Chromosomal duplication is a transient evolutionary solution to stress. Proc Natl Acad Sci USA 109:21010–21015. https://doi.org/10.1073/pnas.1211150109

Zhang W, Zhao G, Luo Z et al (2017) Engineering the ribosomal DNA in a megabase synthetic chromosome. Science 355:eaaf3981. https://doi.org/10.1126/science.aaf3981

The Use of In Silico Genome-Scale Models for the Rational Design of Minimal Cells

Jean-Christophe Lachance, Sébastien Rodrigue, and Bernhard O. Palsson

Abstract Organism-specific genome-scale metabolic models (GEMs) can be reconstructed using genome annotation and biochemical data available in literature. The systematic inclusion of biochemical reactions into a coherent metabolic network combined with the formulation of appropriate constraints reveals the set of metabolic capabilities harbored by an organism, hereby allowing the computation of growth phenotypes from genotype information. GEMs have been used thoroughly to assess growth capabilities under varying conditions and determine gene essentiality. This simulation process can rapidly generate testable hypotheses that can be applied for the systematic evaluation of growth capabilities in genome reduction efforts and the definition of a minimal cell. Here we review the most recent computational methods and protocols available for the reconstruction of genome-scale models, the formulation of objective functions, and the applications of models in the prediction of gene essentiality. These methods and applications are suited to the emerging field of genome reduction and the development of minimal cells as biological factories.

Keywords Computational modeling · Metabolic modeling · Constraint-based modeling · Gene essentiality prediction

J.-C. Lachance · S. Rodrigue
Département de Biologie, Université de Sherbrooke, Sherbrooke, QC, Canada

B. O. Palsson (✉)
Department of Bioengineering, University of California, San Diego, CA, USA

Bioinformatics and Systems Biology Program, University of California, San Diego, CA, USA

Department of Pediatrics, University of California, San Diego, CA, USA

Novo Nordisk Foundation Center for Biosustainability, Technical University of Denmark, Lyngby, Denmark
e-mail: palsson@ucsd.edu

A. R. Lara, G. Gosset (eds.), *Minimal Cells: Design, Construction, Biotechnological Applications*, https://doi.org/10.1007/978-3-030-31897-0_6

1 Introduction

1.1 Engineering Biology

Biologists of the past 200 years have provided a breadth of knowledge on the fundamentals of life on Earth. Current theories and dogmas emerged from a maze of suppositions and hypotheses through the succession of key findings and incremental advances. Nowadays, few molecular functions necessary to support life remain unknown. While biology has considerably matured as a science discipline, we will discuss here how the exhaustive characterization of organisms along with proper modeling frameworks should drive a new era, in which cell engineering will develop into an independent discipline. Because of their lower complexity, microorganisms—particularly minimal bacteria—are expected to play a very important role in this endeavor.

> Scientists investigate that which already is; Engineers create that which has never been.
>
> —Albert Einstein

We discuss here the historical context and key steps leading to the birth of biological engineering. This historical recap should highlight the importance of minimal cell models while providing readers with a perspective on the entire field of biology. We divided it in four stages: *classical biology*, *molecular biology*, *genomics*, and finally *synthetic biology* (Fig. 1).

1.1.1 Classical Biology

In 1859, Darwin published his work entitled "On the Origin of Species by Means of Natural Selection, or, the Preservation of Favoured Races in the Struggle for Life." Less than a decade later, in 1865, Mendel proposed mechanisms for heredity. Both theories used observable phenotypes at the organism level to infer potential mechanisms driving their evolution. While Darwin's work explained the driving forces underlying the emergence of phenotypes and speciation, Mendel's work was focused on a mechanistic explanation of the basic principles of genetics. While not specifically described by Mendel, his conclusions gave birth to the concept of gene. Understanding the chemical basis of the gene and heredity then became the main endeavor of this first era of biology, defined here as the *classical biology* era (Fig. 1). This objective remained one of the grand challenges of biology until, in 1953, Watson and Crick published the structure of deoxyribonucleic acid (DNA) (Watson and Crick 1953). This historically significant finding allowed scientists to ask more intricate questions on the molecular functions sustaining life, marking the beginning of the *molecular biology* era (Waddington 1961).

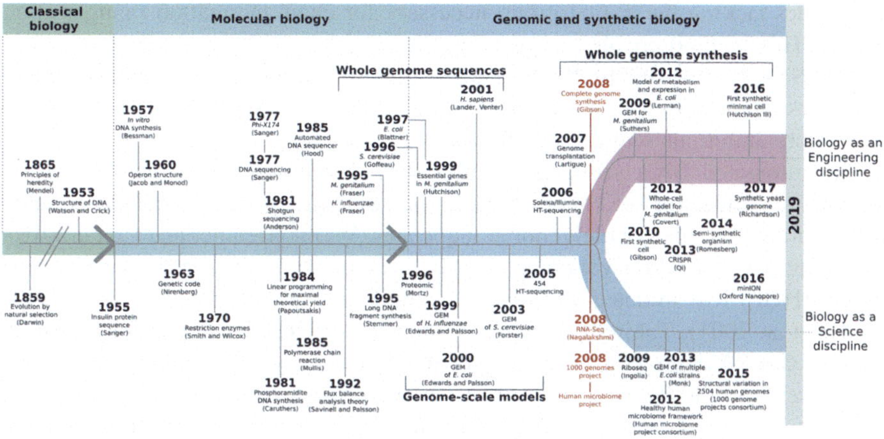

Fig. 1 Synthetic biology and minimal cells: a historical perspective. Elucidating the DNA double helix marked the beginning of the molecular biology era, and it became possible to study molecular mechanisms that underpinned observable phenotypes. DNA sequencing methods improved, leading to whole-genome sequencing at the end of the 1990s. Methods for mathematical cell modeling were developed during the 1980s and 1990s, and genome-scale models of metabolism, as well as computer simulations of metabolic networks, could be reconstructed. A defining moment took place in 2008 (red), with the creation of the first artificial genome that mimicked the genetic information of *M. genitalium*, the smallest genome, free-living, non-synthetic organism known to date. Thanks to developments in next-generation sequencing (NGS) methods, this was paired with the rise of large-scale genome sequencing ventures, such as the Human Microbiome and the 1000 Genomes Projects. Advances in whole-genome synthesis, assembly, and transplantation helped create the first cell living with an entirely synthetic genome shortly after. Altogether, these achievements marked the coming of age for synthetic biology (Reproduced from Lachance et al. 2019a)

1.1.2 Molecular Biology

With the structure of the DNA double helix, the gene concept became tangible, greatly accelerating the pace of discovery. Iconic, groundbreaking findings of the molecular biology era include the cracking of the genetic code (Nirenberg et al. 1965; Holley 1965) and the definition of the operon by Jacob and Monod, which, for the first time, revealed molecular mechanisms underlying gene expression (Jacob et al. 1960). The later discovery of a restriction enzyme (enzyme capable of cutting DNA at a specific sequence) in *Haemophilus influenzae* (Smith and Wilcox 1970) and its application to cut the genome of the human virus SV40 (Danna and Nathans 1971) marked the beginning of DNA manipulations (Roberts 2005). The repurposing of a restriction enzyme provided the first genetic engineering tool, and biologists were now poised to start deciphering the molecular mechanisms that underlie cellular phenotypes.

Cleaving DNA at specific sites is useful, but a pending important challenge was to decode the sequence of genes. Given the determination of the genetic code in 1963 (Nirenberg et al. 1963), DNA sequencing would provide the amino acid sequence of proteins, which in turn mediates its function. In 1977, Frederick Sanger published a method for the sequencing of DNA by random incorporation of radiolabeled nucleotides lacking the 3′-OH group necessary for chain elongation (Sanger et al. 1977a). This method enabled sequencing the complete 5375 bp genome of phage *ϕ*X174 (Sanger et al. 1977b). While Sanger's dideoxy termination method generated sequences ranging from 15 to 200 nucleotides, massive scaling up was necessary to allow more ambitious sequencing efforts.

The automated DNA sequencer (Smith et al. 1986) and the advent of shotgun sequencing (Anderson 1981) considerably increased the capacity of DNA sequencing, resulting in longer whole-genome sequences (WGS) (Heather and Chain 2016). Following the Santa Cruz workshop in 1985 (Sinsheimer 1989), the Human Genome Project (HGP) was initiated and reached completion in 2001 (Venter et al. 2001; Lander et al. 2001). Taking advantage of the technologies developed for the HGP, smaller-scale WGS projects were completed before the new millennium (Fig. 1). In a historic reference to the first type II restriction enzyme isolated, the first WGS of a free-living organism, *Haemophilus influenzae*, was reported in 1995 (Fleischmann et al. 1995). The genome of *Mycoplasma genitalium*, the smallest free-living organism, was published shortly after (Fraser et al. 1995). The more complex model organisms *Saccharomyces cerevisiae* and *Escherichia coli* followed in 1996 and 1997, respectively (Goffeau et al. 1996; Blattner et al. 1997).

1.1.3 Genomic

We arbitrarily defined the beginning of the genomic era with the completion of the first WGS of a free-living organism, the ~1.9 million bp genome of *Haemophilus influenzae* (Fleischmann et al. 1995) (Fig. 1). The number and size of WGS made available following this first sequencing effort steadily increased, eventually including the ~3.2 billion base pairs (bp) haploid human genome (Venter et al. 2001; Lander et al. 2001). Automation and computational tools were further improved to expand the capacity of Sanger sequencing. Nevertheless, the advent of next-generation sequencing (NGS) technologies developed by private companies upon the completion of the Human Genome Project represented a major breakthrough. While the sequencing by synthesis paradigm was preserved between Sanger sequencing and the NGS methods, the ability to parallelize the sequencing within one reaction massively increased the throughput (Heather and Chain 2016).

NGS allowed the elaboration of new initiatives such as the 1000 Genomes Project (Spencer 2008) and the Human Microbiome Project (McGuire et al. 2008), both initiated in 2008. The power of NGS technologies could not be exemplified any better than by considering that, in spite of their much greater scale, these projects reached their primary goals within 4 years (1000 Genomes Project Consortium 2012; Human Microbiome Project Consortium 2012), merely one third of the time required for the HGP. The accessibility of sequencing now contributes to an unprecedented breadth of knowledge that is meant to continue. Recently, the development by Oxford Nanopore (minION) of a portable, benchtop, real-time sequencer (Lu et al. 2016) further expands the applications of NGS for fundamental discovery.

Obtaining the genome sequences of a wide number of species is of paramount importance for understanding their phylogenetic relationships and the potential functions they encode. However, the genetic information encrypted in the DNA of a cell is essentially static and does not reveal the dynamic nature of molecular phenotypes. This reality became evident shortly after the completion of the HGP, when the predicted number of genes in human was found to have been grossly overestimated (Brower 2001). Gladly, the efforts for high-throughput interrogation of other important cellular components started early with the development of untargeted approaches for the sequencing of proteins (Mørtz et al. 1996). More than a decade later, the elaboration of a protocol for high-throughput RNA sequencing using NGS technologies revealed the full transcriptomic profile of yeast (Nagalakshmi et al. 2008). From that point on, the three main macromolecules of the *central dogma* of biology (Crick 1970) could be sequenced at a genome-scale in an untargeted way.

The remaining components of the cell are less ubiquitous, and the application of untargeted methods for organism-wide identification is more complex. The identification of all water-soluble components is termed metabolomic, whereas the hydrophobic content is generally referred to as lipidomic (Riekeberg and Powers 2017). Liquid chromatography followed by mass spectrometry (LC-MS) allows for both metabolomic and lipidomic determination (Riekeberg and Powers 2017; Yang and Han 2016) with the difference in extraction method reflecting the polarity of the compounds. These methods along with others (Ingolia et al. 2009; Lahner et al. 2003; Zamboni et al. 2009) allow the characterization of a dynamic state of the cell that can be leveraged in systems biology (Haas et al. 2017).

1.1.4 Synthetic Biology

The term synthetic biology is closely associated with the application of engineering principles to biological systems. DNA synthesis enabled the generation and assembly of synthetic DNA parts. In turn, these capabilities allowed creating "that which did not exist," hence defining synthetic biology as a field (Andrianantoandro et al. 2006; Heinemann and Panke 2006; Hughes and Ellington 2017). The first attempt at synthesizing DNA happened shortly after the elucidation of its structure. In 1957, Bessman and colleagues used the DNA polymerase from *E. coli* to produce DNA fragments. They noted that the presence of polymerized DNA is necessary for the reaction. This concept was later reused both by Sanger for DNA sequencing (Sanger et al. 1977a) and later for the famous polymerase chain reaction (PCR) (Saiki et al. 1985). The DNA oligonucleotide primers used for the development of PCR were produced using the phosphoramidite method (Matteucci and Caruthers 1981; Beaucage and Caruthers 1981). This chemistry is still currently used in most modern DNA synthesis platforms (LeProust 2016) but is limited by the oligonucleotide length that can be obtained without accumulating undesired mutations. This problem was circumvented by Stemmer in 1995, who first reported a technique to generate a long synthetic DNA fragment (>1000 bp) by assembling oligonucleotides (Stemmer et al. 1995). While the cost of DNA synthesis stayed more or less the same in the last 10 years (Hughes and Ellington 2017), recent progress toward high-throughput DNA synthesis strategies using microarrays may soon overcome this issue (LeProust 2016) and promise to make the synthesis of large DNA fragments an affordable solution for routine molecular biology experiments or industrial strain design (Hughes and Ellington 2017; Bassalo et al. 2016).

The utmost objective of DNA synthesis is the conception and assembly of entire genomes. To reach this goal, the development of robust methods to assemble DNA fragments into larger sequences was necessary. This goal was met in 2008 when a team at the John Craig Venter Institute (JCVI) realized the complete synthesis and assembly of the *Mycoplasma genitalium* genome (Gibson et al. 2008). This achievement was made possible by a hierarchical strategy relying on in vitro recombination of DNA cassettes (Gibson et al. 2009). The assembly of overlapping DNA oligonucleotides to create larger fragments was later shown to be even more effective in vivo using yeast (Gibson 2009). The development of whole-genome synthesis and assembly methods together with that of whole-genome transplantation (Lartigue et al. 2007) enabled the creation of the first cell living with an entirely synthetic genome (Gibson et al. 2010).

Recent years have seen groundbreaking synthetic biology efforts that will undoubtedly have an impact on the future of this field. In 2014, Romesberg and colleagues created a bacteria functioning with an altered DNA containing six different bases (Malyshev et al. 2014), thereby offering an additional base pairing combination. No known living organism contains these synthetic nucleobases, resulting in a new life form on Earth. Following the path of the first free-living organism containing a synthetic genome, the team at JCVI designed and assembled a cell with a greatly reduced gene content, resulting in a working approximation of a minimal cell (Hutchison et al. 2016). Finally, the Sc2.0 project was initiated and in 2017 an international consortium reported the complete de novo synthesis of five entire chromosomes of the yeast *Saccharomyces cerevisiae* (Richardson et al. 2017).

With the advent of NGS, multiple omics methods for the dynamic characterization of the cell, targeted genome editing methods (Qi et al. 2013), and the development of high-throughput DNA synthesis and assembly methods, synthetic biology is now poised to create life forms that will revolutionize many industrial research fields such as microbial drug synthesis, biofuel production, or alternative approaches for disease treatment (Smolke et al. 2018).

1.2 The Minimal Cell Concept

> The hydrogen atom of biology
>
> —Harold J. Morowitz

The idea of a minimal cell was approached by biophysicist Harold J. Morowitz in a guest lecture in 1984 (Morowitz 1984) where he reasoned that a free-living organism would have a lower limit on the number of atoms from which it is composed. Below this number, the necessary functions to support life would not be met. This logical deduction somewhat resembles that of Schrödinger in his famous book *What is life*? (Schrodinger 1967), where the famous physicist questioned the material support of the gene and applied limitations imposed by quantum physics to correctly predict that it would be a molecule that could form a crystal. In his lecture, Morowitz proposed that the mollicutes, a phylogenetic group of bacterium deprived of a cell wall, would be the best candidates to match the constraint and the endeavor of generating what he then defined as a "minimal cell." The choice was made firstly according to their size with the idea that the smaller cell, much like the hydrogen atom in physics, would be the simplest system to study and hence yield fundamental understanding applicable to other, more complex biological systems. The prediction was accurate. The mollicute *Mycoplasma genitalium*, second entirely sequenced free-living organism (Fig. 1) (Fraser et al. 1995), has the smallest gene content of any known naturally occurring organisms. The purpose of studying minimal cells was then clearly stated: defining the basic principles of life (Glass et al. 2017).

As soon as more than one whole-genome sequence was generated, Mushegian and Koonin sought to compare the two phylogenetically distant species in the hope of finding orthologous genes that would be a working approximation of a minimal gene set (Mushegian and Koonin 1996). The initial proposition was that 256 genes would be sufficient to support life. This proposition was later experimentally shown to be a relatively low estimate. Gene essentiality in genome-reduced bacteria probed with random transposon insertion estimated that the number of genes would be between 265 and 350 (Hutchison et al. 1999). With the increasing number of whole-genome sequences available, comparative genomics allowed to deepen the understanding of the concept of minimal gene set. When comparing the eukaryote *Saccharomyces cerevisiae* to its initial proposition, Koonin realized that very few genes were conserved (only 40%) (Koonin 2000). The suggested explanation for this was that non-orthologous gene displacement (Koonin et al. 1996) (NOD) would have a higher frequency than originally anticipated. The definition of NOD states that genes with similar functions can evolve independently. This induced a paradigm shift in the concept of minimal gene set, where the identity of the genes themselves was deferred to a second level, with the functional activity they provide becoming more important. From an engineering standpoint, the minimal set of functions is indeed more interesting than the set of genes (Danchin and Fang 2016). In this context, the various genes become interchangeable parts to accomplish a given function (Fig. 2).

The many progresses in synthetic biology realized by scientists at the JCVI led to the design and synthesis of the first working approximation of a minimal cell: JCVI-syn3.0 (Hutchison et al. 2016). The 463 genes encoded in the chromosome of this cell is a lower number than any other known free-living organism (Glass et al. 2017) but is substantially higher than computationally and experimentally determined minimal gene sets (Mushegian and Koonin 1996; Hutchison et al. 1999; Koonin 2000; Glass et al. 2006). Although essential for cell growth, a significant fraction (149/463, ~30%) of the JCVI-syn3.0 gene set had no proposed function (Hutchison et al. 2016; Glass et al. 2017; Danchin and Fang 2016). Danchin and Fang extensively reviewed these genes in search for a molecular mechanism that would need to be fulfilled (Danchin and Fang 2016) and provided potential functions based on known or projected necessities for 32 of those 84 generic and 65 "unknown unknowns." The validity of these hypotheses has yet to be determined, and therefore the original question raised by Morowitz, seeking for the completeness of molecular biology, remains unanswered.

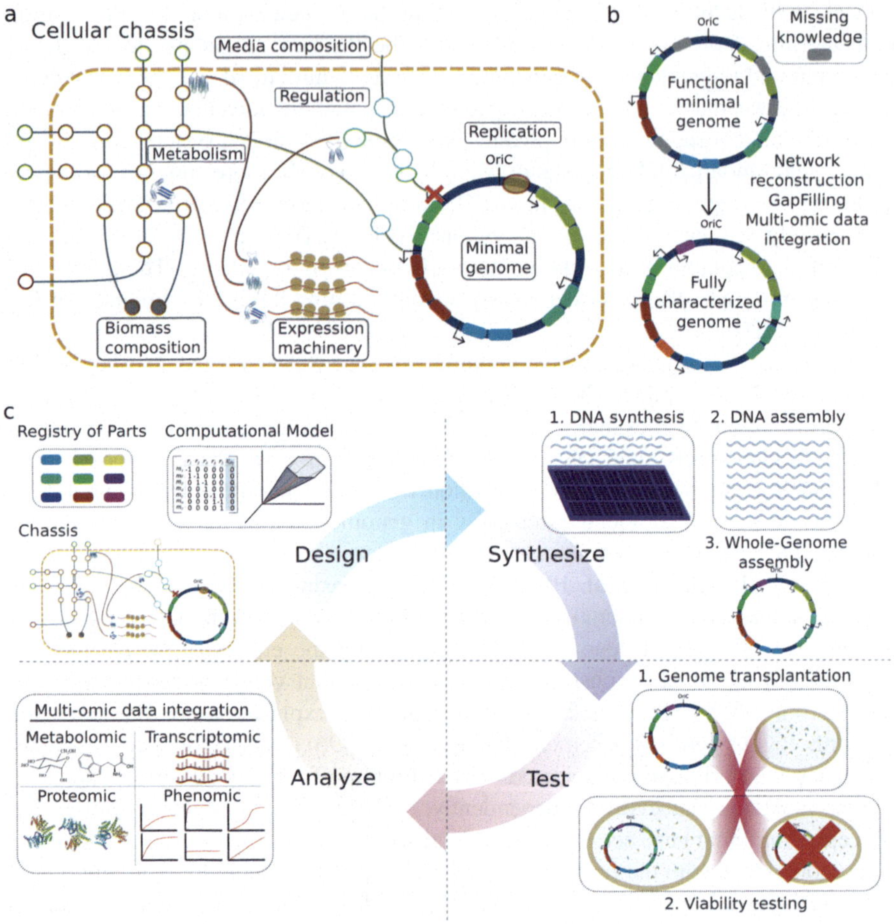

Fig. 2 Design of cells using a computer model. (**a**) Naive representation of a cellular chassis in which all mandatory cellular functions and their interactions are understood and characterized. (**b**) The generation of computer models for minimal cells can accelerate the identification of missing knowledge and facilitate the generation of hypothesis for essential uncharacterized cellular functions. (**c**) A design-build-test-analyze loop for the generation of minimal cells and their improvement toward production strains. Mathematical models are used to predict functional genotypes, and the current DNA synthesis technologies mentioned in the text are used to generate the proposed genome. Cloning of entire genomes in living cells allows to test for viability, and multiple omic datasets are used to characterize the synthetic organism

While absolute minimal cells should inform on the first principles of life, heavily reduced cells entail a strong interest from an engineering perspective. Cho and colleagues reviewed some of the potential advantages of reduced bacteria for strain design (Choe et al. 2016). As mentioned, high-throughput characterization of cellular phenotypes through omic data generation and increase in throughput of DNA synthesis should allow for in vitro fabrication of designed genomes (Fig. 2). We list here some of the advantages that were pointed out.

The first bacterium harboring a synthetic chromosome, JCVI-syn1.0 (Gibson et al. 2010), was reported to represent a 40-million-dollar endeavor (Sleator 2010). A smaller genome obviously results in a reduced DNA synthesis cost. While a decrease in the price per base was announced by companies such as Twist Bioscience, others do not foresee a price reduction that would defeat Moore's law (Smolke et al. 2018). Hence, the economic impact of generating several small genomes would remain significant. From a systems biology or design perspective, a reduced number of genes translate into a lower probability of negative interactions that could affect the desired outcomes of the initiated design. The development of high-throughput and untargeted approaches in the *genomic* era has allowed the rapid characterization of cells, but the outcome of genetic modifications is still not entirely reliable. The idea here is that genome-scale modeling of minimal cells could lead to more accurate model predictions. For instance, efforts have already been invested in reducing the complexity of metabolic models in the attempt of making the generated solutions more human readable (Ataman and Hatzimanikatis 2017). Genome reduction and minimization also allows for the design of biocontainment strategies. These include auxotrophy(ies) or programmed cell death, which will be highly beneficial as synthetic biology becomes more common in commercial applications. Finally, for more complex organisms, the deletion of genomic sections could accelerate genome replication while potentially increasing genomic stability through the removal of duplicated elements.

2 Constraint-Based Modeling

In the last section, we reviewed how biology developed from a pure science discipline at its inception to a mechanistic and engineering discipline in our times. The advent of high-throughput characterization methods for organisms together with biological reductionism that entails mechanistic description of processes sustaining life led to the birth of synthetic biology as a field. In this context, we reviewed the idea of a minimal cell, which should be a functional chassis for the design of production strains or a platform for fundamental understanding of biology (Danchin 2012). As mentioned, the current status of minimal cell research, with the 149 genes with no function associated with the synthetic organism JCVI-syn3.0 (Hutchison et al. 2016), requires further characterization of molecular functions to reach a complete understanding of every molecular functions necessary to sustain life. This biological reductionism approach should feed into a computational framework geared toward integrative analysis where in silico simulations based on mathematical models take advantage of the high-throughput methods to generate predictions. In this section we describe flux balance analysis (FBA) (Orth et al. 2010), a mathematical approach that allowed the generation of genome-scale models from whole-genome sequences around the new millennium (Edwards and Palsson 1999, 2000). This modeling approach is a solid basis on which minimal cells can be designed in silico.

2.1 Concept of Constraints in Metabolism

Flux balance analysis (FBA) arose from an attempt at generating simple coarse-grained models for the fermentation of the chemical industry's feedstocks in bacterial hosts (Papoutsakis 1984). One initial model suggested by Papoutsakis relied on the assumption that the fermentative process could be resumed in a single stoichiometric equation where elemental balance is conserved. Interestingly, the definition of the so-called fermentation equation utilized the known stoichiometry of reactions involved in the fermentation of butyric acid. The stoichiometry of biochemical reactions in a metabolic network was later used by Majewski and Domach in an attempt to establish a theoretical understanding for acetate overflow metabolism in *Escherichia coli* cells grown under aerobic conditions (Majewski and Domach 1990). The model presented for the acetate overflow entailed many key elements of FBA. The proposed hypothesis was that a flow network with a given objective could represent and explain the shift in metabolic state of *E. coli* responsible for the excretion of acetate.

The problem was summarized as a linear optimization problem on which network constraints would apply. Fixing the objective as to maximize the production of ATP and applying two constraints, (1) limiting the amount of reducing equivalents that can be produced by the electron transport chain and (2) assuming that a given enzyme of the Krebs cycle is limiting hereby limits the flux through a given reaction, the authors demonstrated that linear programming could correctly predict a bacterial metabolic state.

The use of a metabolic flux network optimized with linear programming served as a basis for the development of mathematical formalism for FBA (Savinell and Palsson 1992a, b). The concept was extended with the definition of a stoichiometric matrix (***S***). In this matrix, each column represents a reaction in the metabolic network, and each row is a different metabolite (Fig. 3). The mathematical formulation of the metabolite concentration over time using the ***S*** matrix then becomes:

$$\frac{d\boldsymbol{X}}{dt} = S \cdot \boldsymbol{v} \tag{1}$$

where $\boldsymbol{X}$ is the vector of metabolites and $\boldsymbol{v}$ is the vector of fluxes. FBA assumes that the metabolic network will reach a steady state. In this case, the concentration of metabolites over time should be in equilibrium where the inputs are equal to the outputs so that:

$$0 = S \cdot v \tag{2}$$

FBA has the advantage of requiring only the stoichiometry of the reactions to operate. The details of thermodynamics for each reaction are not necessary. Nevertheless, reaction directionality can be obtained from thermodynamics, hereby adding another set of constraint on the system. A physiologically meaningful objective (***Z***) can be defined in order to simulate the desired metabolic phenotype.

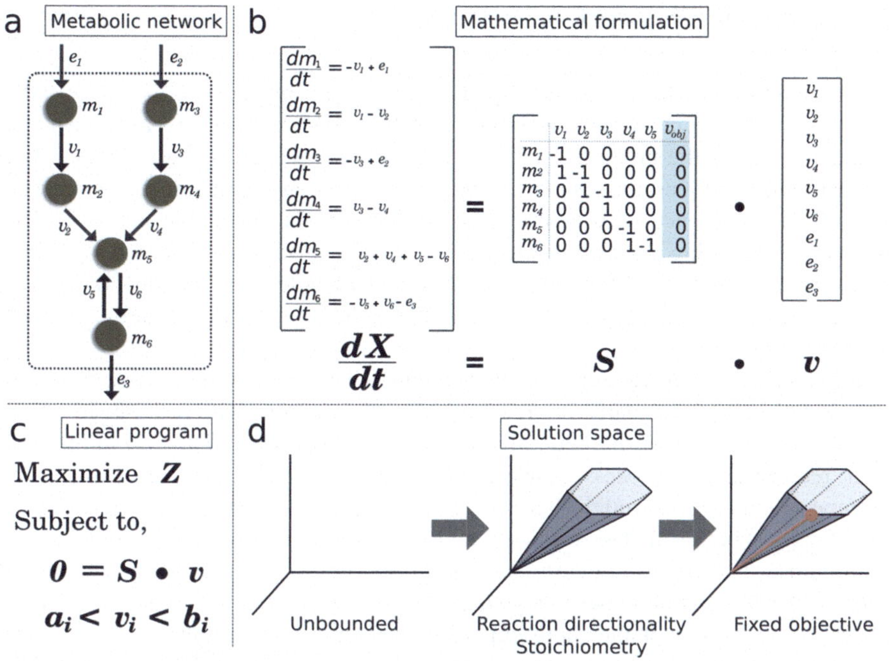

Fig. 3 Constraint-based modeling using linear programming. (**a**) A given metabolic network composed of metabolites (nodes) and reactions (links) can be represented in the form of a stoichiometric matrix ***S***. (**b**) In this matrix, each row represents a metabolite while each column is associated with a reaction. The variation of metabolite concentration over time $\left(\frac{dX}{dt}\right)$ can then be represented as the matrix-vector product of ***S*** by ***v***, the vector of fluxes for each reaction in the network. (**c**) Defining a physiologically meaningful objective ***Z***, the optimal solution for the metabolic network can be represented as a linear optimization problem with given flux constraints on metabolic reactions, and, at steady state, the variation in metabolite concentration is equal to 0. (**d**) The application of constraints on the optimization problem limits the solution space, while applying a proper objective allows finding the line of optimality within that bounded convex solution space

$$\begin{gathered}\text{maximize } Z,\\ 0 = S \cdot v\\ a_i < v_i < b_i\end{gathered} \tag{3}$$

This mathematical formulation can be solved using linear programming and allows finding the optimal solution of a given metabolic network at steady state. We will now review how this formulation allows for the generation of genome-scale models and how the objective function can be tailored to represent specific physiological states.

2.2 Metabolic Network Reconstruction

The completion of whole-genome sequences in the genomic era (Fig. 1) allowed the generation of genome-scale metabolic models (GEMs). In most cases, genome annotation yields the predicted function of proteins encoded by an organism. For the metabolic enzymes, the annotation together with extensive literature research can link a DNA sequence to a biochemical reaction in the metabolic network. The process of extracting a maximum number of reactions from the genome is termed reconstruction and has been reviewed in detail (Thiele and Palsson 2010). We explain the key steps in the reconstruction of a stoichiometric matrix at the genome scale (Fig. 4).

First, a draft reconstruction must be generated. The process of building this draft can be performed manually or automatically. The automated methods for draft reconstruction of metabolic networks are reviewed in Sect. 3. The generation of a draft reconstruction process consists in extracting biochemical reactions from genome annotation. Through this process, the stoichiometry of every reaction in the network is obtained. The reactions can be fetched from annotated EC numbers or gene names, and the candidate metabolic genes are linked to a reaction of the $\boldsymbol{S}$ matrix. The association between a gene and its reaction is key for future predictions generated by the model and should therefore be evaluated carefully.

Second, the initial draft is examined more closely through a refinement step. The key elements of this step are the examination of the mass-balance conservation for each reaction; that is the number of atoms in the reactants should be equal to the number of atoms in the products. The same rationale goes for the charge of the reactions. The balanced equations should have a neutral charge. These assumptions are linked to the fundamental principles of chemistry, hereby ensuring that no mass or energy is created in any reaction of the metabolic network. The gene-protein-reaction (GPR) association is then verified for all reactions, and a confidence score is attributed that facilitates further evaluation of the results once the model simulations are compared to experimental data.

Non-gene-associated reactions are then added. Spontaneous reactions are reactions for which no gene is associated and represent the natural occurrence of a reaction that is thermodynamically favorable without the need for a gene-encoded catalyst (enzyme). Other non-gene-associated reactions are exchange, sinks, and demands. These reactions represent the environment/culture media of the cell. They are not mass-balanced or charge-balanced by default since they represent the uptake/dumping of metabolites from/to the media. They are nevertheless necessary for the simulation of growth phenotypes under a given environment. Finally, a biomass reaction and ATP maintenance (ATPM) are added. The idea of a biomass reaction is to force the model to produce metabolites necessary for the growth of the organism, and its potential in simulating growth will be discussed later. The ATPM reaction is an ATP hydrolysis reaction that allows modelers to set a certain rate of ATP consumption for a growing cell. Knowing the experimental energy requirements hereby allows for more precise growth rate predictions.

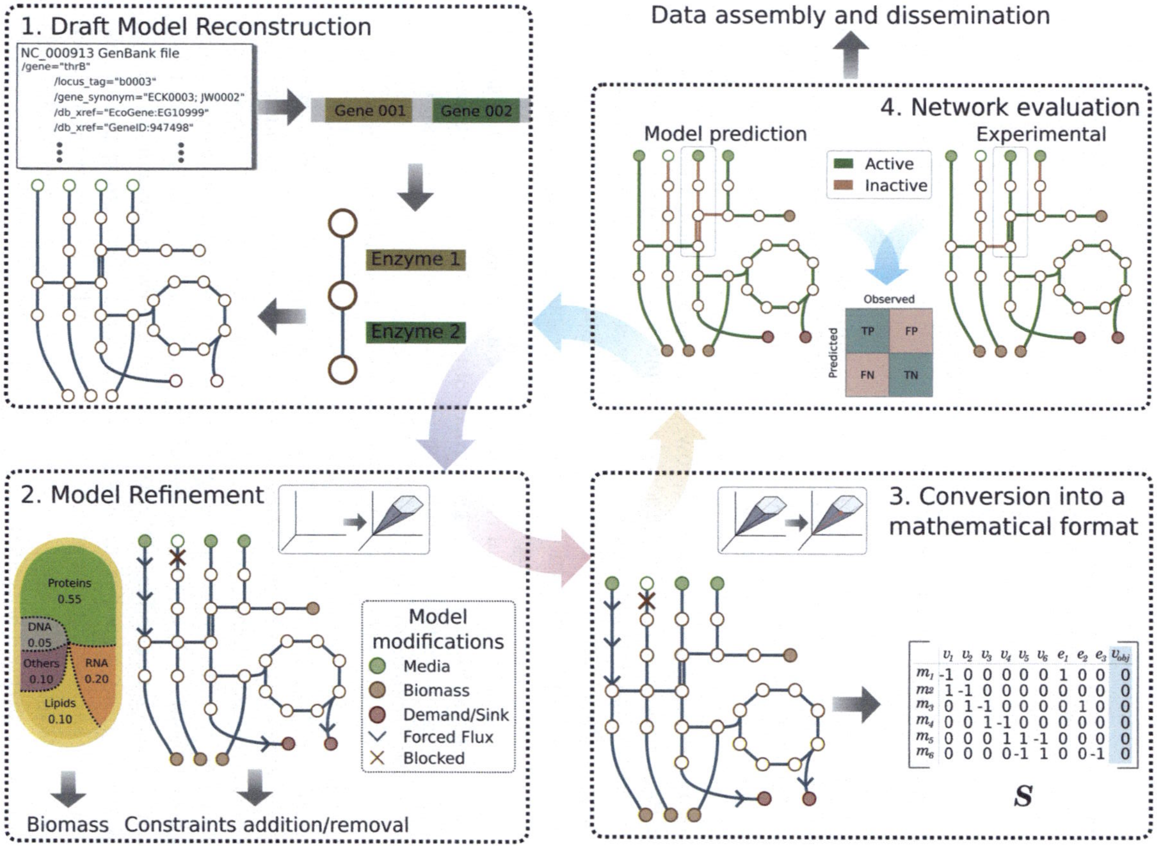

Fig. 4 The four main stages of metabolic network reconstruction and simulation. (1) A draft reconstruction is generated from genome annotation. (2) The draft reconstruction is refined by generating a biomass equation and applying constraints. (3) The reconstruction is converted into an optimization problem allowing the simulation of cellular phenotype. (4) Model predictions are compared to experimental observations. The discrepancies between predictions and observations are used to improve the model in an iterative process

Finally, the reconstructed network is ready for validation and simulation. Setting simulation constraints together with a defined objective allows the formulation of predictions that can be tested. The model may or may not yield a feasible solution. In the latter case, extensive unit testing may be required to address issues in the formulated reconstruction (e.g., metabolite accumulation). Iteratively fixing those issues allows the generation of a functional model that can be used for simulation. The comparison of the formulated predictions with readily available experimental data can either confirm how the system is expected to function or reveal potential gaps in knowledge.

2.3 Objective Function

In linear programming, the objective function is the numerical value to maximize or minimize. The significance of the value to be optimized is dependent on the situation that the modeler wishes to simulate. For instance, Papoutsakis (Papoutsakis 1984) generated a model for butyrate production. In that case, the numerical value is the amount of butyrate (a feedstock of the chemical industry) that can be produced over time. To simulate and explain acetate overflow, Majewski and Domach (Majewski and Domach 1990) maximized the ATP production by the network. Finally, the FBA red blood cell (RBC) model (Bordbar et al. 2011) maximizes the flux through the Na+/K+ ATPase pump. The choice of the objective function hence reflects the physiological situation and is key for the predictions generated by the model. Since the RBC cannot duplicate itself, it is assumed that the actual biological objective of the cell is to maintain a proper gradient of sodium and potassium, a task that requires the production of energy in the form of ATP. This proper objective definition along with the integration of high-throughput experimental data allowed the identification of biomarkers for RBC degradation upon storage (Yurkovich et al. 2017).

A common objective for modelers is to predict a growth phenotype. In this case, a biomass objective function (BOF) is defined that contains every metabolite necessary for the doubling of the cell (Feist and Palsson 2010). The BOF is modeled through the addition of an extra reaction (column) of the stoichiometric matrix (***S***). The proportion of each element within the cell is given as stoichiometric coefficients in the reaction. In order to provide an estimation of growth rate, a basis is given (Varma and Palsson 1993) such that the product of cell weight by time is equal to 1 g of cellular dry weight per hour (gDW/h). While the metabolite composition of the BOF may vary from a species to another, many components are shared across prokaryotes that are necessary for growth (Xavier et al. 2017). The proper integration of biomass components effectively present along with the stoichiometric coefficients that reflect the experimental composition of the cell (Beck et al. 2018) in species changes the accuracy of the model predictions (Lachance et al. 2018). The definition of the BOF is therefore crucial to generate gene essentiality predictions, a key topic for the endeavor of generating minimal cells in silico through genome-scale modeling with FBA.

2.4 Conversion into a Mathematical Format and Evaluation

Genome-scale models have the power to simulate the organism's metabolic capabilities. Converting the reconstruction into a mathematical format through the establishment of proper objective (e.g., precise definition of the biomass objective function) and constraints (e.g., media definition, internal flux bounds, uptake and secretion rates, etc.) provides the model with that potential. The model can then be used to formulate predictions of the metabolic state of the organism. The predictions formulated by the model and the datasets used to validate them vary based on the scientific objective of the conducted research.

A common objective used to enhance the quality of the model is growth (maximize flux through the biomass reaction). The direct prediction is the growth rate, which can be matched with the experimentally determined value. Getting a correct doubling time is dependent on correctly determining the cellular energy expenses and the stoichiometric coefficients of biomass precursors included in the biomass reaction. Optimizing for biomass production can also be used to determine gene essentiality by iteratively removing single genes and solving the model, a common measure of a model's quality, and will be covered in more detail later (see Sect. 4). Finally, FBA provides a flux state with the given solution. While FBA finds a unique optimal solution for the given objective function, many flux states may lead to it. Different methods have been developed that study the variability of the flux states that will be covered later (Gudmundsson and Thiele 2010). Modelers can then sample and study the variability of the flux state to identify fluxes that are out of biologically feasible ranges and apply supplementary constraints that improve the model's quality.

Compliance with experimental data can then be assessed. As mentioned, the gene essentiality prediction of the model is commonly used as a reference for a model's general quality since it accounts for the quality of the assigned GPRs together with the network topology, biomass, and media composition. A Punnett matrix is often used to visualize the predictions formulated by the model with all four combinations of true/false positive/negative represented. A metric such as accuracy or Matthews correlation coefficient can be used to quantify the quality of the model's prediction in a single number.

3 Computational Methods for Genome-Scale Reconstruction

With whole-genome sequences available for a greater number of species, the number of genome-scale metabolic models (GEMs) developed over the last two decades increased steadily (Monk et al. 2014). The number of computational tools tailored for the reconstruction of biological metabolic networks as well as the analysis and integration of omics data in these models has been developed accordingly (Lewis et al. 2012). In this section we review the methods and databases used for the reconstruction of the stoichiometric matrix (***S***), filling reaction gaps in the network and objective definition.

3.1 Tools for Network Reconstruction

The reconstruction of a GEM begins with the reconstruction of the stoichiometric matrix of reactions and metabolites (Fig. 4). Careful inspection of the genome annotation allows to link a gene and its sequence to a particular function in the network. In order to connect these elements together, modelers can use the many publicly available databases of pathways and biochemical reactions that are specifically designed to provide the association between genes, biochemical reactions, and/or the metabolic pathways (Kanehisa et al. 2017; Artimo et al. 2012; Placzek et al. 2017; Wattam et al. 2017; Aziz et al. 2008; Devoid et al. 2013; Fabregat et al. 2018; King et al. 2016; Caspi et al. 2008).

The identification of metabolic candidates in the reference genome is the first step of genome-scale reconstruction. To do so, modelers can either obtain enzyme commission numbers (EC) from specialized software (Nursimulu et al. 2018) or extract the information contained in the publicly available databases. In both cases, the standardization of metabolite and reaction identifier is key for the consistency and readability of the model. Since these identifiers vary considerably from one database to another, draft reconstructions may not be readable in another format. This type of issue has been addressed and can potentially be overcome by the use of MetaNetX (Moretti et al. 2016) or BiGG (King et al. 2016). MetaNetX is a web-based platform that attempts to centralize the identification of metabolite and reactions while also providing methods for automated genome-scale reconstructions. The main focus of the BiGG database is to list GEMs formulated in the BiGG nomenclature. Nevertheless, reactions and metabolites stored on BiGG are linked to other commonly used databases such as Reactome, KEGG, SEED, CHEBI, BioCyc, and MetaNetX. Choosing an identification system and ensuring the conversion from an annotation system to another is therefore key for the establishment of the draft reconstruction of the model.

The network reconstruction can be executed in different frameworks based on modeler's preferences. The SEED (Devoid et al. 2013) and Merlin (Dias et al. 2015) both allow for the automated generation of GEMs. While these functional models provide predictions, exhaustive literature search and model fine-tuning are usually necessary before a model is released (Thiele and Palsson 2010). The Open COBRA (Constraint-Based Reconstruction and Analysis) suite is designed to include every step of the process and is currently available under three different programming languages: Python (COBRApy, Ebrahim et al. 2013), MatLab (COBRA Toolbox 3.0, Schellenberger et al. 2011), and Julia (COBRA.jl, Heirendt et al. 2017). Implemented in MatLab, the RAVEN toolbox (Agren et al. 2013) is another option for reconstruction that also entails the visualization of the metabolic networks. The sybil toolbox allows R users to operate FBA, MOMA (Segrè et al. 2002), and ROOM (Shlomi et al. 2005) in their preferred language (Gelius-Dietrich et al. 2013). While Open COBRA does not specifically entail visualization, the generated network can be visualized by building a metabolic map with Escher (King et al. 2015).

3.2 *Tools for Network Analysis*

The main functionalities of the reconstruction toolboxes mentioned above are to allow the creation of foundational elements of models (metabolite, reaction, and gene objects) and store them into a model object that can be saved or imported in the desired format(s). These toolboxes also include basic model simulation functionalities such as the definition of objective and a bridge to the solver interface necessary to optimize the model. These preliminary simulation functionalities are useful for the conversion of the model into a mathematical format which can later be used for more intensive simulation processes and the evaluation of the organism's metabolic capabilities. We cover here some of the algorithms that have been developed to increase the quality of models before they are used for simulation.

3.2.1 Gaps in Network

In order to reveal biological capabilities, the network needs to be maximally functional, that is, flux can go through as many reactions as possible. As discussed, the mathematical formulation of FBA (steady-state assumption) does not allow for the accumulation of metabolites. This means that for a given linear pathway, a single missing reaction would block flux through every upstream and downstream reactions. The entire pathway would then be considered unfunctional, a hypothesis of debatable biological value that should be handled with care by modelers.

Several algorithms have been developed that aim at identifying problematic metabolites and reactions, solving gaps in the biological network, finding reactions that could fill those gaps and eventually proposing genes that could catalyze the suggested reaction(s) (Orth and Palsson 2010; Pan and Reed 2018). As mentioned, the general framework of these algorithms first identifies dead-end metabolites, that is, metabolites that cannot be produced or consumed in the metabolic network. Solving a gap in the network may be accomplished by adding one or many reactions. To find candidate reactions, these algorithms usually query larger reaction databases such as those contained in KEGG (Kanehisa et al. 2017) or MetaCyc (Caspi et al. 2008). The value of adding a given specific set of reactions can only be measured by the relatedness of this proposed mechanism to the actual species being studied. Therefore, the third step aims at identifying the best possible genes that can associate with those reactions.

The first gap filling algorithm (Satish Kumar et al. 2007) did not include this third task, but subsequent versions incorporated different ways to input experimental data along with the suggested reactions. GlobalFit (Hartleb et al. 2016) and ProbAnnoPy (King et al. 2018) are good examples of gap filling methods attempting to improve a metabolic model based on experimental data. For a deeper coverage of the available methods, interested readers can consult this review by Pan and Reed (Pan and Reed

2018). GlobalFit was used to increase the quality of two GEMs, *Escherichia coli* *i*JO1366 and *Mycoplasma genitalium* *i*PS189. It uses a bi-level optimization problem to minimize the gap between predicted gene essentiality and the experimental data and allows the incorporation of new reactions within the model or new exchange reactions (media components) as well as biomass precursors (metabolites of the BOF). ProbAnno (Web and Py) attributes a probability based on BLASTp search e-value to rank the reactions used to fill the gaps in the network.

Such approaches are relevant in the current context of minimal cell research and design. While a minimal cell has already been generated experimentally, the number of genes it contains for which a precise function could not be attributed is a significant portion of the complete genome (149/463). An ideal cellular chassis should have no unknown properties (Danchin 2012) since it would serve as a blueprint for further design. Hence, reconstructing metabolic networks and using gap filling algorithms that provide functional annotation are systematic ways to address the fulfillment of missing knowledge.

3.2.2 Objective Functions

The metabolic objectives of the cells can be summarized in a reaction of the stoichiometric matrix and set as an objective: the biomass objective function (BOF). The identification of key components necessary for a cell to grow is nevertheless a daunting task. This process can be accomplished in a biased way, which attempts at incorporating as much of the current knowledge of the organism's composition as possible or in an unbiased way where experimental data is utilized to infer the cellular objectives. A worthy effort at summarizing the current knowledge on prokaryotic biomass composition was accomplished by Rocha and colleagues (Xavier et al. 2017). In this extensive study, the biomass composition of 71 manually curated models available in the BiGG database (King et al. 2016) was compared along with the phylogenetic distance of the species that they represent. Swapping the BOF from a model to another showed that reaction essentiality prediction is sensitive to the BOF composition. Further studying the impact of biomass composition on gene essentiality predictions of several species, the authors found a set of universally essential cofactors in prokaryotes. This foundational knowledge highlights the importance of accurate BOFs for gene essentiality prediction by GEMs and provides an important resource for future modeling work.

Using previously established essential cell components, modelers can partly define the BOF for their organism of interest. Nevertheless, the remaining part of the BOF is species-specific and can be completed using an unbiased approach. Much like gap filling, cellular objective search can be performed algorithmically. Historically, most algorithms developed for that purpose have used metabolic flux analysis (MFA) data together with various optimization methods (Burgard and Maranas 2003; Gianchandani et al. 2008; Zhao et al. 2016). While MFA is a particularly well-suited type of data for flux models, the state-of-the-art number of fluxes generated by the method does not scale to the number of reactions included in GEMs. Recently developed algorithms attempt to use other types of data to find cellular objectives. BOFdat (Lachance et al. 2019b) uses a genetic algorithm to find the biomass compositions that provide the best match between predicted and experimental gene essentiality. The metabolites identified by the algorithm are then clustered based on their relative distance in the metabolic network to form clusters of metabolic objectives that can be interpreted by modelers. Another approach called BIG-BOSS integrates multiple omics data types to formulate the cellular objective by using a proteome constrained model, with a bi-level optimization problem similar to BOSS (Gianchandani et al. 2008). By augmenting MFA for a subset of fluxes with proteomics, the biomass composition was recovered more accurately than using just one data type alone.

4 Data Integration and Phenotypic Predictions

Once a GEM is reconstructed, converted into a mathematical format, and validated with experimental data, systematic model-driven hypothesis generation can take place that will guide the design of the desired strain. Much like the design of a production strain, the realization of a minimal cell requires in-depth knowledge of the organism that can be acquired through the generation of extensive high-throughput data. The integration of such data is made possible by GEMs, and a plethora of software has been written that helps modelers in this task. Here we cover available methods for the integration of high-throughput data as well as strain design algorithms that can be leveraged for the design of synthetic minimal cells (Fig. 6).

4.1 Cellular Objectives and Gene Essentiality Prediction

A key concept for the design of minimal cells is the identification of removable content. That is: "what genes are non-essential under laboratory conditions?". To formulate such prediction in silico, one must first determine the requirements for growth (Fig. 6). As we mentioned, those are represented by the BOF in GEMs. The definition of the BOF is tightly linked to the evolutionary pressure applied on the strain, which is in turn function of its growth environment. For instance, given an *E. coli* cell suddenly shifted from aerobic to anaerobic conditions, the instantaneous modification in phenotype is the result of chemical and physical properties, i.e., utilization of new substrate, shift in metabolic state, changes in gene expression, etc. This rapid adaptation can be termed proximal causation (Palsson 2015). Its counterpart, termed distal causation, happens over time and is the result of evolutive adaptation. Distal causation is proper to biological systems and entails a modification of the genotype to fit the constraints imposed by the environment under which the species is grown. Since the biomass composition of a cell is a result of its evolution, each species entails different metabolite requirements for growth with some essential components shared across a wide range of organisms (Xavier et al. 2017) (Fig. 6).

4.1.1 Gene Essentiality Prediction

GEMs can be used to formulate reaction or gene essentiality predictions (Fig. 6). To formulate that prediction, all reactions in the model are individually removed, and every time, the model is optimized for growth (Suthers et al. 2009a; Joyce and Palsson 2008). An appropriate threshold is necessary to discriminate between viable and nonviable phenotypes. This allows determining which reactions are essential to carry flux through the biomass reaction in the model. The proper definition of the BOF is therefore critical for the accurate prediction of gene essentiality. The qualitative definition of the BOF defines the growth requirements of the organism, and the pathways that lead to the production of these metabolites are then activated. Other constraints such as the growth media, uptake rates of the main carbon sources, and/or oxygen also impact gene essentiality predictions.

The added value of GEMs is that most reactions within the framework are associated with one or many genes. This association between gene and reactions is termed GPR and accounts for reactions catalyzed by a single gene or multiple genes in a complex, symbolized by an "and" rule, as well as isozymes, symbolized by an "or" rule (Fig. 6). Whole-model gene essentiality can be generated easily using the reconstruction toolboxes mentioned previously since they include an implementation of this function.

It is noteworthy that GEMs are very efficient at predicting gene essentiality. Highly curated models like that of *E. coli* have achieved essentiality predictions on different growth conditions with accuracies up to 93.4% (Monk et al. 2017). The quality of the prediction relies both on the high level of biochemical information included in the *E. coli* reconstruction and the precise knowledge of the growth conditions. These limitations will be discussed later.

4.1.2 Beyond Single Gene Deletion

An advantage of GEMs is the ability to formulate predictions of synthetic lethality (SL) (Fig. 6). The phenomenon was reported early in the classical era of biology in an attempt to describe the observation that the combination of observable traits did not yield viable descendants (Bridges 1922). At the gene level, SL is known as the observation that simultaneously knocking out two genes yields a lethal phenotype when their independent individual knockout provided a viable phenotype (Fig. 6). Experimentally studying SL at the systems level is complex since it involves screening several combinations of gene knockouts. For an organism containing N individually nonessential genes, the number of combinations is the binomial coefficient: $\frac{n!}{k!(n-k)!}$. Obtaining all possible SL combinations for an organism implies generating a library of knockouts on top of a knockout library. This task has been accomplished for heavily studied organisms such as *Saccharomyces cerevisiae* in which gene editing methods are commonplace (Goodson et al. 1996; Deutscher et al. 2006) but is generally too demanding to be generated for most species.

The computation of SL genes in FBA models is computationally expensive but still orders of magnitude faster than generating the data experimentally. Using this approach can guide the design of minimal genomes since it adds a level of information that could not be otherwise fetched from the genome or single knockout libraries. GEMs also provide the possibility to expand the SL study to more than gene pairs and include triple or quadruple knockouts (Suthers et al. 2009a), an undeniable advantage over the strictly experimental approach.

An interesting usage of SL analysis is the MinGenome algorithm written by Wang and Maranas (2018). This algorithm takes as input the genome sequence of the organism of interest, a GEM, genome-scale in vivo essentiality data, operon and promoter sites, and transcription factor information. Using this information, MinGenome iteratively finds the largest section of DNA that can be removed without killing the cell. The operon structure along with the promoters and transcription factor information are used to keep regulatory elements in place which should increase the probability that the suggested minimal genome is functional in vivo.

4.2 Multiple Omics Dataset Integration

As previously mentioned, the genomic era has enabled the generation of high-throughput data ("omic") for many different types of molecules. The integration of these sizeable datasets into comprehensive biological knowledge requires a proper framework. Metabolic models have been shown to provide a systematic way for the integration of multiple omic datasets for mechanistic understanding (Monk et al. 2014; Bordbar et al. 2014). Ralser and colleagues discussed the integration of seven types of omic datasets: genomic, transcriptomic, proteomic, lipidomic, metabolomic, ionomic, and phenomic (Haas et al. 2017). The approach used to incorporate this multi-omics information in GEMs will be discussed below.

GEMs use genomic information to extract biological functions of metabolic genes. While the regulation of gene expression is not accounted in metabolic models, the transcriptomic and proteomic datasets can be used to apply supplementary constraints on the model. The flux bound can be limited based on the level of expression or simply shut down when the genes are not expressed so that the reaction(s) associated with these genes cannot carry flux (Fig. 6). The concept of a minimal cell assumes a very specialized cell with reduced metabolic capabilities. Integrating the gene expression datasets in models hence has the potential to generate context-specific models that meet the expectation of highly specialized minimal cells.

Other datasets characterize molecules outside the central *dogma* of biology (Crick 1970). Metabolite concentrations themselves are not included in standard FBA, but a variant termed uFBA (Bordbar et al. 2017) allows the incorporation of time-course metabolomics into GEM resulting in more accurate predictions of the metabolic state of the cell. Lipidomic and ionomic results are useful to determine the composition of the cell, valuable information for the definition of the BOF.

The integration of multiple omic datasets with genome-scale models provides mechanistic explanation of the organism's phenotype under different environments (Lewis et al. 2010). Using multiple omic datasets, Lewis et al. showed that *E. coli* strains evolved under different conditions modify their pattern of gene expression in a manner that is consistent with a variant of FBA termed parsimonious FBA (pFBA). pFBA uses a bi-level linear programming approach to minimize the enzyme-associated flux while maximizing biomass production. The flux state generated using pFBA was consistent with the differential gene expression across conditions. These findings provided support for the biological relevance of FBA. The implication for the design of minimal cells is that generating an FBA-based model for such a cell would allow to design its optimal state ahead of conception.

5 Systems Biology of Minimal Cells

Since the proposition by Morowitz that minimal cells would allow understanding the basic principles of life (Morowitz 1984), many efforts have been driven toward the identification of theoretical minimal gene sets through comparative genomics (Mushegian and Koonin 1996), gene-wide essentiality probing (Glass et al. 2006), and a combination of these approaches (Baby et al. 2018). Genome reduction in complex bacteria has also been attempted experimentally for several complex bacteria (Choe et al. 2016), and ultimately, nearly 10 years of groundbreaking efforts led to the realization of a working approximation of a minimal cell in vitro (JCVI-syn3.0) (Hutchison et al. 2016; Sleator 2010).

We covered how the use of GEMs, which are mathematically structured knowledge bases of metabolism, provides phenotypic predictions from genomic information and thus can be leveraged for the rational design of minimal cells (Wang and Maranas 2018). We will now review GEMs for some naturally occurring near-minimal bacteria from the class of mollicutes and then cover expansion of modeling methods beyond metabolism.

5.1 Available GEMs for Naturally Occurring Minimal Organisms

Mollicutes have been the object of much research since they were proposed as the smallest free-living organisms (Morowitz and Tourtellotte 1962). Extensive knowledge of the particular metabolism (Miles 1992) of these species allowed the generation of GEMs for the most studied of them. The first GEM for a mollicute was reconstructed for the human urogenital pathogen *Mycoplasma genitalium* (Suthers et al. 2009b). This model includes 189 genes, 168 gene-associated reactions, and

274 metabolites. Using the experimental essentiality data (Glass et al. 2006), the model was consistent with 87% of essential genes and 89% of nonessentials. While this model prediction may be accurate, several approximations were used for the reconstruction. The biomass composition and the growth and non-growth-associated maintenance costs that can be calculated from substrate uptake rate and secretion rates were estimated from *E. coli*. Since there is no defined media for *M. genitalium*, the growth media was also estimated.

Formerly known as Eaton's agent, *Mycoplasma pneumoniae* is associated with atypical pneumonia in humans (Dajani 1965; Lind 1966). Multiple efforts at characterizing *M. pneumoniae* have been undertaken providing genome re-annotation (Dandekar et al. 2000), and the transcriptome (Güell et al. 2009), proteome (Kühner et al. 2009), and metabolism (Yus et al. 2009) have been studied in-depth. This allowed the generation of a quantitative model for *M. pneumoniae* (Wodke et al. 2013). The amount of experimental data available allowed modelers to compare predicted sugar utilization and obtain the energy utilization throughout the growth phases. Constraining the model with this data allowed dissecting the pathway usages at different growth stages.

The predictions formulated by the *M. pneumoniae* model revealed that a substantial amount of ATP is not directed toward biomass production but rather toward cell maintenance functions such as chaperone-assisted protein folding, DNA maintenance, and posttranslational modifications. Strikingly, the ATPase was responsible for most of the energy usage (57–80%) in order to maintain intracellular pH and a favorable proton gradient across the membrane. The authors suggested that four factors may impact the overall energy usage: the topology of the metabolic network, the growth rate, the environmental conditions, and the cell size. These findings are particularly interesting as they show that using a systems biology approach such as GEMs for the design of bacteria can go beyond gene essentiality prediction and reveal intrinsic properties affecting cellular energetics. These factors could hardly be predicted without the integration of experimental data into a mathematically structured knowledge base.

5.2 Genome-Scale Modeling of Synthetic Minimal Organisms

Recently, modeling efforts were dedicated to JCVI-syn3.0, a synthetic working approximation of a minimal cell (Breuer et al. 2019). The metabolic reconstruction was generated using the gene annotation of the parental strain JCVI-syn1.0 (*Mycoplasma mycoides*) for which much information is available. Collecting the breadth of knowledge into a single computational format is a significant step forward in order to define the functional metabolic requirements of a minimal cell. As discussed, GEMs can be used to formulate phenotypic predictions such as gene essentiality and integrate high-throughput data such as gene expression (see Sect. 4). Breuer et al. recently provided a dataset of high-density transposon mutagenesis operated on JCVI-syn3.0 as well as a quantitative proteomic dataset. The gene essentiality data allowed identifying discrepancies between model predictions and observations.

Together with the reconstruction process, the authors were able to formulate several hypotheses on the remaining gene functions that could not be removed but nevertheless unknown.

Using proteomic data allowed contextualizing the activity of the expressed proteins in JCVI-syn3.0, but the analysis is nevertheless limited. Indeed, while the resulting GEM for this organism is the first and closest representation of a synthetic minimal cell, more accurate model predictions would have required the detailed biomass composition of this bacterium along with a chemically defined medium. Including these parameters within the model should expand its predictive capabilities.

6 Perspectives on the Use of Models for Minimal Cell Design

A key objective of minimal cell research is to gather exhaustive understanding of the cell. The FBA framework allows to generate multiple predictions on the metabolic state of the cell, but the scope is limited to metabolism. Other approaches have been developed that allow including constraints from various cellular functions such as the expression machinery, regulatory network, enzyme kinetics, and thermodynamics. We propose here to extend the definition originally proposed by Morowitz for "the completeness of molecular biology" which entailed that every element in the cell should be characterized.

> What I do not understand I cannot create
>
> —Richard Feynman

6.1 Expanding the Scope of Models Beyond Metabolism

6.1.1 Modeling Gene Expression

Using the constraint imposed by the stoichiometry of reactions was key for the development of flux balance analysis (Kauffman et al. 2003) and later to genome-scale models of metabolism. In an attempt to expand the scope of models beyond metabolism, Thiele et al. reconstructed the expression matrix for *E. coli* (Thiele et al. 2009). The reconstruction of this matrix, named E-matrix by opposition to the M-matrix for metabolism, was executed using the same protocol that was mentioned above (Thiele and Palsson 2010). All reactions necessary for RNA transcription and protein translation are included in the E-matrix. Interestingly, every element necessary for the synthesis of proteins is considered as a metabolite in the network. For instance, transfer RNA (tRNA) and ribosomal RNA (rRNA) are both metabolites that can be produced from the transcription reactions. The tRNA are then charged and used in another reaction which synthesizes proteins. While the number of genes included in the E-matrix (423 genes) was smaller than that of the M-matrix [1515 genes (Monk et al. 2017)], the number of reactions is significantly higher (13,694 reactions in E vs 2719 reactions in M). The large size of the E-matrix is due to the high number of similar reactions catalyzed by the expression machinery.

Much like the M-matrix, the stoichiometry imposed by the E-matrix can be used as a constraint, and the reconstruction can be converted into a mathematical format by applying reaction bounds and fixing an objective. In this case, the uptake rates of amino acids and nucleotides need to be fixed as they are the necessary metabolites for the production of every downstream metabolites. The production of ribosome by the model can then be optimized for different growth rates since ribosome production is key for cell growth. Refining the constraints allowed the model to generate a number of ribosomes matching the experimental data. This work demonstrated the applicability of FBA to systems other than metabolism.

In order to couple the machinery of gene expression to the metabolism of the cell and generate a unified model for cellular growth, additional constraints were needed. Termed "coupling constraints," these equations are a function of the organism's doubling time and account for the dilution of material in doubling cells while providing upper limits on enzyme expression (Lerman et al. 2012; Lloyd et al. 2018; Thiele et al. 2012; O'Brien et al. 2013). These new constraints are both integer and linear and therefore define a mixed integer linear programming (MILP) problem. This type of problem is computationally more intensive than the regular linear programming problem solved in FBA and also requires more specific solvers (Yang et al. 2016).

6.1.2 Simulating with ME Models

An ME-model links metabolism to gene expression and can be used to generate experimentally testable predictions such as growth rate, substrate uptake and secretion rates, metabolic fluxes, and gene product expression levels (O'Brien et al. 2013). This last property is important as it simplifies comparison with experimental gene expression levels, which can now be routinely generated under many different environments. The ease of integration of multiple omics data in ME models has allowed the identification of key biological regularities (Ebrahim et al. 2016). Experimental proteomic data can provide absolute protein counts within a cell, which can be used to constrain the amount of protein in the ME model. Fluxomic data can also be used as a constraint since it provides the flux through a certain number of reactions. Combining these two types of data into ME-model simulations allowed to generate turnover rates (k_{eff}) for enzymes in the model, an example of model-driven generation of knowledge.

Simulating ME models over 333 different environmental conditions, Yang et al. identified genes consistently essential for optimal growth in *E. coli* (Yang et al. 2015). The formulated model-driven prediction of the core proteome was also found to be consistent with non-differentially expressed genes. Obviously, the functional incorporation of expression subsystems provided by the ME matrix allows the identification of more functional categories [COG (Koonin et al. 2004)] when determining a minimal gene set. This was further exemplified by the fact that DNA replication and repair mechanisms, functional categories absent from the ME model, were not represented in the core proteome. The further expansion of the ME-model to include other cellular systems such as a constraint-based approach of the regulatory machinery would provide a working approximation of a whole-cell model requiring fewer experimental parameters than what has been previously generated (Karr et al. 2012).

Potentially because of the size of the E-matrix, the reconstruction of entire ME-models has been contained to only two species so far, namely, *Thermotoga maritima* and *E. coli* (Lerman et al. 2012; O'Brien et al. 2013). Much like the generation of M-models is eased by the existence of toolboxes, the reconstruction of ME-models could be widespread by the recent publication of COBRAme, a Python framework for the reconstruction of ME-models (Lloyd et al. 2018).

6.2 Perspectives on the Use of Models to Design Minimal Cells

We delved into the historical evolution of biology and highlighted the possibility that a part of the discipline could turn into a field of engineering, in which the concept of a minimal cell would play a central role. The main idea surrounding this minimal cell concept is that of biological reductionism (Glass et al. 2017), which entails the complete description of every molecular functions harbored by a free-living cell (Lachance et al. 2019a). Reaching this level of knowledge is of paramount importance for the establishment of key design rules for organisms. With the advent of DNA synthesis techniques and whole-genome assembly, the creation of entirely new organisms is within reach. Such an example has been achieved with JCVI-syn3.0 (Hutchison et al. 2016), completing the first functional in vitro approximation of a minimal cell.

JCVI-syn3.0 reveals the state of the art in the design of minimal cells. Cutting-edge methods together with extensive work over many years have been put in place in order to produce this framework. The amount of labor necessary is met with the high-throughput capabilities of our days and age, both in terms of DNA synthesis and cloning and assembly, but the limiting factor remains the predictability of a given design. This struggle, relevant for both academic and industrial researchers, is one of the grand challenges that lays ahead in synthetic biology, and it is understood that laboratories which possess the best predictive power may outcompete those with high-throughput production and analysis capabilities.

In this context, the development of models for minimal cells is of paramount importance. We have reviewed the standard FBA approach for the genome-scale modeling of metabolism (Figs. 3, 4, and 5) and its applications for high-throughput data integration and the formulation of phenotypic predictions such as the flux

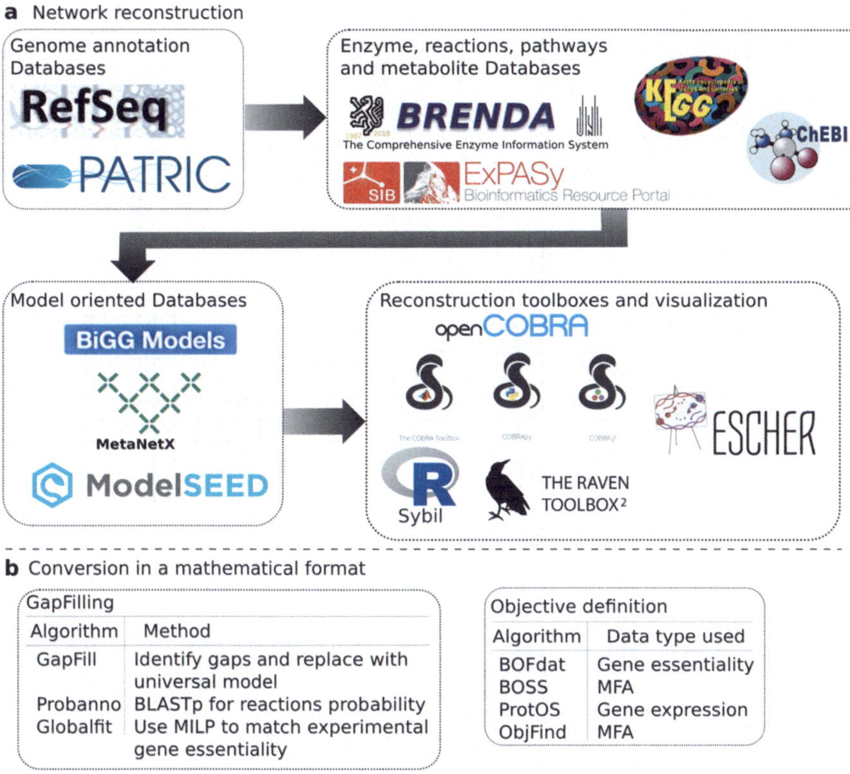

Fig. 5 Tools for genome-scale reconstruction and analysis. (**a**) Non-exhaustive list of computational tools and databases for the reconstruction of metabolic networks. First, querying annotation databases allows the identification of metabolic gene candidates (RefSeq, PATRIC). These genes can be associated to reactions by consulting reaction databases (KEGG, Brenda, ExPaSy, Chebi). The reactions and metabolites are associated to model specific identifiers using model-oriented databases (BiGG, MetaNetX, ModelSEED). The reconstruction toolboxes are designed to facilitate the creation of reaction, metabolite, gene and model objects programmatically (Open COBRA, Sybil, Raven). (**b**) Non-exhaustive list of computational tools to facilitate the identification of gaps in the network and cellular objectives

through metabolic reactions and gene essentiality (Fig. 6) (Suthers et al. 2009a; Zomorrodi and Maranas 2010). Integrating this knowledge into a single framework is important to offer a systematic way of addressing knowledge gaps (Orth and Palsson 2010; Pan and Reed 2018) as demonstrated by Breuer et al. in their GEM of JCVI-syn3.0 (Breuer et al. 2019).

What lays ahead is up for debate. Further development of models for mollicutes will require more exhaustive biomass and growth media definition to impose relevant constraints on the system. Given their small genomes, the number of biochemical studies needed before exhaustive characterization is reached is reduced and, with the help of models, could be addressed rather quickly (Danchin and Fang 2016). A recently developed algorithmic method allows to generate a minimal genome sequence from transcriptional architecture and an ME-model (Wang and Maranas 2018) which could help in reducing genomes of more elaborate organisms

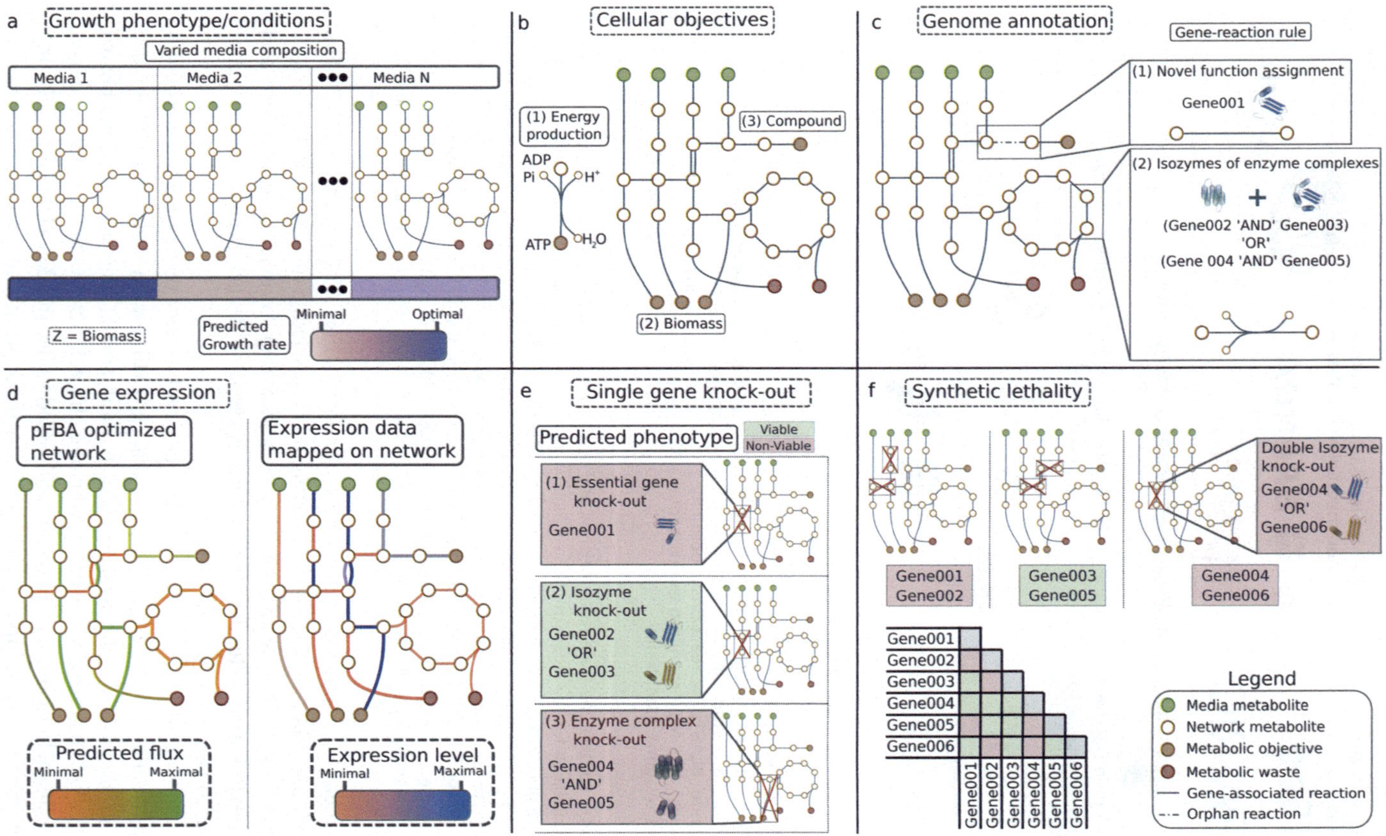

Fig. 6 The multiple uses of genome-scale models. (**a**) Growth media composition. The opened exchange reactions determine the in silico media composition. The impact of modifying the media composition is twofold. First, the necessary metabolites for growth are identified, hereby defining the minimal media. Second, the growth capabilities of the organism are explored, that is, the identification of nitrogen and carbon sources allowing growth. (**b**) Cellular objectives.

that are already used as production strains such as *E. coli* and *S. cerevisiae*. As we just discussed, constraint-based approaches can be expanded beyond metabolism, allowing the generation of models of metabolism and expression, ME-models. These models have already been employed to generate an in silico prediction of the core proteome by simulating on a wide array of different environments (Yang et al. 2015). With one of the main conclusions of this study being that the inclusion of more cellular systems be important for accurate predictions of a minimal gene set, it is interesting to consider that the expansion of modeling methods beyond metabolism and expression may be key for the rational design of minimal cells.

Finally, in silico writing of functional genome should be the following step. The integration of software tools for the conception of genomes is underway with the "Autocad" for genome recently published (Bates et al. 2017) as well as a genetic circuit compiler (Waites et al. 2018). Such tools are inspired by the experience acquired in the field of engineering, and the interest spurred by the community suggests a widespread application for the future of biology. For now, no organism is fully characterized, and hence the proposed completeness of biology (Morowitz 1984) is yet to be achieved. The use of genome-scale models together with genome writing tools might accelerate this process, and once a well-understood minimal cell chassis is described, strain design will reach a new paradigm.

References

1000 Genomes Project Consortium, Abecasis GR, Auton A, Brooks LD, DePristo MA, Durbin RM et al (2012) An integrated map of genetic variation from 1,092 human genomes. Nature 491:56–65

Agren R, Liu L, Shoaie S, Vongsangnak W, Nookaew I, Nielsen J (2013) The RAVEN toolbox and its use for generating a genome-scale metabolic model for Penicillium chrysogenum. PLoS Comput Biol 9:e1002980

Anderson S (1981) Shotgun DNA sequencing using cloned DNase I-generated fragments. Nucleic Acids Res 9:3015–3027

Andrianantoandro E, Basu S, Karig DK, Weiss R (2006) Synthetic biology: new engineering rules for an emerging discipline. Mol Syst Biol 2:2006.0028

Fig. 6 (continued) The cellular objective can be modified to suit modeling needs. The production of energy, of biomass, or of a relevant metabolite can be studied. (**c**) Genome annotation. The reactions in the network are associated with a gene via the gene-reaction rule. The metabolic model provides a dynamic context in which to situate the gene annotation and review a gene's function. (**d**) Gene expression. Parsimonious FBA (pFBA) is used to generate optimal flux state assuming an optimal enzyme usage by the cell. The flux through each reaction can then be compared to the gene expression level. (**e**) Gene essentiality. Genes and reactions are associated together through the gene-reaction rule. A gene is deemed essential if its deletion identifies it as the single determinant for the production of a biomass precursor. (**f**) Synthetic lethality. Performing simultaneous knockouts is possible with GEMs, an important asset for genome reduction and the production of synthetic minimal cells

Artimo P, Jonnalagedda M, Arnold K, Baratin D, Csardi G, de Castro E et al (2012) ExPASy: SIB bioinformatics resource portal. Nucleic Acids Res 40:W597–W603

Ataman M, Hatzimanikatis V (2017) lumpGEM: systematic generation of subnetworks and elementally balanced lumped reactions for the biosynthesis of target metabolites. PLoS Comput Biol 13:e1005513

Aziz RK, Bartels D, Best AA, DeJongh M, Disz T, Edwards RA et al (2008) The RAST Server: rapid annotations using subsystems technology. BMC Genomics 9:75

Baby V, Lachance J-C, Gagnon J, Lucier J-F, Matteau D, Knight T et al (2018) Inferring the minimal genome of Mesoplasma florum by comparative genomics and transposon mutagenesis. mSystems 3. https://doi.org/10.1128/mSystems.00198-17

Bassalo MC, Garst AD, Halweg-Edwards AL, Grau WC, Domaille DW, Mutalik VK et al (2016) Rapid and efficient one-step metabolic pathway integration in *E. coli*. ACS Synth Biol 5:561–568

Bates M, Lachoff J, Meech D, Zulkower V, Moisy A, Luo Y et al (2017) Genetic constructor: an online DNA design platform. ACS Synth Biol 6:2362–2365

Beaucage SL, Caruthers MH (1981) Deoxynucleoside phosphoramidites—a new class of key intermediates for deoxypolynucleotide synthesis. Tetrahedron Lett 22:1859–1862

Beck AE, Hunt KA, Carlson RP (2018) Measuring cellular biomass composition for computational biology applications. Processes 6:38

Blattner FR, Plunkett G III, Bloch CA, Perna NT, Burland V, Riley M et al (1997) The complete genome sequence of *Escherichia coli* K-12. Science 277:1453–1462

Bordbar A, Jamshidi N, Palsson BO (2011) iAB-RBC-283: a proteomically derived knowledge-base of erythrocyte metabolism that can be used to simulate its physiological and patho-physiological states. BMC Syst Biol 5:110

Bordbar A, Monk JM, King ZA, Palsson BO (2014) Constraint-based models predict metabolic and associated cellular functions. Nat Rev Genet 15:107–120

Bordbar A, Yurkovich JT, Paglia G, Rolfsson O, Sigurjónsson ÓE, Palsson BO (2017) Elucidating dynamic metabolic physiology through network integration of quantitative time-course metabolomics. Sci Rep 7:46249

Breuer M, Earnest TM, Merryman C, Wise KS, Sun L, Lynott MR et al (2019) Essential metabolism for a minimal cell. Elife:8. https://doi.org/10.7554/eLife.36842

Bridges CB (1922) The origin of variations in sexual and sex-limited characters. Am Nat 56:51–63

Brower V (2001) Proteomics: biology in the post-genomic era: companies all over the world rush to lead the way in the new post-genomics race. EMBO Rep 2:558–560

Burgard AP, Maranas CD (2003) Optimization-based framework for inferring and testing hypothesized metabolic objective functions. Biotechnol Bioeng 82:670–677

Caspi R, Foerster H, Fulcher CA, Kaipa P, Krummenacker M, Latendresse M et al (2008) The MetaCyc Database of metabolic pathways and enzymes and the BioCyc collection of Pathway/Genome Databases. Nucleic Acids Res 36:D623–D631

Choe D, Cho S, Kim SC, Cho B-K (2016) Minimal genome: worthwhile or worthless efforts toward being smaller? Biotechnol J 11:199–211

Crick F (1970) Central dogma of molecular biology. Nature 227:561–563

Dajani AS (1965) Experimental infection with mycoplasma pneumoniae (Eaton's agent). J Exp Med 121:1071–1086

Danchin A (2012) Scaling up synthetic biology: do not forget the chassis. FEBS Lett 586:2129–2137

Danchin A, Fang G (2016) Unknown unknowns: essential genes in quest for function. Microb Biotechnol 9:530–540

Dandekar T, Huynen M, Regula JT, Ueberle B, Zimmermann CU, Andrade MA et al (2000) Re-annotating the Mycoplasma pneumoniae genome sequence: adding value, function and reading frames. Nucleic Acids Res 28:3278–3288

Danna K, Nathans D (1971) Specific cleavage of simian virus 40 DNA by restriction endonuclease of Hemophilus influenzae. Proc Natl Acad Sci U S A 68:2913–2917

Deutscher D, Meilijson I, Kupiec M, Ruppin E (2006) Multiple knockout analysis of genetic robustness in the yeast metabolic network. Nat Genet 38:993–998

Devoid S, Overbeek R, DeJongh M, Vonstein V, Best AA, Henry C (2013) Automated genome annotation and metabolic model reconstruction in the SEED and Model SEED. Methods Mol Biol 985:17–45

Dias O, Rocha M, Ferreira EC, Rocha I (2015) Reconstructing genome-scale metabolic models with merlin. Nucleic Acids Res 43:3899–3910

Ebrahim A, Lerman JA, Palsson BO, Hyduke DR (2013) COBRApy: COnstraints-based reconstruction and analysis for python. BMC Syst Biol 7:74

Ebrahim A, Brunk E, Tan J, O'Brien EJ, Kim D, Szubin R et al (2016) Multi-omic data integration enables discovery of hidden biological regularities. Nat Commun 7:13091

Edwards JS, Palsson BO (1999) Systems properties of the Haemophilus influenzaeRd metabolic genotype. J Biol Chem 274:17410–17416

Edwards JS, Palsson BO (2000) The *Escherichia coli* MG1655 in silico metabolic genotype: its definition, characteristics, and capabilities. Proc Natl Acad Sci U S A 97:5528–5533

Fabregat A, Jupe S, Matthews L, Sidiropoulos K, Gillespie M, Garapati P et al (2018) The reactome pathway knowledgebase. Nucleic Acids Res 46(D1):D649–D655. https://doi.org/10.1093/nar/gkx1132

Feist AM, Palsson BO (2010) The biomass objective function. Curr Opin Microbiol 13:344–349

Fleischmann RD, Adams MD, White O, Clayton RA, Kirkness EF, Kerlavage AR et al (1995) Whole-genome random sequencing and assembly of Haemophilus influenzae Rd. Science 269:496–512

Fraser CM, Gocayne JD, White O, Adams MD, Clayton RA, Fleischmann RD et al (1995) The minimal gene complement of Mycoplasma genitalium. Science 270:397–403

Gelius-Dietrich G, Desouki AA, Fritzemeier CJ, Lercher MJ (2013) Sybil--efficient constraint-based modelling in R. BMC Syst Biol 7:125

Gianchandani EP, Oberhardt MA, Burgard AP, Maranas CD, Papin JA (2008) Predicting biological system objectives de novo from internal state measurements. BMC Bioinf 9:43

Gibson DG (2009) Synthesis of DNA fragments in yeast by one-step assembly of overlapping oligonucleotides. Nucleic Acids Res 37:6984–6990

Gibson DG, Benders GA, Andrews-Pfannkoch C, Denisova EA, Baden-Tillson H, Zaveri J et al (2008) Complete chemical synthesis, assembly, and cloning of a Mycoplasma genitalium genome. Science 319:1215–1220

Gibson DG, Young L, Chuang R-Y, Venter JC, Hutchison CA III, Smith HO (2009) Enzymatic assembly of DNA molecules up to several hundred kilobases. Nat Methods 6:343–345

Gibson DG, Glass JI, Lartigue C, Noskov VN, Chuang R-Y, Algire MA et al (2010) Creation of a bacterial cell controlled by a chemically synthesized genome. Science 329:52–56

Glass JI, Assad-Garcia N, Alperovich N, Yooseph S, Lewis MR, Maruf M et al (2006) Essential genes of a minimal bacterium. Proc Natl Acad Sci U S A 103:425–430

Glass JI, Merryman C, Wise KS, Hutchison CA, Smith HO (2017) Minimal cells—real and imagined. Cold Spring Harb Perspect Biol 9(12). https://doi.org/10.1101/cshperspect.a023861

Goffeau A, Barrell BG, Bussey H, Davis RW, Dujon B, Feldmann H et al (1996) Life with 6000 genes. Science 274(546):563–567

Goodson HV, Anderson BL, Warrick HM, Pon LA, Spudich JA (1996) Synthetic lethality screen identifies a novel yeast myosin I gene (MYO5): myosin I proteins are required for polarization of the actin cytoskeleton. J Cell Biol 133:1277–1291

Gudmundsson S, Thiele I (2010) Computationally efficient flux variability analysis. BMC Bioinf 11:489

Güell M, van Noort V, Yus E, Chen W-H, Leigh-Bell J, Michalodimitrakis K et al (2009) Transcriptome complexity in a genome-reduced bacterium. Science 326:1268–1271

Haas R, Zelezniak A, Iacovacci J, Kamrad S, Townsend S, Ralser M (2017) Designing and interpreting "multi-omic" experiments that may change our understanding of biology. Curr Opin Syst Biol 6:37–45

Hartleb D, Jarre F, Lercher MJ (2016) Improved metabolic models for *E. coli* and *Mycoplasma genitalium* from GlobalFit, an algorithm that simultaneously matches growth and non-growth data sets. PLoS Comput Biol 12:e1005036

Heather JM, Chain B (2016) The sequence of sequencers: the history of sequencing DNA. Genomics 107:1–8

Heinemann M, Panke S (2006) Synthetic biology—putting engineering into biology. Bioinformatics 22:2790–2799

Heirendt L, Thiele I, Fleming RMT (2017) DistributedFBA.jl: high-level, high-performance flux balance analysis in Julia. Bioinformatics 33:1421–1423

Holley RW (1965) Structure of an alanine transfer ribonucleic acid. JAMA 194:868–871

Hughes RA, Ellington AD (2017) Synthetic DNA synthesis and assembly: putting the synthetic in synthetic biology. Cold Spring Harb Perspect Biol 9. https://doi.org/10.1101/cshperspect.a023812

Human Microbiome Project Consortium (2012) Structure, function and diversity of the healthy human microbiome. Nature 486:207–214

Hutchison CA III, Chuang R-Y, Noskov VN, Assad-Garcia N, Deerinck TJ, Ellisman MH et al (2016) Design and synthesis of a minimal bacterial genome. Science 351:aad6253

Hutchison CA, Peterson SN, Gill SR, Cline RT, White O, Fraser CM et al (1999) Global transposon mutagenesis and a minimal Mycoplasma genome. Science 286:2165–2169

Ingolia NT, Ghaemmaghami S, Newman JRS, Weissman JS (2009) Genome-wide analysis in vivo of translation with nucleotide resolution using ribosome profiling. Science 324:218–223

Jacob F, Perrin D, Sanchez C, Monod J (1960) The operon: a group of genes whose expression is coordinated by an operator. C R Seances Acad Sci 250:1727–1729

Joyce AR, Palsson BØ (2008) Predicting gene essentiality using genome-scale in silico models. Methods Mol Biol 416:433–457

Kanehisa M, Furumichi M, Tanabe M, Sato Y, Morishima K (2017) KEGG: new perspectives on genomes, pathways, diseases and drugs. Nucleic Acids Res 45:D353–D361

Karr JR, Sanghvi JC, Macklin DN, Gutschow MV, Jacobs JM, Bolival B Jr et al (2012) A whole-cell computational model predicts phenotype from genotype. Cell 150:389–401

Kauffman KJ, Prakash P, Edwards JS (2003) Advances in flux balance analysis. Curr Opin Biotechnol 14:491–496

King ZA, Dräger A, Ebrahim A, Sonnenschein N, Lewis NE, Palsson BO (2015) Escher: a web application for building, sharing, and embedding data-rich visualizations of biological pathways. PLoS Comput Biol 11:e1004321

King ZA, Lu J, Dräger A, Miller P, Federowicz S, Lerman JA et al (2016) BiGG Models: a platform for integrating, standardizing and sharing genome-scale models. Nucleic Acids Res 44:D515–D522

King B, Farrah T, Richards MA, Mundy M, Simeonidis E, Price ND (2018) ProbAnnoWeb and ProbAnnoPy: probabilistic annotation and gap-filling of metabolic reconstructions. Bioinformatics 34:1594–1596

Koonin EV (2000) How many genes can make a cell: the minimal-gene-set concept. Annu Rev Genomics Hum Genet 1:99–116

Koonin EV, Mushegian AR, Bork P (1996) Non-orthologous gene displacement. Trends Genet 12:334–336

Koonin EV, Fedorova ND, Jackson JD, Jacobs AR, Krylov DM, Makarova KS et al (2004) A comprehensive evolutionary classification of proteins encoded in complete eukaryotic genomes. Genome Biol 5:R7

Kühner S, van Noort V, Betts MJ, Leo-Macias A, Batisse C, Rode M et al (2009) Proteome organization in a genome-reduced bacterium. Science 326:1235–1240

Lachance J-C, Monk JM, Lloyd CJ, Seif Y, Palsson BO, Rodrigue S et al (2018) BOFdat: generating biomass objective function stoichiometric coefficients from experimental data [Internet]. bioRxiv:243881. https://doi.org/10.1101/243881

Lachance J-C, Rodrigue S, Palsson BO (2019a) Minimal cells, maximal knowledge. Elife 8. https://doi.org/10.7554/eLife.45379

Lachance J-C, Lloyd CJ, Monk JM, Yang L, Sastry AV, Seif Y et al (2019b) BOFdat: generating biomass objective functions for genome-scale metabolic models from experimental data [Internet]. PLoS Comput Biol:e1006971. https://doi.org/10.1371/journal.pcbi.1006971

Lahner B, Gong J, Mahmoudian M, Smith EL, Abid KB, Rogers EE et al (2003) Genomic scale profiling of nutrient and trace elements in Arabidopsis thaliana. Nat Biotechnol 21:1215–1221
Lander ES, Linton LM, Birren B, Nusbaum C, Zody MC, Baldwin J et al (2001) Initial sequencing and analysis of the human genome. Nature 409:860–921
Lartigue C, Glass JI, Alperovich N, Pieper R, Parmar PP, Hutchison CA III et al (2007) Genome transplantation in bacteria: changing one species to another. Science 317:632–638
LeProust EM (2016) Rewriting DNA synthesis. Chem Eng Prog 2016:30–35
Lerman JA, Hyduke DR, Latif H, Portnoy VA, Lewis NE, Orth JD et al (2012) In silico method for modelling metabolism and gene product expression at genome scale. Nat Commun 3:929
Lewis NE, Hixson KK, Conrad TM, Lerman JA, Charusanti P, Polpitiya AD et al (2010) Omic data from evolved *E. coli* are consistent with computed optimal growth from genome-scale models. Mol Syst Biol 6:390
Lewis NE, Nagarajan H, Palsson BO (2012) Constraining the metabolic genotype-phenotype relationship using a phylogeny of in silico methods. Nat Rev Microbiol 10:291–305
Lind K (1966) Isolation of mycoplasma pneumoniae (Eaton agent) from patients with primary atypical pneumonia. Acta Pathol Microbiol Scand 66:124–134
Lloyd CJ, Ebrahim A, Yang L, King ZA, Catoiu E, O'Brien EJ et al (2018) COBRAme: a computational framework for genome-scale models of metabolism and gene expression. PLoS Comput Biol 14:e1006302
Lu H, Giordano F, Ning Z (2016) Oxford Nanopore MinION sequencing and genome assembly. Genomics Proteomics Bioinformatics 14:265–279
Majewski RA, Domach MM (1990) Simple constrained-optimization view of acetate overflow in *E. coli*. Biotechnol Bioeng 35:732–738
Malyshev DA, Dhami K, Lavergne T, Chen T, Dai N, Foster JM et al (2014) A semi-synthetic organism with an expanded genetic alphabet. Nature 509:385–388
Matteucci MD, Caruthers MH (1981) Synthesis of deoxyoligonucleotides on a polymer support. J Am Chem Soc 103:3185–3191
McGuire AL, Colgrove J, Whitney SN, Diaz CM, Bustillos D, Versalovic J (2008) Ethical, legal, and social considerations in conducting the Human Microbiome Project. Genome Res 18:1861–1864
Miles RJ (1992) Catabolism in mollicutes. J Gen Microbiol 138:1773–1783
Monk J, Nogales J, Palsson BO (2014) Optimizing genome-scale network reconstructions. Nat Biotechnol 32:447–452
Monk JM, Lloyd CJ, Brunk E, Mih N, Sastry A, King Z et al (2017) iML1515, a knowledgebase that computes *Escherichia coli* traits. Nat Biotechnol 35:904–908
Moretti S, Martin O, Van Du Tran T, Bridge A, Morgat A, Pagni M. MetaNetX/MNXref--reconciliation of metabolites and biochemical reactions to bring together genome-scale metabolic networks. Nucleic Acids Res 2016;44: D523–D526.
Morowitz HJ (1984) Special guest lecture the completeness of molecular biology. Isr J Med Sci 2
Morowitz HJ, Tourtellotte ME (1962) The smallest living cells. Sci Am 206:117–126
Mørtz E, O'Connor PB, Roepstorff P, Kelleher NL, Wood TD, McLafferty FW et al (1996) Sequence tag identification of intact proteins by matching tanden mass spectral data against sequence data bases. Proc Natl Acad Sci U S A 93:8264–8267
Mushegian AR, Koonin EV (1996) A minimal gene set for cellular life derived by comparison of complete bacterial genomes. Proc Natl Acad Sci U S A 93:10268–10273
Nagalakshmi U, Wang Z, Waern K, Shou C, Raha D, Gerstein M et al (2008) The transcriptional landscape of the yeast genome defined by RNA sequencing. Science 320:1344–1349
Nirenberg MW, Jones OW, Leder P, Clark BFC, Sly WS, Pestka S (1963) On the coding of genetic information. Cold Spring Harb Symp Quant Biol 28:549–557
Nirenberg M, Leder P, Bernfield M, Brimacombe R, Trupin J, Rottman F et al (1965) RNA codewords and protein synthesis, VII. On the general nature of the RNA code. Proc Natl Acad Sci U S A 53:1161–1168
Nursimulu N, Xu LL, Wasmuth JD, Krukov I, Parkinson J (2018) Improved enzyme annotation with EC-specific cutoffs using DETECT v2. Bioinformatics 34:3393–3395

O'Brien EJ, Lerman JA, Chang RL, Hyduke DR, Palsson BØ (2013) Genome-scale models of metabolism and gene expression extend and refine growth phenotype prediction. Mol Syst Biol 9:693

Orth JD, Palsson BØ (2010) Systematizing the generation of missing metabolic knowledge. Biotechnol Bioeng 107:403–412

Orth JD, Thiele I, Palsson BØ (2010) What is flux balance analysis? Nat Biotechnol 28:245–248

Palsson BØ (2015) Systems biology: constraint-based reconstruction and analysis. Cambridge University Press, Cambridge

Pan S, Reed JL (2018) Advances in gap-filling genome-scale metabolic models and model-driven experiments lead to novel metabolic discoveries. Curr Opin Biotechnol 51:103–108

Papoutsakis ET (1984) Equations and calculations for fermentations of butyric acid bacteria. Biotechnol Bioeng 26:174–187

Placzek S, Schomburg I, Chang A, Jeske L, Ulbrich M, Tillack J et al (2017) BRENDA in 2017: new perspectives and new tools in BRENDA. Nucleic Acids Res 45:D380–D388

Qi LS, Larson MH, Gilbert LA, Doudna JA, Weissman JS, Arkin AP et al (2013) Repurposing CRISPR as an RNA-guided platform for sequence-specific control of gene expression. Cell 152:1173–1183

Richardson SM, Mitchell LA, Stracquadanio G, Yang K, Dymond JS, DiCarlo JE et al (2017) Design of a synthetic yeast genome. Science 355:1040–1044

Riekeberg E, Powers R (2017) New frontiers in metabolomics: from measurement to insight. F1000Res 6:1148

Roberts RJ (2005) How restriction enzymes became the workhorses of molecular biology. Proc Natl Acad Sci U S A 102:5905–5908

Saiki RK, Scharf S, Faloona F, Mullis KB, Horn GT, Erlich HA et al (1985) Enzymatic amplification of beta-globin genomic sequences and restriction site analysis for diagnosis of sickle cell anemia. Science 230:1350–1354

Sanger F, Nicklen S, Coulson AR (1977a) DNA sequencing with chain-terminating inhibitors. Proc Natl Acad Sci U S A 74:5463–5467

Sanger F, Air GM, Barrell BG, Brown NL, Coulson AR, Fiddes JC et al (1977b) Nucleotide sequence of bacteriophage φX174 DNA. Nature 265:687

Satish Kumar V, Dasika MS, Maranas CD (2007) Optimization based automated curation of metabolic reconstructions. BMC Bioinf 8:212

Savinell JM, Palsson BO (1992a) Optimal selection of metabolic fluxes for in vivo measurement. I. Development of mathematical methods. J Theor Biol 155:201–214

Savinell JM, Palsson BO (1992b) Optimal selection of metabolic fluxes for in vivo measurement. II. Application to *Escherichia coli* and hybridoma cell metabolism. J Theor Biol 155:215–242

Schellenberger J, Que R, Fleming RMT, Thiele I, Orth JD, Feist AM et al (2011) Quantitative prediction of cellular metabolism with constraint-based models: the COBRA Toolbox v2.0. Nat Protoc 6:1290–1307

Schrodinger E (1967) What is life? The physical aspect of the living cell and mind and matter; mind and matter. Cambridge University Press, Cambridge

Segrè D, Vitkup D, Church GM (2002) Analysis of optimality in natural and perturbed metabolic networks. Proc Natl Acad Sci U S A 99:15112–15117

Shlomi T, Berkman O, Ruppin E (2005) Regulatory on/off minimization of metabolic flux changes after genetic perturbations. Proc Natl Acad Sci U S A 102:7695–7700

Sinsheimer RL (1989) The Santa Cruz Workshop—May 1985. Genomics 5:954–956

Sleator RD (2010) The story of Mycoplasma mycoides JCVI-syn1.0: the forty million dollar microbe. Bioeng Bugs 1:229–230

Smith HO, Wilcox KW (1970) A restriction enzyme from Hemophilus influenzae. I. Purification and general properties. J Mol Biol 51:379–391

Smith LM, Sanders JZ, Kaiser RJ, Hughes P, Dodd C, Connell CR et al (1986) Fluorescence detection in automated DNA sequence analysis. Nature 321:674–679

Smolke C, Lee SY, Nielsen J, Stephanopoulos G (2018) Synthetic biology: parts, devices and applications. Wiley

Spencer G (2008) International consortium announces the 1000 Genomes project. See http://www.1000genomes.org/bcms/1000_genomes/Documents/1000Genomes-NewsRelease pdf
Stemmer WP, Crameri A, Ha KD, Brennan TM, Heyneker HL (1995) Single-step assembly of a gene and entire plasmid from large numbers of oligodeoxyribonucleotides. Gene 164:49–53
Suthers PF, Zomorrodi A, Maranas CD (2009a) Genome-scale gene/reaction essentiality and synthetic lethality analysis. Mol Syst Biol 5:301
Suthers PF, Dasika MS, Kumar VS, Denisov G, Glass JI, Maranas CDA (2009b) Genome-scale metabolic reconstruction of Mycoplasma genitalium, iPS189. PLoS Comput Biol 5(2): e1000285. https://doi.org/10.1371/journal.pcbi.1000285
Thiele I, Palsson BØ (2010) A protocol for generating a high-quality genome-scale metabolic reconstruction. Nat Protoc 5:93–121
Thiele I, Jamshidi N, Fleming RMT, Palsson BØ (2009) Genome-scale reconstruction of *Escherichia coli*'s transcriptional and translational machinery: a knowledge base, its mathematical formulation, and its functional characterization. PLoS Comput Biol 5:e1000312
Thiele I, Fleming RMT, Que R, Bordbar A, Diep D, Palsson BO (2012) Multiscale modeling of metabolism and macromolecular synthesis in *E. coli* and its application to the evolution of codon usage. PLoS One 7:e45635
Varma A, Palsson BO (1993) Metabolic capabilities of *Escherichia coli*: I. synthesis of biosynthetic precursors and cofactors. J Theor Biol 165:477–502
Venter JC, Adams MD, Myers EW, Li PW, Mural RJ, Sutton GG et al (2001) The sequence of the human genome. Science 291:1304–1351
Waddington CH (1961) Molecular biology or ultrastructural biology? Nature 190:184
Waites W, Mısırlı G, Cavaliere M, Danos V, Wipat A (2018) A genetic circuit compiler: generating combinatorial genetic circuits with web semantics and inference. ACS Synth Biol. https://doi.org/10.1021/acssynbio.8b00201
Wang L, Maranas CD (2018) MinGenome: an in silico top-down approach for the synthesis of minimized genomes. ACS Synth Biol 7:462–473
Watson JD, Crick FHC (1953) Others. Molecular structure of nucleic acids. Nature 171:737–738
Wattam AR, Davis JJ, Assaf R, Boisvert S, Brettin T, Bun C et al (2017) Improvements to PATRIC, the all-bacterial bioinformatics database and analysis resource center. Nucleic Acids Res 45: D535–D542
Wodke JAH, Puchałka J, Lluch-Senar M, Marcos J, Yus E, Godinho M et al (2013) Dissecting the energy metabolism in Mycoplasma pneumoniae through genome-scale metabolic modeling. Mol Syst Biol 9:653
Xavier JC, Patil KR, Rocha I (2017) Integration of biomass formulations of genome-scale metabolic models with experimental data reveals universally essential cofactors in prokaryotes. Metab Eng 39:200–208
Yang K, Han X (2016) Lipidomics: techniques, applications, and outcomes related to biomedical sciences. Trends Biochem Sci 41:954–969
Yang L, Tan J, O'Brien EJ, Monk JM, Kim D, Li HJ et al (2015) Systems biology definition of the core proteome of metabolism and expression is consistent with high-throughput data. Proc Natl Acad Sci U S A 112:10810–10815
Yang L, Ma D, Ebrahim A, Lloyd CJ, Saunders MA, Palsson BO (2016) solveME: fast and reliable solution of nonlinear ME models. BMC Bioinf 17:391
Yurkovich JT, Yang L, Palsson BO (2017) Biomarkers are used to predict quantitative metabolite concentration profiles in human red blood cells. PLoS Comput Biol 13:e1005424
Yus E, Maier T, Michalodimitrakis K, van Noort V, Yamada T, Chen W-H et al (2009) Impact of genome reduction on bacterial metabolism and its regulation. Science 326:1263–1268
Zamboni N, Fendt S-M, Rühl M, Sauer U (2009) 13C-based metabolic flux analysis. Nat Protoc 4:878
Zhao Q, Stettner AI, Reznik E, Paschalidis IC, Segrè D (2016) Mapping the landscape of metabolic goals of a cell. Genome Biol 17:109
Zomorrodi AR, Maranas CD (2010) Improving the iMM904 S. cerevisiae metabolic model using essentiality and synthetic lethality data. BMC Syst Biol 4:178

From Minimal to Minimized Genomes: Functional Design of Microbial Cell Factories

Paul Lubrano, Antoine Danchin, and Carlos G. Acevedo-Rocha

Abstract The minimal genome is a theoretical concept asking what is the minimal gene set that defines life under a given environment. Experimental efforts show that stripping off most non-essential genes results in fragile organisms with "minimal genomes". By contrast, eliminating cryptic genes and mobile DNA results in strains with "minimized genomes" suitable for biotechnological applications because they display enhanced productivity, robust growth and upscalability. While it is believed that a minimal genome could be used to plug in "metabolic modules", we argue that there is no universal "chassis" because different organisms are suited to different environments. A further issue with the minimal genome is that it places DNA at the top of the hierarchy that led to the origin of life, ignoring metabolism and supporting a gene-centric view of evolution. This hardly accommodates the fact that the invention of nucleotides must have been a late event in prebiotic evolution. In this work, we take a "metabolism first" approach to describe the emergence of the first cells and the evolution of selected metabolic pathways that provided different solutions to the same problem. Understanding such processes provides insights for developing platform strains in metabolic engineering and industrial biotechnology.

Keywords Maxwell demons · Origin of life · LUCA · Evolution · Metabolic engineering · Industrial biotechnology · Synthetic biology

---

P. Lubrano
Unité Molécules de Communication et Adaptation des Microorganismes (MCAM), Muséum National d'Histoire Naturelle (MNHN), Centre National de la Recherche Scientifique (CNRS), Paris, France

A. Danchin
Department of Infection, Immunity and Inflammation, Institute Cochin, INSERM U1016, CNRS UMR8104, Université Paris Descartes, Paris, France

C. G. Acevedo-Rocha (✉)
Biosyntia, Copenhagen, Denmark
e-mail: car@biosyntia.com

A. R. Lara, G. Gosset (eds.), *Minimal Cells: Design, Construction, Biotechnological Applications*, https://doi.org/10.1007/978-3-030-31897-0_7

1 Introduction

Synthetic biology (SB) brings together professionals from various disciplines with different methods, languages and goals for making biology more predictable and useful to humankind (Acevedo-Rocha 2016). One such goal is the creation of synthetic cells that can perform predefined tasks such as the sustainable production of food, medicines and chemicals. The idea of using synthetic cells as chemical machines is centuries-old and dates back to Jacques Loeb (Porcar and Peretó 2018). With the advent of contemporary SB, bioengineers aim to develop not only "microbial cell factories" for the manufacturing of goods but also "minimal cells" to primarily answer the question: what is the minimal gene set allowing cellular life? (Juhas et al. 2011) To answer this question, two main experimental approaches have emerged, both using microbes with a small genome size and suitable for cultivation in the laboratory: top-down and bottom-up engineering. The former approach uses massive random combinatorial or directed single-gene mutagenesis techniques to remove all non-essential genes, resulting in minimal gene sets of 200–1000 genes (Acevedo-Rocha et al. 2013). Some of these gene sets can be useful to identify essential genes from pathogenic bacteria to develop new antimicrobial targets (Juhas et al. 2012). But the gene number depends on the organism, the gene mutagenesis method used in the construction and the environment—usually rich media where many metabolites are externally provided in a context where cells cannot produce their one building block to construct macromolecules (e.g. amino acids). In minimal media, which is a prerequisite for industrial applications, an additional number of 1000 "essential" genes are needed to provide de novo most basic metabolites (Danchin 2012).

By contrast, the bottom-up approach, which belongs to the field known as "synthetic genomics" (Schindler et al. 2018), harnesses high-throughput gene synthesis to build up entire chromosomes of reduced size that can be transplanted into cells. This remarkable feat has been accomplished in *Mycoplasma mycoides* using not only a slightly modified genome of 1000 kb called "JCVI-syn1.0" (Gibson et al. 2010) but also a half reduced version thereof dubbed "JCVI-syn3.0" (Hutchison et al. 2016). Surprisingly, almost one-third of the genes in that construct were apparently coding for unknown functions (Coyle et al. 2016). More recently, the chemical synthesis of larger and recoded genomes including reduced versions of *Caulobacter ethensis-2.0* (Venetz et al. 2019) and *Escherichia coli MDS42* has been reported (Fredens et al. 2019). The ultimate goal of synthetic genomics is to build from simple chemicals the genomes of eukaryotic cells including yeasts (Pretorius and Boeke 2018) and humans (Boeke et al. 2016) for understanding genomes and for developing applications in biotechnology.

A third nonexperimental approach is based on in silico comparative genomics, which exploits advanced phylogenetic methods of sequence similarity to infer protein-coding genes present across a series of microbes. However, as the number of genomes increases (from not only free-living *Bacteria* and *Archaea* but also non-free-living microbes such as symbionts and parasites), the number of conserved genes decreases down to zero (Acevedo-Rocha et al. 2013). This is because functions must be preserved while different structures can perform the same function.

Such genes code for specific protein primary, secondary and tertiary structures, which can be extremely variable, yet the function of these proteins is conserved (Mazumder and Vasudevan 2008). The absence of gene homologs in closely related species, even if combined with gene mutagenesis experiments, is a common finding (Baby et al. 2018), indicating the existence of minimal genomes suited to different niches.

Several years back, we suggested that, in order to be productive, the concept of gene essentiality should be replaced by "gene persistence" to be able to find universal gene functions that can be determined by nonuniversal protein-coding structures (Acevedo-Rocha et al. 2013). Persistent genes are microbial protein-coding genes that tend to be conserved in many genomes of widely different clades. They are mainly located on the leading DNA strand and expressed at a high level. Using a "functional approach" based on lists of persistent genes, it was possible to identify universal functions that are needed for life: biogenesis of the cell envelope (membrane, and cell wall when relevant), energy production and conversion, transport and metabolism of building blocks (organic acids, coenzymes, amino acids, lipids, carbohydrates, nucleotides) as well as information transfer (replication, recombination and repair) (Fang et al. 2005, 2008; Forterre et al. 2007). More recently, a similar functional approach was used to assign function to many of the unknown genes of *Mycoplasma mycoides* JCVI-syn3.0, allowing the identification of one third of genes involved in protein translation, stress response, small molecule transport as well as in RNA and DNA metabolism (Danchin and Fang 2016).

Obviously, a functional approach alone won't identify all functions needed to sustain life because important functions may be overlooked despite careful functional analysis of the cell's behaviour. To try and identify unexpected functions, it is important to combine data from large-scale experiments such as proteomics or metabolomics in order to develop in silico models that can enhance our understanding of apparently unidentified functions within a minimal cell. For example, an in silico flux balance analysis (FBA) model based on 338 metabolic reactions and 304 metabolites was elaborated and compared with in vivo data for JCVI-syn3A (Hutchison et al. 2016), revealing the function of genes mostly related to RNA modification (Breuer et al. 2019). To go further, we need to integrate many other factors such as isoenzymes and media optimization to improve predictability (Jean-Christophe Lachance et al. 2019).

Theoretically, a minimal cell might provide a clearer metabolic environment and reduce potential interactions between the endogenous genome and a targeted metabolic pathway. However, a major issue of minimal cells is that their minimal genomes are usually adapted to constant temperature and rich media. Moreover, the elimination of stress-responsive genes and restriction/methylase or toxin/antitoxin systems that are dispensable under ideal conditions render fragile cells to fluctuations of temperature and nutrient availability (Acevedo-Rocha et al. 2013). In addition, the elimination of two non-essential genes can lead to a lethal phenotype, a phenomenon known as synthetic lethality and related to nonadditive epistatic effects (Butland et al. 2008). Finally, another factor that is often not considered is growth rate, for instance, JCVI-syn3A has a very long doubling time of 2 h in rich media. Hence, these minimal cells are not suitable for industrial applications.

In parallel to the "minimal genome" projects during the past two decades, other efforts emerged to develop applications with model microbes in Europe (1998 EU *Bacillus subtilis* factory project) and Japan (2001 minimum genome *E. coli* factory project). These efforts resulted in a better understanding in the processes of protein expression and secretion in *B. subtilis* and in *E. coli* strains more suitable for the production of the amino acid threonine (Chi et al. 2019). In the USA, besides the "tour-de-force" works developed in *Mycoplasma*, other efforts reduced rationally the genome of *E. coli*. For instance, it was possible to reduce the genome of *E. coli* strain MG1655 from 6 up to ca. 39% (Pósfai et al. 2006), resulting in the *E. coli* MDS strain series that lack mostly transposons, mobile genetic elements and cryptic genes (Fehér et al. 2007; Umenhoffer et al. 2010; Csörgo et al. 2012). Recently such genetic modifications were transferred to the *E. coli* BL21 strain, which is commonly used to produce recombinant proteins (Draskovits et al. 2017).

Since then, many organisms like *B. subtilis*, *Pseudomonas putida*, *Streptomyces* and yeast have been added to the list (Chi et al. 2019). All those strains can be defined as cells with "minimized genomes". Notably, such engineered strains display features that are superior to produce chemicals and proteins compared to their ancestors. In some cases, however, some features have been compromised, indicating that reverse engineering is needed to fix the issues. For example, *E. coli* MSD42 showed metabolic instability in chemostats (Couto et al. 2018). Likewise, the slow growth rate of *E. coli* MS56 in minimal media was improved after applying adaptive laboratory evolution (Choe et al. 2019). In the case of *B. subtilis*, the removal of almost 40% of the chromosome resulted in strains that can grow only in rich media (Müller et al. 2017; Reuß et al. 2016; Suárez et al. 2019).

In summary, the removal of "junk" sequences has resulted in cells with "minimized genomes" that are more robust for certain biotechnological applications. This contrasts with cells with "minimal genomes" that are fragile and thus not suitable for practical applications (Zhang et al. 2010).

In the next section, we examine how persistent functions may have arisen at the onset of life. We then provide selected examples showing how enzymes were invented at least twice to solve the same problem but using different approaches by microorganisms adapting to different environments. Our main goal is to understand better metabolic design principles for designing microbial cell factories, thus filling a gap between top-down targeted genome reduction and metabolic engineering efforts.

2 From Minerals to Metabolism to Encoding of Information in Genomes

2.1 Origins of Life

The origin of life has always been a matter of debate. The main cause of discrepancy is the divergent preconceived views that scholars from different disciplines have

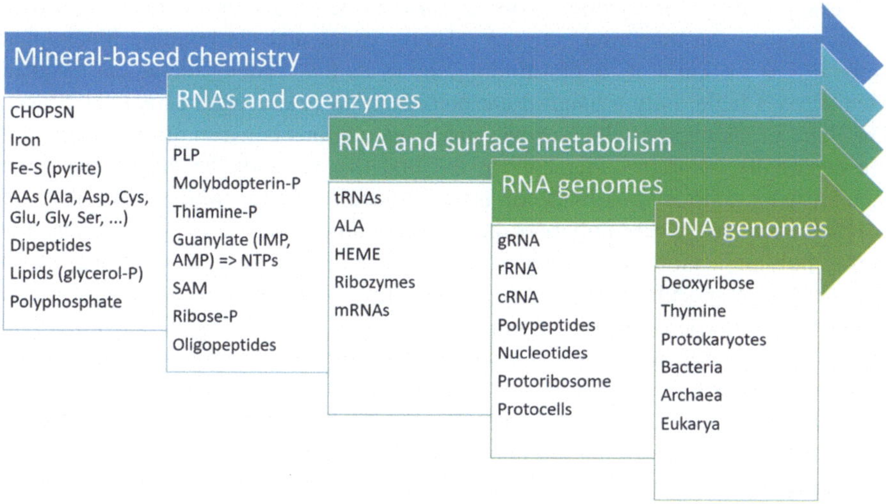

Fig. 1 A possible scenario of the origin of cellular life based on metabolism. *CHOPSN* carbon hydrogen oxygen phosphate sulphur nitrogen, *AAs* amino acids, *ALA* aminolevulinic acid, *RNA* ribonucleotide acid, *tRNA* transfer RNA, *gRNA* genomic RNA, *rRNA* ribosomal RNA, *cRNA* complementary RNA

when embarking in such an elusive problem. On the one hand, prebiotic chemists study the emergence of molecules from conditions that might not have been relevant early on. On the other hand, other scholars pose RNA at the onset of life and place the origin when DNA appeared, but this molecule is quite stable and could not have been present without a wealth of other pre-existing metabolic processes. There will always be opposite, yet sometimes complementary ideas on the origin or origins of life on Earth. The scenario below integrates physics, physicochemistry, organic chemistry, surface chemistry, enzymology, protein evolution, genomics and bioinformatics to study the evolution of ions to minerals, cofactors and coenzymes, followed by peptides, RNAs, protein machineries like the proto-ribosome and finally DNA. In this scenario, protocells gave birth to protokaryotes, of which *Bacteria* and *Archaea* arose, followed by *Eukarya*, as summarized in Fig. 1.

2.1.1 From Minerals to RNAs

The scenario presented here is at odds with mainstream views, yet it is built on a rational exploration of constraints that must have existed on Earth when life began to develop. It is based on the idea, proposed by Freeman Dyson, that life existence required at least two origins (Dyson 1985). It also assumes that the early cells are likely to have been made of fairly large entities. Present-day cells have been optimized and miniaturized for several billions of years, in a path reminding us how our awkward and big computers gave birth to smart and small cell phones.

Cells are built up from atoms. Any scenario of the origins of life must begin with accounting for the transition between relevant atoms and the building blocks that make life as we know it. A first finding shows us that, among the atoms listed in Mendeleev's table, only a handful are present in all extant organisms (CHOPSN). The logic of chemistry, which drives stable interactions between atoms, made that life essentially comprises elements of the first two rows of the table, plus some metallic ions. Life is also made of macromolecules. It encompasses a subset of polymer chemistry where polymers are formed with production of a water molecule during polymerization of monomers. Polymerization is a remarkable process as it requires that a specific entropy component allows monomers to get together in a stable subset of conformations. Finally, life is developing in water, and water is a highly unusual solvent, precisely because it manages a huge entropy component due to the ubiquitous hydrogen bond-driven interactions between water molecules.

If we are to account for the formation of macromolecules, then it is hardly possible to allow them to happen in a purely liquid water phase, as hydrolysis would prevail. The only way out is to benefit from (not fight against!) the entropy moiety of the energy involved in making polymers. This suggests that the most likely process would require that polymerization happened on surfaces. This was proposed by many explorers of the origins of life, among whom Bernal (1951), Granick (1957), Cairns-Smith (1982) and Wächtershäuser (1988). This would liberate a water molecule free to move in the environment, hence associated with a large entropy increase, while fitting with a further constructive view of a primitive metabolism. This metabolism would use surfaces as templates for local concentration and selective generation of a subset of carbon-based molecules, which would otherwise have generated an unlimited set of varied compounds, poisoning the environment (Cairns-Smith 1982; Wächtershäuser 1988; Danchin 1989).

A key feature of extant metabolites is that they are often negatively charged. Remarkably, their charge is not involved in their function, suggesting that the solid surfaces that selected the first metabolites comprised metal ions, iron in particular, diluting out uncharged molecules in the surrounding environment (Wächtershäuser 2007). During this first period, a subset of the amino acids present in extant cells would be formed on relevant surfaces, with formation of a variety of peptides. In addition, this peptide-based metabolism would also create lipids, essential for the synthesis of the membrane of primitive cells, in a process that involved thioester bonds, sometimes deemed essential for the origin of life (de Duve 1990). Such a mixture would lead to some reproduction of a subset of the molecules, via a process named "graded autocatalysis replication domain" (GARD) as described by Lancet and co-workers (Segré et al. 1998).

The corresponding chemistry has been documented in various experimental setups (Miller and Urey 1959; Lohrmann and Orgel 1968; Shapiro 2000; Costanzo et al. 2007; Kim and Benner 2015; Gibard et al. 2018; Xu et al. 2019), rarely matching plausible prebiotic environments and often ignoring the role of surfaces, so that the main questions asked for scenarios of origins remain open. How were coenzymes formed—they are the true effectors of biological catalysis—and what about nucleotides? Their chemical build-up is a difficult challenge, because ribonucleotides, which are essential besides polypeptides for building up the first ancestors of the life's effector structures, are extremely unstable. Any relevant scenario must therefore propose a steady-state process that would continuously generate ribonucleotides. A further difficulty comes from the fact that nucleotides are nitrogen-rich, so that, in a redox-neutral atmosphere, nitrogen fixation had to be discovered very early on.

Several scenarios have been proposed to answer either of these questions, but it must be acknowledged that it is difficult to see how they could have been actually implemented. A possibility is that an alternation of humid conditions followed by desiccation allowed the formation of polyphosphates and ribonucleotide polymers (Ross and Deamer 2016). But, in order to proceed to life, this must have happened in environments where a specific nitrogen-fixing process had already developed. A chemical scenario that makes use of a process related to non-ribosomal peptide synthesis triggering a reaction reversing that now acting in the synthesis of the ubiquitous molybdopterin cofactor would generate a guanylate derivative in a single step involving formic acid assimilation, ending up in the direct creation of a ribose moiety linked to the nitrogen-rich guanine base without asking for a separate, independent, generation of ribose (Danchin 1989).

At this point in the development of prebiotic surface chemistry, the environment of iron-containing surfaces would be enriched in mixtures of peptides and ribonucleotides. Polymerization of the latter would result in the formation of oligonucleotides, which, only if associated with peptides—yet another feature showing that peptides must have been early actors of prebiotic chemistry—would form the correct 3′–5′ bonds, not the 2′–5′ bonds that a spontaneous polymerization also generates (Wieczorek et al. 2013).

2.1.2 RNA Metabolism

When primitive metabolic GARD aggregates reached this level, they comprised short RNA molecules which, carrying phosphate bonds, could bind metal ions and progressively substitute to surfaces, where they carried over surface metabolism. Present-day transfer RNAs are likely to be descendants of these primitive metabolism-supporting RNAs. Less than 80 nucleotide long, and made of two similar halves (Hopfield 1978), parents of transfer RNAs are plausible candidates for an RNA metabolism step. Indeed, in extant organisms, these molecules are heavily modified (Machnicka et al. 2012), showing that they still interact strongly with metabolic pathways not necessarily related to the process of translation. tRNA molecules would then behave as "handles" maintaining selected molecules to be modified by reactive groups, in a process that has been named "homeotopic transformation" (Danchin 1989). We still recognize this process in the modification of the first residue of protein, a methionine that is formylated, or in the formation of glutamine on glutaminyl-tRNA loaded by glutamate, as found in *Firmicutes* and related bacterial clades (Feng et al. 2004). Naturally, the very process of peptidyl transfer in the ribosome belongs to this category of functions (Leung et al. 2011).

Finally, further witnessing their role as support of metabolism, tRNA molecules are still used in a variety of cells as a handle in the synthesis of the heme precursor aminolevulinate (Feng et al. 2004), or in the synthesis of some bacterial envelope components (Kamiryo and Matsuhashi 1969).

It is therefore plausible to think that metabolism was organized around synthesis of peptides, nucleotides, coenzymes and lipids, with tRNA ancestors playing the role that minerals had played previously. This created a novel GARD system centered on RNA metabolism. Among the first ribonucleotide bases likely to have been present during this early period, one expects to find, besides guanine derived from the nitrogen-fixing process, adenine, inosine, uracil and cytosine (note that pyrimidines, today, derive from amino acid metabolism a feature that may have been present very early on). The RNA molecules generated with these bases tended to fold on themselves. This allowed them to discover that A could pair with U and G with C or U. This uncovered a first type of coding, based on the general law of complementarity, in the form stressed by Pauling and Delbrück as early as 1940 (Pauling and Delbrück 1940). Folding generated stems and loops, and more complicated shapes such as pseudo-knots, that allowed RNA molecules to be able to develop enzyme activities, becoming ribozymes. Among those, formation of peptide bonds, with an ancestor of ribosomal RNAs stabilized by relevant peptides, associated ancestors of tRNAs acting as amino acid carrying handles. Progressively, the thioester-based synthesis of peptides was replaced by an RNA-based synthesis. Initially, the overall specificity of the reaction remained limited, forming more or less random assemblies of amino acids into varied peptides.

It is expected that peptides were not equivalent, with some peptides playing a critical role in the primitive cell. This entails that the discovery of how to favour their synthesis became a key function. At times, discovered from early RNA synthesis, the law of complementarity between RNA bases made it possible to ensure that the peptidyl transfer would become ordered rather than random, driving selection of a precise sequence of the amino acids in the neosynthesized peptide. This required that tRNAs, each loaded with a specific amino acid, could be aligned with some reference RNA, using complementarity with part of their sequence. Then the presence of that guiding sequence, playing the role of a mentor RNA (mRNA), would behave as an indirect template for building up the peptide sequence. In this process, tRNA acquired a new function besides that of support of metabolism. It became an adaptor between the mRNA sequence and the peptide sequence. This function would obviously be subjected to positive selection, in a way fitting well with the GARD model of primitive evolution.

The complementarity between two classes of RNAs introduced, for the first time, a law of coding, creating a cipher—a primitive genetic code—that allowed correspondence between a string of nucleotides and a string of amino acids (Danchin 1983). This code associated a short mRNA sequence of nucleotides—a codon—to a specific amino acid, witnessing the emergence of a totally new function of RNA molecules, based on management of information.

In summary, besides its role as *substrate* in its metabolic functions (presentation of a metabolite for a reaction and ribozyme activity), the RNA molecular class, under the form of mRNAs, now acted as *templates* carrying the memory of a chemical reaction (here, the ordered sequence of specific amino acids to form particular peptides). This brought to light an *information*, embedded in the arbitrary symbolic link between a memory (the nucleic acid) and a function (a peptide contributing to the survival of the protocell). The increasing specification of this relationship between memory and function developed into the extant rule known as the genetic code.

At this point, the cell harboured a functional trio: (1) tRNAs, which we can now truly call "transfer" RNAs; (2) template mRNAs, precursors of what will become messenger RNAs and genes; and (3) a family of ribozymes, precursors of ribosomes (peptidyl transferases), catalyzing the reaction required to form peptide bonds and of other activities. When cells reached this stage, they gradually went through a whole series of improvements. The RNAs of this world of RNA metabolism kept evolving. They recruited new chemical abilities, through the evolution of cofactors of enzyme catalysis, the elimination of parasitic molecules, and the improvement of exchanges between the inside and the outside of the cell. In parallel, an ever more efficient management of energy channelled electron transfers and handled polyphosphates and osmotic pressure via the synthesis of large molecules, including complex sugars, that behaved as matter and energy storage compounds. How did these primitive cells store the memory of the most relevant metabolic ensembles? An obvious challenge is expected here to have evolved some stable solution: indeed, because RNA is so unstable, a critical question that arose was that of its durability.

2.1.3 RNA Genomes

The protocells containing the RNA trio kept splitting and merging together, sometimes engulfing one another, propagating the most efficient metabolic setups. They reproduced as described for GARD systems, in parallel with the associated peptide-coding system. Starting as short peptides, the length of coded peptides increased progressively with the increase in size of the ribozyme endowed with peptidyl transferase activity, the future ribosome, and peptides began to further evolve enzymatic activities. Among the functions invented by ribozymes emerged that which allowed RNA replication (Lau and Ferré-D'Amaré 2016). This function derived from the discovery that complementarity between the sequences of nucleotides of a ribozyme or any other RNA allowed that a complimentary copy (a cRNA) could be maintained faithfully, replicated. This cRNA resulted from a simple coding rule: A<=>U and G<=>C (likely with some ambiguity allowing G<=>U). The replication process led to the formation of an RNA double helix, an RNA gene (gRNA), that comprised two strands in opposite directions. The strand coding the structure of ribozymes (rRNA) and tRNAs, but also that of the templates ensuring the generation of peptides of defined sequence (mRNA), was associated with a complementary strand (cRNA). We must note here that the phosphodiester bond forming the RNA molecule is not symmetrical: in three dimensions 5′–3′ bonds differ from 3′–5′ bonds. This entails that the copy is orientated in the opposite direction to that of the copied strand, and this must be considered by any machinery using RNA strands. We expect here that some peptides were involved as cofactors of this primitive RNA replication process, thus interlocking the replication of RNA with the synthesis of peptides.

Alas, implementing a replication process could not be straightforward because the double-helix RNAs (gRNAs) form a linear molecule, making that their ends (telomeres) are free. Now, the replicative machinery had to bind to these ends, preventing replication of these extremities. This led to shortening of the RNA genes at each replicative cycle making them shrink and then progressively disappear. This constraint favoured discovery of an RNA-based way out. In extant replication (that operates in extant cells on DNA, a variant of nucleic acids that is discussed below) an enzyme, telomerase, fulfills this function. Remarkably, today it still makes use of an RNA template to extend the extremities of the (DNA) molecule to be replicated (Wu et al. 2017). This makes it plausible to assume that this patching-up process is an archive of the primitive RNA replication system that operated in the very first cells. Finally, the replication process unveiled yet another obvious functional asymmetry. Indeed, while essential RNAs (mRNAs, tRNAs and rRNAs) were required in substantial amounts in the cell to perform their function, their complement (the cRNAs) had only a role in storing a replicable memory. This did not require the presence of multiple copies. To solve this asymmetry, a new function became indispensable. Analogous to replication, but asymmetrical and naturally derived from ribozymes, it consisted in repeatedly transcribing only one of the two strands of the replicable memory.

Another critical feature of life is likely to have appeared at this stage, especially when the peptide bond-forming ribosomal RNA increased in size by accretion of RNA fragments allowing the ribosome to properly accommodate tRNAs deciphering an mRNA. Indeed, RNA is made of only four nucleotides, and long RNA molecules may fold in a very large number of structures, only one of which retaining a proper function. There was an absolute need for channelling folding into the right shapes in a process discriminating between functional and nonfunctional entities. This information-loaded step requires specific agents, likely to be polypeptides, that load a source of energy, allowing them to selectively identify the functional structures, letting the other ones to be either refolded or degraded. This step is reversible, but the agents need to be reset to their original state, and this requires dissipating the energy from the source they had associated to for their discriminating role. Remarkably this two-step behaviour (reversible information identification followed by energy dissipation for reset) follows exactly the principle discovered by Rolf Landauer in 1961 (Landauer n.d.). That agents displaying related functions are essential for life is obvious when remarking that minimal genome code at least for 50 such agents (Boël et al. 2019). Despite its key importance, this family of functions has been entirely overlooked in all previous explorations of the origins of life.

At this point, protocells contained sets of double-helix RNAs of various sequences, as seen today in the genome of some RNA viruses. Transcription produced the mRNA templates, the tRNAs and the proto-ribosome (rRNA), whereas replication increased the copy number of both strands of the ancestral genes. Transcription, being intrinsically asymmetric, is functionally distinct from replication. This made that the management by telomerases of the ends of the many individual double-stranded RNA genes was wasteful and difficult to coordinate with transcription. This hurdle must have led early on to the splicing of several of the gRNAs into longer double-stranded RNA coding simultaneously for several functions as in extant chromosomes.

Remarkably, we note that RNA splicing is therefore likely to have been a primitive cell function, not a late discovery. Still, the former function, transcription, required large amounts of RNA, while the latter, clustering memory fragments within RNA genomes (gRNA), required much less. Consequently, transcription and replication evolved separately. A first function associated with transcription was, for example, the separate recognition of the beginning of proto-genes, what we name promoters today. Furthermore, it seems natural that the processes preceding formation of the final products of the cell metabolism—transcription and replication—were confined to a given compartment. This would make a proto-nucleus. Yet, these protocells must have depended upon a very active metabolism, ensuring the continuous synthesis of ribonucleotides and polyribonucleotides, chemically unstable molecules. This process is expected to have developed in a significantly larger volume than the proto-nucleus, making the first cells fairly large contrivances. Protocells must have been made of a large compartment, where metabolism was evolving, and a small compartment, where transcription and replications operated.

2.1.4 DNA Genomes and the First Stable Cells

In a primitive cell, therefore, two types of compartments were expected: a cytoplasm, including most of the RNA metabolism, and a nucleus, where the elements forming the RNA genome and the nucleic acids wielding machinery operated, allowing transcription in most of the cell's lifetime and replication in a fraction of the time. This situation remained unstable because of the fragility of RNAs, asking that their synthesis be continuously ensured. The merging, engulfment, and splitting of protocells within large populations made it possible to maintain some permanence. However, these processes missed the selective advantage that would be brought about by the stabilization, over time, of the memory of past successes. This was not allowed by the initial memory molecule, the RNA genome, as it was composed of very fragile nucleotides.

The expected selective stabilization was enabled by the discovery of deoxyribose, a molecule much more stable than ribose, which solved the problem. Extant forensic techniques based on DNA, as well as the study of the genomes of organisms that disappeared millennia ago, are a present-day vivid proof of DNA stability. DNA appeared only at the very end of a long evolution process that progressively improved and stabilized the memory that orchestrated the recipes driving metabolism. This new molecule—with a few further changes, such as the discovery of thymine (an analogue of uracil, U) that allowed it to be told from the rest of the cell as much as possible as well as generated a clock for ongoing replication—separated the memory from the general functioning of the cell. Transcription evolved to recognize DNA and produce RNA, while replication was essentially acting on DNA. However, we still have an archive of the previous RNA genomes in the fact that DNA replication initiates with RNA primers (Bell 2019) and with tRNA primers in reverse transcription of some viruses (Jin and Musier-Forsyth 2019).

In summary, the first cells, protokaryotes, were born during a long transition, with a multiplicity of roots. It can be expected that these fairly large cells formed a population associating, in a single cell, several compartments with different but complementary destinies. These cells split, merged and engulfed one another. This allowed exchanges that kept enriching the evolution of metabolic pathways. Predatory organisms, they behaved as rather large cannibals (like many extant protists are). Yet, this situation was unstable. It created an asymmetry between predators and preys. This inevitably led to the emergence of a novel function, that which allowed preys to resist being engulfed by predators. Two solutions were explored: either discovery of a metabolic pathway allowing the cell to enclose itself in a protecting envelope, resisting engulfment, or ensure that the cell's membrane and its various functions are not assimilated in a productive functional fusion with that of the predator cell. This led to two very different domains of life, both evolving towards miniaturization.

Bacteria developed the first solution, losing their nuclear compartment and forming small cells surrounded by a strong wall. *Archaea* (cells without nucleus like *Bacteria*, thus prokaryotes) discovered how to envelope themselves with a membrane-escaping predation. To this aim their envelope lipids were based on structures that mirror, in three dimensions, those of the protokaryotes' predators. This discovery may look unlikely. Indeed, until very recently, the transition from membranes based on a particular lipid symmetry to the opposite chiral form was perceived as far-fetched. Remarkably, the construction of a bacterium genetically modified to produce a membrane composed of both forms gives considerable weight to this view (Caforio et al. 2018). This form of evolution led to a particular autonomy of *Archaea*, which then lost the possibility of becoming pathogenic and acquired in parallel that of occupation of extreme environments.

Bacteria, whose membranes had remained identical to that of the predatory ancestral cells, subsequently developed another way to evade predation, replacing it by symbiosis. We can see here how eukaryotes were born (Hartman 1984). They were the progeny of protokaryotes that recruited certain bacteria as symbionts. This happened multiple times. The mitochondria—whose central role is not the management of energy, but the formation of iron-sulphur clusters (see below), an archive of the mineral origins of life—are cases in point, displaying a significant variety of origins (Stairs et al. 2015). Later, after oxygen had been contaminating the Earth's atmosphere, cyanobacteria were engulfed in *Eukarya*, leading to a variety of protists as well as plants. *Bacteria*, *Archaea* and *Eukarya* further evolved. The former miniaturized all their metabolic functions while losing all but their introns, and *Eukarya* evolved into multicellular organisms, keeping introns as spacers and timers. These processes match those we are witnessing today in the way our computers have evolved into several families of objects, such as very large computers or mobile phones, more or less incompatible with one another.

In conclusion, the initial steps of life are likely to have been made from a collection of cells that preyed on one another, fused and split. With this view there cannot be a last universal common ancestor (LUCA) that would proceed into all life forms, but a last ancestral cell ensemble (LACE) where horizontal gene transfer was the norm (Koonin and Wolf 2008). The origin of cells looks more like the reticulated trunk of a banian tree in tropical countries than a straight trunk, as usually represented. We are aware that this is an unorthodox scenario, but we must remember that during the first 3 billion years of life's development, the Earth underwent cataclysmic changes, with very hot and very cold periods. This must have created a series of bottlenecks that are doomed to have obscured the scenario of origins (Danchin 2007), so that we must refrain from being driven by poorly supported mainstream views (see Cavalier-Smith (2006) for the reasoning supporting a view that differs from the present one).

3 Origin and Evolution of Selected Metabolic Pathways

Based on the consideration of LACE instead of LUCA at the onset of cellular life, we favour a "metabolism first" view on the origins of life (Danchin 2017; Koc and Caetano-Anolles 2017). This means that we consider reproduction of metabolism and biochemical reactions as the first step of cellular evolution, followed by the invention of DNA replication as a memory storage. The evolution of cellular reproduction and chromosome replication resulted in the immense biodiversity that we can appreciate nowadays.

3.1 Reproduction Versus Replication

Contrary to their unfortunate name, GARD structures are not replicating structures, but structures that make similar, not exact copies, of what they are: they reproduce. How did the transition from reproduction to replication develop? Extant cells can be compared to programmable factories. They can be designed to produce ad hoc chemicals and behaviours. However, by contrast with man-made factories, cells evolve. This results from the separation of two processes in the cell: reproduction of the cell's machinery and replication of the genetic programme. While accumulating errors, reproduction may improve over time, generating replicative processes. In the course of evolution, since the beginnings of life, cells harnessed processes meant to accumulate information from whatever source, so that they could produce a progeny which is systematically fitter than its ageing parents. The requirement of a transition from reproduction to replication has been demonstrated by Freeman Dyson who showed that this entails that life had at least two origins, combining reproduction and replication (Dyson 1985).

3.2 *The First "Metabolic Pathways"*

The first steps of biochemical reactions and metabolism were most innovative in terms of chemistry. To explore new niches and adapt to various environments, cells needed to exploit the available resources and use them for their reproduction and afterwards their replication. These processes are ensured by organisms through the operation of various key functions. To carry on such functions, cells relied on processes that evolved into enzymes and metabolic pathways.

Non-enzymatic chemical reactions predated their improvement when enzymes progressively introduced three-dimensional selectivity and lowered the activation energy of cognate reactions, according to the laws of thermodynamics (Pascal et al. 2013) and organic chemistry (Danchin and Sekowska 2015). For example, a non-enzymatic origin of primary metabolic pathways such as the Embden-Meyerhof-Parnas (EMP) and the pentose phosphate pathway (PPP) has been demonstrated experimentally, using Fe(II) as a catalyst in temperature conditions ranging from 40 to 70 °C (Keller et al. 2014). In a more recent example, it was demonstrated that 9 out of the 11 TCA cycle intermediates could be formed from two simple carbon sources (pyruvate and glyoxylate) in the presence of ferrous iron too, an abundant catalyst in prebiotic chemistry (Muchowska et al. 2019b). Interestingly, inspired by existing pathways (reductive TCA cycle, Wood-Ljungdahl), CO_2 fixation has been recently achieved by using transition metals (Fe, Zn, Ni) as catalysts under high temperature and pressure conditions (Muchowska et al. 2019a). Notably, CO_2 fixation, the TCA, EMP or PPP cycles are essential for most modern organisms directly or indirectly because of the functions they accomplish including, inter alia, energy storage, respiration, synthesis of intermediates used by other pathways, etc.

3.3 *Same Function But Different Process*

Studies of the origin of metabolic pathways are still in their infancy, yet thanks to modern analytical methods, we are gaining insights on the fascinating complexity and diversity of metabolism. In this section we look at some examples of different pathways that produce molecules that provide similar functions but using a different and unrelated process (i.e. enzymes that are not necessarily conserved). We illustrate some of these functions below (summarized in Fig. 2).

3.3.1 Fe-S Clusters and Isoprenoids

As described at the beginning of this article, among mineral surfaces that are likely to have been critical for generating the first reactions of life, pyrite-like iron-sulphur (Fe-S) complexes appear to be omnipresent. In fact, it was hypothesized that the first chemical reactions were catalysed by Fe-S clusters (Wächtershäuser 1988). Recent

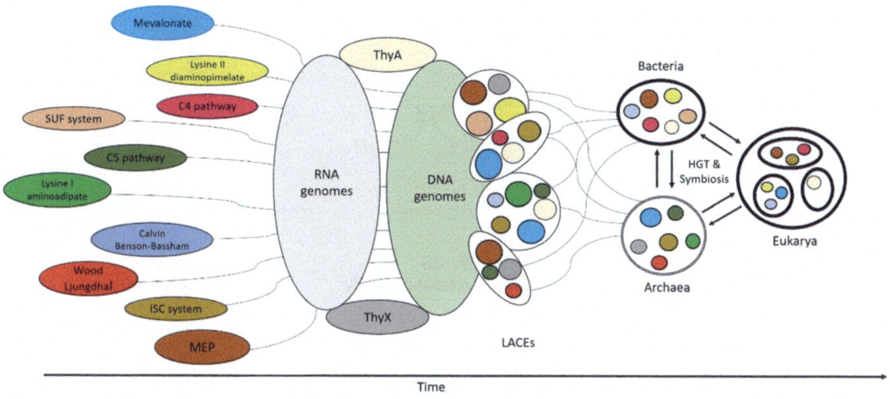

Fig. 2 Emergence and evolution of selected biosynthetic metabolic pathways. *LACEs* Last Ancestral Cell Ensemble, *HGT* Horizontal Gene Transfer

phylogenetic studies suggest that the first unicellular microorganism would have contained proteins with Fe-S clusters and S-adenosyl methionine (SAM) as cofactors (Weiss et al. 2016). Fe-S proteins are involved in electron transfer, gene regulation, sulphur donation, respiration and DNA repair, etc. (Py and Barras 2010). These proteins are a relic of the primordial use of Fe-S clusters for various functions in primordial cells (Hall et al. 1971; Boyd et al. 2017), yet modern cells use sophisticated protein machineries called ISC (iron-sulphur cluster), SUF (sulphur assimilation) and NIF (nitrogen fixation) systems to insert Fe-S cluster into target proteins. Fe-S clusters come in many forms, but the most common forms are rhombic [2Fe-4S] and cubane [4Fe-4S] (Fig. 3). For example, ferredoxins have the former forms, while the latter type of cluster is ubiquitous in radical SAM enzymes. These enzymes utilize a [4Fe-4S] cluster and SAM to initiate a diverse set of radical reactions involved in the biosynthesis of vitamins (thiamine, biotin), cofactors (lipoic acid, F420), amino acids (pyrrolysine), complex metal clusters (FeMoCo, H-cluster), antibiotics (gentamicin, aminoglycosides), tetrapyrroles (heme, cobalamin) as well as tRNA and rRNA modifications. Since SAM can be spontaneously produced from methionine, adenosine and ATP (Laurino and Tawfik 2017), it has been suggested that the first cells used Fe-S clusters and radical SAM chemistry for processes such as CO_2 and N_2 fixation, energy generation and biosynthesis of complex molecules (Weiss et al. 2016).

Surface-exposed cubane-type Fe-S clusters, however, are extremely sensitive to oxygen (Giel et al. 2006), compared to rhombic clusters, which are very stable. Thus, rhombic clusters could have evolved from cubane clusters to avoid their inactivation by oxygen, and/or some proteins evolved to protect clusters from inactivation (Dai et al. 2014). If this process was not possible, Fe-S cluster proteins were replaced by other enzymes using alternative catalytic mechanisms. In *E. coli*, for example, haem biosynthesis depends on a crucial catalytic step carried by the radical SAM protein HemN, but there is also an oxygen-dependent enzyme called HemF that converts coproporphyrinogen III to Protoporphyrinogen IX using different chemistries

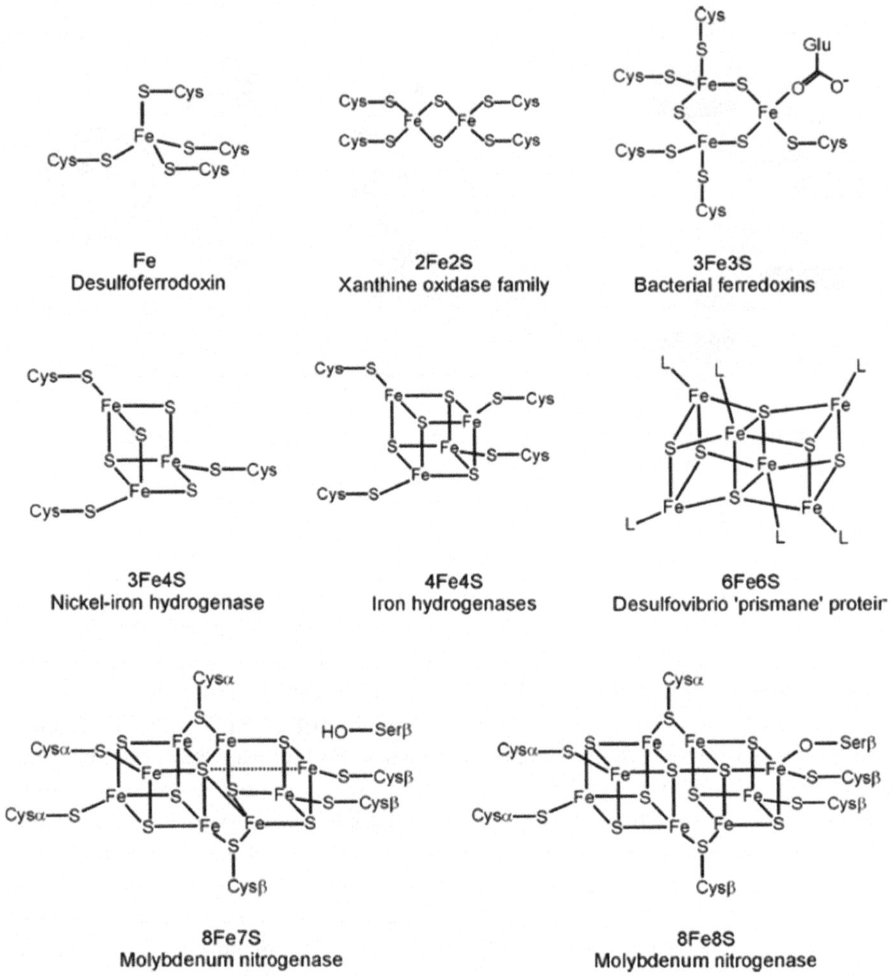

Fig. 3 Different types of Fe-S clusters (Source: Brzóska et al. 2006)

(Schobert and Jahn 2002). Despite the different processes, the substrate and product of both enzymes are the same (the function is conserved).

Another example is the methylerythritol phosphate (MEP) pathway for isoprenoid biosynthesis that relies on two Fe-S cluster proteins IspG and IspH that are mostly found in *Bacteria* (and a few plants and parasites). An alternative to this pathway, the mevalonate pathway (MVA), which is present mostly in *Eukarya* and *Archaea* (and a few bacteria), does not use Fe-S cluster chemistry. Yet, these two pathways start from close precursors (pyruvate and acetyl-CoA) and end with the same products, isopentenyl diphosphate (IPP) and dimethylallyl diphosphate (DMAPP), which are used to build various isoprenyl chains and thousands of different compounds (Frank and Groll 2017). Another variation in the MVA pathway is the use of three enzymes instead of two for the biosynthesis of IPP from mevalonate in extreme acidophiles (Vinokur et al. 2016). Figure 4 shows the MEP and MVA pathways.

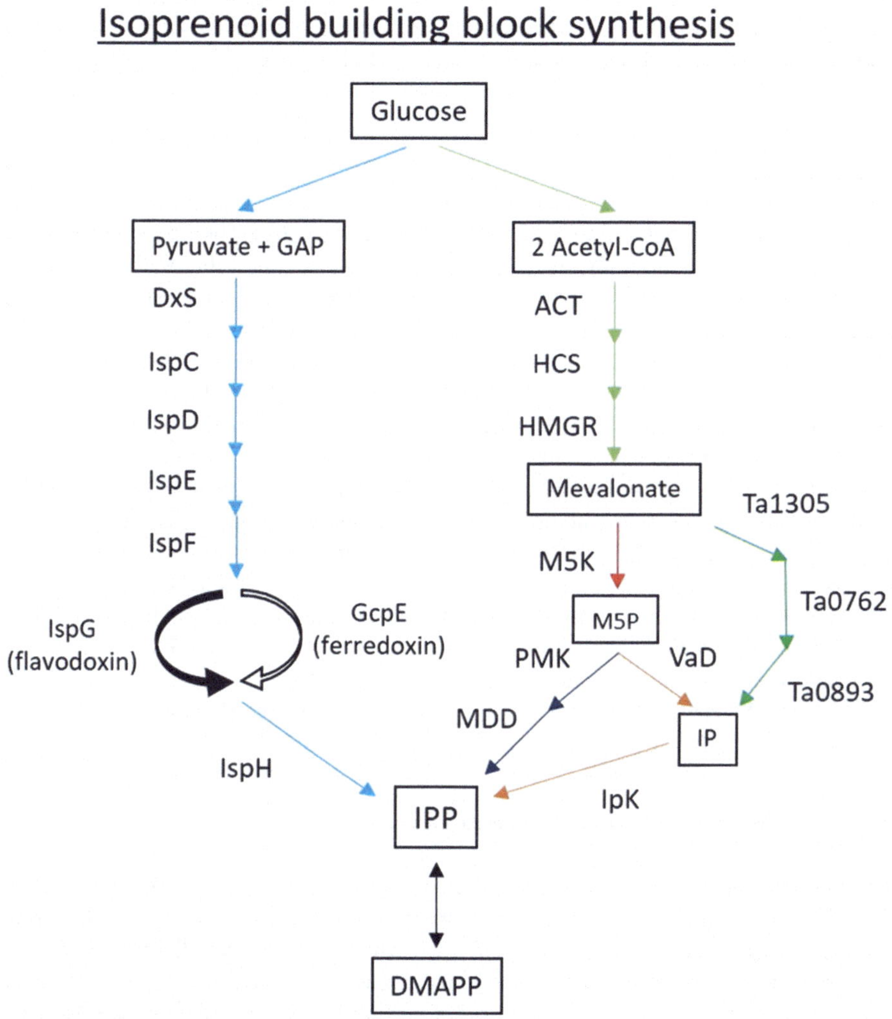

Fig. 4 Alternative pathways for the biosynthesis of the universal isoprenoid building blocks IPP and DMAPP. The pathway on the left corresponds to the MEP pathway, and the MVA pathway can be found on the right. Sub-pathway variations in the MVA pathway are also represented, notably using IP as an intermediate. Arrows have similar colour when they represent a single pathway. Important branching intermediates are named and within a square. Circular arrows represent a catalytic step that is carried by alternative enzymes with different catalytic mechanisms. Enzyme names are in black. *GAP* glyceraldehyde 3-phosphate, *IPP* isopentenyl diphosphate, *IP* isopentenyl phosphate, *DMAPP* dimethylallyl diphosphate, *M5P* mevalonate 5-phosphate, *MEP* methylerythritol phosphate, *MVA* mevalonate

3.3.2 Phospholipids

Among them are notably the composition of bacterial and archaeal cell walls as discussed above. Cell walls accomplish the same function (compartmentalization). Archaeal cell walls are mostly made of *sn*-glycerol-1-phosphate (G1P) linked with an isoprenoid carbon chain, whereas bacterial cell walls are made of *sn*-glycerol-3-phosphate (G3P) linked with fatty acids (Koga et al. 1998).

3.3.3 Lysine

L-Lysine is an essential amino acid and has the particularity of being synthesized by two different metabolic pathways. The first pathway is called the diaminopimelate (DAP) pathway, and the second pathway is called the L-2 aminoadipate (L2A) pathway (Xu et al. 2006). The DAP pathway uses L-aspartate as a starting metabolite, whereas the L2A branches out of the TCA cycle through 2-oxoglutarate and acetyl-CoA. Both pathways use significantly different intermediates and enzymes (Kanehisa and Goto 2000). There are also variations within the pathways themselves. Four different methods for the conversion from (S)-2,3,4,5-tetrahydrodipicolinate to meso-diaminopimelate have been reported for the DAP pathway, relying on different enzymatic mechanisms, namely, dehydrogenation, succinylation, amine group transfer or acetylation (Con Dogovski 2012). It is notable that *Corynebacterium glutamicum,* among other microorganisms, have adopted more than one variation of the DAP pathway in their metabolism (succinylation and dehydrogenation) (Schrumpf et al. 1991). Two variants of the L2A pathway have also been reported, both of which share common steps and differ in the methods used to convert L-2 aminoadipate to L-lysine (Nishida et al. 1999).

3.3.4 Aminolevulinic Acid and Tetrapyrroles

Tetrapyrroles are complex cofactors that play crucial roles in energy generation, oxygen transfer and catalysis in most organisms (haem, chlorophyll, cobalamin, F430 cofactor). The main precursor of tetrapyrroles is aminolevulinic acid (ALA), for which two biosynthetic pathways exist (Dailey et al. 2017). The C4 pathway depends on a single step, the conversion of succinyl-CoA and glycine into ALA by ALA synthase (ALAS), whereas the C5 pathway relies on the conversion of L-glutamate to ALA via three enzymes (GltX, HemA and HemL), with a notable implication of a tRNA for the first committed step (Zhang et al. 2015). Alternative pathways for the NADPH-dependent conversion of 2-oxoglutarate to L-glutamate also exist in *E. coli.* The nitrogen atom can either be acquired through ammonium by the enzyme GdhA (Veronese et al. 1975) or through L-glutamine by the [4Fe-4S] cluster enzyme GltBD (Reitzer 2004). The different enzymatic solutions offered by HemF and HemN in haem biosynthesis were mentioned above. Their product, protoporphyrinogen IX, can be converted to protoporphyrin also by two alternative enzymes, oxygen-dependant (HemY as in *B. subtilis*) or independent (HemG as in *E. coli*) (Layer et al. 2010). An alternative pathway for haem synthesis can be found in *Archaea* and denitrifying bacteria, involving the precursor precorrin-2, obtained after methylation of the C2 and C7 carbons of uroporphyrinogen III by SUMT (S-adenosyl-L-methionine-dependent uroporphyrinogen III C methyltransferase) enzymes. Precorrin-2 is the common precursor of cobalamin, haem d1, sirohaem and F430 cofactor (Fig. 5). The biosynthesis of sirohaem from precorrin-2 requires three steps, all of which are ensured by a multifunctional enzyme CsyG with a SUMT domain in *E. coli*, but differently in other organisms (Vévodová et al. 2004). Synthesis of haem from precorrin-2 is proposed to involve the conversion of sirohaem into haem via the enzymes AhbABCD, with AhbC and AhbD being two radical SAM enzymes (Bali et al. 2011). Other alternatives, like the coproporphyrin-dependent pathway, exist in this pathway and have been reviewed in detail (Dailey et al. 2017). Notably, the biosynthesis of other tetrapyrroles can be achieved through different pathways, as shown by the oxygen-dependent and oxygen-independent cobalamin biosynthesis route (with the use of the Fe-S protein CobG in the former) (Schroeder et al. 2009; Fang et al. 2017).

3.3.5 DNA Biosynthesis

As discussed above, DNA was likely first composed of uracil (U) alongside adenosine (A), cytosine (C) and guanidine (G), as a relic of its RNA ancestry. The discovery of DNA as a backbone in which U is replaced by thymine (T) is assumed to have been carried on by methylation of the U backbone. Two unrelated enzymes, ThyA and ThyX, are in charge of such reaction in modern organisms and are essential for de novo DNA biosynthesis (Forterre et al. 2007). Both enzymes do not carry reductive methylation of deoxyuridine monophosphate (dUMP) in the same fashion but use the same substrate, CH_2H_4-folate. While ThyX uses CH_2H_4-folate as a methyl donor and NADPH/FADH as reductants, ThyA uses CH_2H_4-folate both as a methyl donor and as a reductant (Cho et al. 2012). Organisms using both ThyA and ThyX have been described like *Mycobacterium tuberculosis* (Hunter et al. 2008).

3.3.6 Other Examples

How proton gradient systems differ between acetogens and methanogens is also another interesting example (Pomiankowski and Lane 2014). Whereas ATPases are shared and conserved by both, Na^+ pumping, required for ATP synthesis, is ensured by unrelated antiporters (A-type for methanogens and F-type for acetogens).

This section aimed to show the versatility of metabolic pathways exploring functional space. Countless of other examples exist, and the list of metabolic processes used by different organisms to achieve similar functions is non-exhaustive (Pereira et al. 2013). The main question is how metabolic pathways of similar functions but arising from different processes can be harnessed for the

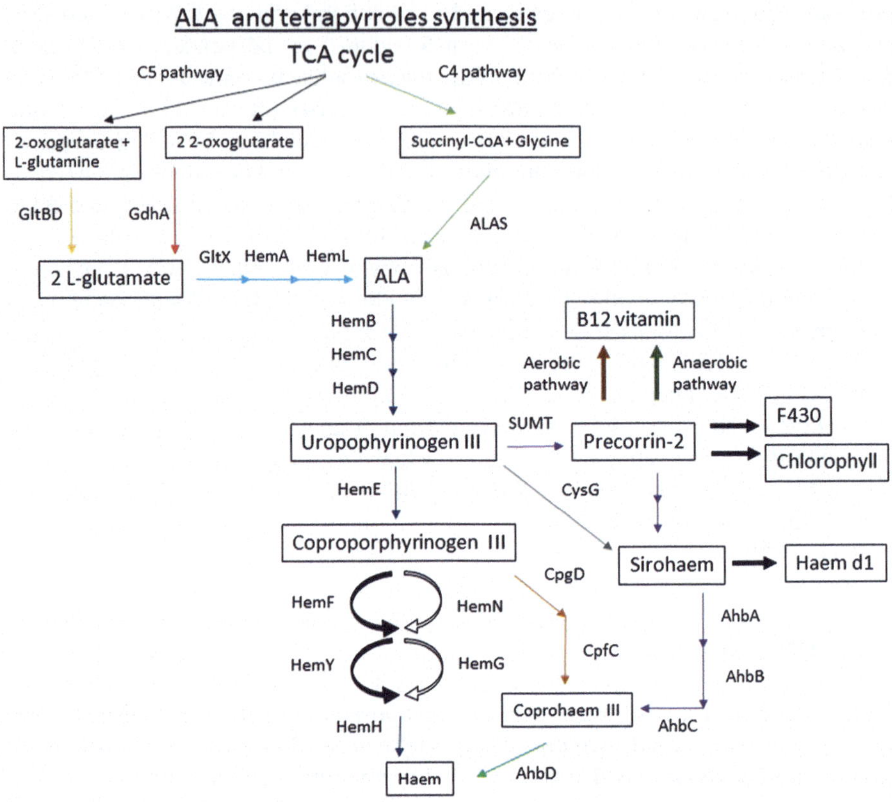

Fig. 5 Alternative pathways for the synthesis of aminolevulinic acid (ALA) and tetrapyrroles with emphasis on haem and sirohaem. The two ALA alternative pathways, C5 (left) and C4 (right), are represented. Various branchings from important intermediate are shown. The two enzymes carrying the transformation of precorrin-2 to sirohaem have not been named yet. Arrows with similar colour represent a single pathway. Important branching intermediates are named and within a square. Bold arrows correspond to pathways not shown in detail. Circular arrows represent a catalytic step that is carried by alternative enzymes with different catalytic mechanisms. Enzyme names are in black

design of microbial cell factories for industrial biotechnology. In the section below we provide a few selected examples.

4 A Functional Approach to Design Cell Factories for White Biotech

Industrial or white biotechnology is an area of research aiming at producing chemicals, drugs, food ingredients, biofuels and biomaterials (to name a few) by using enzymes and/or cells as biocatalysts (Lorenz and Zinke 2005; Heux et al. 2015; Straathof et al. 2019). The overall goal is to replace current industrial processes that rely on nonrenewable sources such as petroleum-derived compounds and that require toxic chemicals and/or high energy demand compared to bioprocesses that are sustainable because they rely on renewable feedstocks and are more environmentally friendly (Nielsen and Keasling 2016; Yu et al. 2019).

To develop a circular bio-economy, enzymes and microorganisms are powerful allies. Processes that rely on both enzymes (Choi et al. 2015; Sheldon and Woodley 2018) and organisms (Jullesson et al. 2015; Nielsen and Keasling 2016) are already found at an industrial scale in different companies. However, enzymes do not often show the desired characteristics for a bioprocess (activity, selectivity or stability), but these can be generally solved by directed evolution (Reetz 2013). In the case of microorganisms, these do not produce large quantities of the targeted molecules to permit their cost-effective manufacture. To address this issue, metabolic engineering optimizes the metabolism of microbial hosts to turn them into "cell factories" by redirecting fluxes and tuning up enzyme levels and genetic regulators. Examples of microbial cell factories have been reported for antibiotics (Weber et al. 2015; van Tilburg et al. 2019); biofuels (Zargar et al. 2017); bioplastics (Murphy 2012; Choi et al. 2019); chemicals (Fact et al. 2019); flavours, fragrances and pharmaceuticals (Zhang et al. 2017; Cravens et al. 2019); natural products (Zhou et al. 2014; Liu et al. 2017; Cravens et al. 2019; Nielsen 2019); pigments (Tolborg et al. 2017; Frandsen et al. 2018); and vitamins (Acevedo-Rocha et al. 2019).

It is important to consider certain engineering aspects to develop successful microbial cell factories. First, the right starting strain needs to be selected depending on the application. Second, this strain should be ideally converted into a platform strain. Third, specific metabolic pathways need to be designed and optimized in iterative cycles of design, building and testing. Finally, the strain should have an optimal performance during upscaling. Ideally, microbial cell factories require simple nutritional requirements, their metabolism is versatile and robust to be adapted to produce different compounds, their genetic modification is straightforward and there is a substantial knowledge and understanding about their metabolism, etc. Cell factories must have not only certain ideal characteristics (Table 1) but also excel in performance during large-scale cultivation to enable large-scale production (Foley and Shuler 2010).

4.1 Choosing the Chassis

In SB jargon, a so-called chassis is a strain that can be used as a "motherboard" for various applications (e.g. a cell factory). There are chassis (*E. coli, B. subtilis, Saccharomyces cerevisiae*) that have proven to be industrial cell factories, but there are also emerging strains that have the potential to become industrial workhorses such as *Pseudomonas putida* (Calero and Nikel 2019). Microbial species are diverse and show interesting characteristics for applications in metabolic engineering, for example, *P. putida* is well adapted to harsh environmental conditions, making it an ideal chassis for producing chemicals from feedstocks that other

Table 1 Requisites for an ideal industrial microbial cell factory

Robust cell growth	Robust cell envelope	Dynamic stability	Metabolic control
• Assimilation of cheap feedstocks • Fast growth rate • Simple chromosome segregation and cell division mechanisms • High levels of viable biomass	• Hydrodynamic reliance to shear forces • Export of target compound in large amounts • Low product toxicity or absence thereof	• Low genetic drift and evolutionary potential • Stable mismatch repair systems • Upscalability to large fermentation volumes	• Simple controls for replication, transcription and translation of target pathway • Metabolic proofreading • Predictable metabolic interactions

organisms would not withstand such as oil spill or lignin (Nikel and de Lorenzo 2018). Other pathways could be tailored for electricity production in *Shewanella oneidensis*, biofuel production in *Clostridium acetobutylicum* (Kim et al. 2016) or long-chain fatty acid production in the oleaginous yeast *Yarrowia lipolytica* (Ledesma-Amaro and Nicaud 2016). More recently, the fast growing *Vibrio natriegens* could be used for protein expression (Lee et al. 2019) or the bacterium *Halomonas campaniensis* for biopolymers (Ling et al. 2019). The list is ever expanding (Kim et al. 2016; Nielsen 2019), and it is only limited by our capacities to characterize new chassis and develop molecular tools. It foreshadows also that the microbial biodiversity can be used as a tool for human needs.

4.2 Platform Strains

An important concept in the design of cell factories is that of "platform strain" (Jullesson et al. 2015). A platform strain has the desired characteristics of a robust cell with a minimized genome (as discussed in Sect. 1). However, a platform cell goes beyond a robust cell factory because it allows the introduction of different metabolic pathways that are compatible with the host in a short time and cost-effective manner. In other words, a platform strain is a chassis that has an assigned function in which it excels. Platform strains can use different feedstock (e.g. sugars, waste, CO_2), and they are robust for operation under industrial conditions (stress tolerance, fast growth, performance). The usage of platform strains is important in the industry to reduce the costs for developing bioprocesses. Since metabolic pathways are very complex and highly interconnected, it usually takes time to engineer each pathway for specific molecules. Yet it is common to find in any organism a metabolite that is at the source of several metabolic pathways of industrial relevance.

First-generation platform strains are already used in the industry (Jullesson et al. 2015), but second-generation strains are in the horizon. This is well exemplified with chorismate that allows the biosynthesis of aromatic compounds (phenylalanine,

2-aminobenzoate), catechol (Balderas-Hernández et al. 2014) or muconic acid (Noda et al. 2016), which is a precursor of several bioplastics (Curran et al. 2013). A platform strain optimized to produce chorismate is essential to produce cost-effectively chorismate-related compounds. A second example is an ALA platform strain, which is at the root of numerous tetrapyrrole biosynthetic pathways (Fig. 5) and can produce various valuable industrial products including ALA itself (Kang et al. 2011; Zhang et al. 2015), haem production via the C4 or C5 pathways (Zhao et al. 2018) and vitamin B_{12} (Fang et al. 2018). Another example is the ubiquity of SAM as a cofactor for C–C bond formation and/or methylation of a considerable repertoire of chemical compounds (Struck et al. 2012; Yokoyama and Lilla 2018). This cofactor is useful for the production of vanillin (Kunjapur et al. 2016), anthocyanin (Cress et al. 2017) or methyl anthranilate (Luo et al. 2019). The various alternative SAM biosynthetic pathways (Sekowska et al. 2019) could also provide valuable knowledge for the design of an efficient platform strain for the production of SAM-related compounds.

The development of platform strains requires consideration of both the chassis and available metabolic pathways. With the increasing discovery of new enzymes and metabolic pathways, many examples of alternative pathways for the same function are becoming available. Such knowledge is of considerable value when designing platform strains because it offers many solutions for a single problem. For example, there are at least six metabolic pathways for CO_2 fixation, each offering high malleability in their catalytic and metabolic mechanisms (Fuchs 2011). Using this information, it was possible to implement the most efficient CO_2 fixation pathway in vitro—the CETCH cycle (Schwander et al. 2016). Table 2 provides further examples of next-generation platform strains that could exploit alternative solutions to the same problem by using a functional approach.

4.3 Resource Optimization

Metabolic pathway optimization usually takes considerable time and effort. It is crucial to balance enzyme levels to achieve a high flux, which is usually done in multicopy systems (plasmids). However, the availability of enzyme cofactors and resources of the chassis needs to be likewise considered. For example, some organisms like *E. coli* have a natural overabundance of enzymes (Donati et al. 2018) to increase its robustness against varying environmental conditions (Mori et al. 2017). Thus, decreasing enzyme levels for unnecessary functions has the capacity to improve productivity, as recently shown for aromatic amino acids in *E. coli* (Sander et al. 2019). Another approach is to reduce enzyme promiscuity and thus metabolic noise. Proteins show often unwanted side activities (Nielsen and Moon 2013). The notion of metabolic proofreading consists in using auxiliary enzymes in a module that would convert unwanted metabolites back into intermediates used in such module (Schwander et al. 2016).

Table 2 Examples of platform strain development strategies

Platform name	Starting metabolite	End metabolite	Industrial chassis example	Number of enzymes in host de novo pathway	Cofactors	Examples of alternative pathways and enzymes	Potential associated downstream biosynthetic pathway	References
MEP pathway	Pyruvate and G3P	IPP/ DMAPP	*E. coli*	8	NADPH, CTP, ATP, Fe-S cluster	MVA pathway and sub-pathways	Q6/8/10, zeaxanthin, resveratrol	Frank and Groll (2017) Niu et al. (2017)
Chorismate pathway	E4P	Chorismate	*E. coli*	7	PEP, NADPH, ATP	Salicylate or catechol synthetic pathways	Muconic acid, PABA, Q6/8/10	Noda et al. (2016) Curran et al. (2013)
Fatty acid metabolism	Acetyl-CoA	Free fatty acid (nC)	*S. cerevisiae*	3	ACP, HCO3-, ATP	Xylose fermentation, inverted beta-oxidation, ACP engineering	DHA, EPA, biotin, octanoic acid	Tran Nguyen Hoang et al. (2018) Curran et al. (2013) Dellomonaco et al. (2011) Hayashi et al. (2016)
SAM biosynthesis	L-aspartate	SAM	*E. coli*	8	ATP, NADPH, succinyl-CoA, L-cysteine, CH3-THF	MTA deaminase (salvage), MetA and CysE engineering	Vitamin B_{12}, haem, vanillin, anthocyanin, RIPPs	Cress et al. (2017) Kunjapur et al. (2016) Danchin and Sekowska (2015)

ALA bio-synthesis (C-5)	2-ketoglutarate	ALA	*E. coli*	5	NADPH, ATP, t-RNA (glu), NH4+	GltBD, C4 pathway	Cobalamin, sirohaem, haem, p450 expression	Zhao et al. (2018) Kang et al. (2011)
CO_2 fixation (Calvin cycle)	CO_2 and D-ribulose-1,5-bisphosphate	3-PG	*Synechocystis* sp. *PCC 6803*	13	ATP, NADPH, G3P	CETCH cycle, reductive citric acid cycle, 3-Hydroxypropionate bicycle, other alternative CO_2 fixating pathways	Biofuels, polyketides, fatty acids, beta-carotene	Schwander et al. (2016) Fuchs (2011) Calero and Nikel (2019)

Level 1 pathways give the name to their respective platform strains. They correspond to metabolic pathways synthesizing precursors usable by various sub-pathways (level 2) that lead to the synthesis of industrially relevant compounds. Various alternative methods, supplementing the ones mentioned in Sect. 3, are listed. Abbreviations (from top left to bottom right): *MEP* methylerythritol phosphate, *SAM* S-adenosyl-L-methionine, *ALA* aminolevulinic acid, *G3P* glycerol 3-phosphate, *E4P* D-erythrose 4-phosphate, *IPP* isopentenyl diphosphate, *DMAPP* dimethylallyl diphosphate, *3-PG* 3-phosphoglycerate, *NADPH* nicotinamide adenine dinucleotide phosphate, *CTP*, cytidine triphosphate *ATP* adenosine triphosphate, *PEP* phosphoenolpyruvate, *ACP* acyl carrier protein, *CH3-THF* methyltetrahydrofolate, *MVA* mevalonate, *MTA* methylthioadenosine, *CETCH* crotonyl-coenzyme A (CoA)/ethylmalonyl-CoA/hydroxybutyryl-CoA, *DHA* docosahexaenoic acid, *EPA* eicosapentaenoic acid

4.4 *The Challenge of Upscaling*

A microbial cell factory should be able to excel at performance in small- and large-scale conditions; otherwise, it is not possible to develop a bioprocess. Certain cells are more sensitive to others when it turns to physical stresses including oxygen availability, product toxicity and mechanical forces. For instance, microbes are 1000-fold more resilient to shear stress than mammalian cells (Foley and Shuler 2010). There are also genetic factors that could affect upscaling. For example, compared to the parent strain, the minimized genome *E. coli* strain MDS42 displays a more stable growth during a simulation of a large-scale fermentation after 60–80 generations that are needed to reach population sizes of 10^{20} in a 200 m^2 fed-batch reactor (Rugbjerg et al. 2018b). Likewise, a stable fermentation can be achieved by coupling biosynthetic production genes to essential genes (Rugbjerg et al. 2018a). Finally, discriminating proper substrates against similar compounds entails that cells code for information-loaded agents that behave as Maxwell's demons would (Boël et al. 2019). Reflection on this requirement for upscaling is just beginning.

5 Conclusions and Perspectives

Although the concept of the minimal genome is useful when trying to understand minimal life, other approaches like the use of cells with "minimized genomes" might offer alternative solutions when looking for applications in industrial biotechnology. For this reason, we aimed to describe a possible scenario for the origin of life to indicate that we should consider metabolism and biodiversity for the design of platform strains. The biodiversity available to us in any place of this planet is a gold mine to find new enzymes and metabolic pathways that can solve challenges in the production of food, medicine and chemicals. It has been estimated that the number of microbial species inhabiting Earth ranges approximately from 10^{11} to 10^{12} (Locey and Lennon 2016). Even though the vast majority of living organisms are not cultivable in laboratory settings (Bernard et al. 2018), we can look at new technologies such as functional genomics to discover new enzymes and pathway components (Van Der Helm et al. 2018). Microbial cell factories will help to reach the sustainable development goals (O'Toole and Paoli 2017) and contribute to economic growth and employment creation (Timmis et al. 2017). Although industrial biotechnology still faces some technical issues (Straathof et al. 2019), perhaps the biggest challenge lies in economics and politics when it turns to replace chemical processes with greener and more sustainable bioprocesses (Ramos and Duque 2019). To address this issue, we will need to increase the literacy in microbiology of our fellow citizens (Timmis et al. 2019).

References

Acevedo-Rocha CG (2016) The synthetic nature of biology. In: Ambivalences of creating life. Springer, Cham, pp 9–53. https://doi.org/10.1007/978-3-319-21088-9_2

Acevedo-Rocha CG et al (2013) From essential to persistent genes: a functional approach to constructing synthetic life. Trends Genet 29(5):273–279. https://doi.org/10.1016/j.tig.2012.11.001

Acevedo-Rocha CG et al (2019) Microbial cell factories for the sustainable manufacturing of B vitamins. Curr Opin Biotechnol 56:18–29. https://doi.org/10.1016/j.copbio.2018.07.006

Baby V et al (2018) Inferring the minimal genome of *Mesoplasma florum* by comparative genomics and transposon mutagenesis. mSystems 3(3). https://doi.org/10.1128/mSystems.00198-17

Balderas-Hernández VE et al (2014) Catechol biosynthesis from glucose in *Escherichia coli* anthranilate-overproducer strains by heterologous expression of anthranilate 1,2-dioxygenase from Pseudomonas aeruginosa PAO1. Microb Cell Factories 13(1):1–11. https://doi.org/10.1186/s12934-014-0136-x

Bali S et al (2011) Molecular hijacking of siroheme for the synthesis of heme and d1 heme. Proc Natl Acad Sci U S A 108(45):18260–18265. https://doi.org/10.1073/pnas.1108228108

Bell SD (2019) Initiating DNA replication: a matter of prime importance. Biochem Soc Trans 47 (1):351–356. https://doi.org/10.1042/BST20180627

Bernal J (1951) The physical basis of life. Routledge and Paul, London. https://www.worldcat.org/title/physical-basis-of-life/oclc/1495902. Accessed 1 June 2019

Bernard G, Pathmanathan JS, Lannes R, Lopez P, Bapteste E (2018) Microbial dark matter investigations: how microbial studies transform biological knowledge and empirically sketch a logic of scientific discovery. Genome Biol Evol 10(3):707–715. https://doi.org/10.1093/gbe/evy031

Boeke JD et al (2016) The Genome Project-Write. Science 353(6295):126–127. https://doi.org/10.1126/science.aaf6850

Boël G et al (2019) Omnipresent Maxwell's demons orchestrate information management in living cells. Microb Biotechnol 12(2):210–242. https://doi.org/10.1111/1751-7915.13378

Boyd ES et al (2017) Origin and evolution of Fe-S proteins and enzymes. Biochem Biosynth Hum Dis 2:445–461. https://doi.org/10.1515/9783110479850-017

Breuer M et al (2019) Essential metabolism for a minimal cell. eLife 8:e36842. https://doi.org/10.7554/eLife.36842

Brzóska K, Meczyńska S, Kruszewski M (2006) Iron-sulfur cluster proteins: electron transfer and beyond. Acta biochim Pol 53(4):685–691. Accessed 31 May 2019

Butland G et al (2008) eSGA: *E. coli* synthetic genetic array analysis. Nat Methods 5(9):789–795. https://doi.org/10.1038/nmeth.1239

Caforio A et al (2018) Converting *Escherichia coli* into an archaebacterium with a hybrid heterochiral membrane. Proc Natl Acad Sci U S A 115(14):3704–3709. https://doi.org/10.1073/pnas.1721604115

Cairns-Smith AG (1982) Genetic takeover and the mineral origins of life. Cambridge University Press, Cambridge

Calero P, Nikel PI (2019) Chasing bacterial chassis for metabolic engineering: a perspective review from classical to non-traditional microorganisms. Microb Biotechnol 12(1):98–124. https://doi.org/10.1111/1751-7915.13292

Cavalier-Smith T (2006) Cell evolution and Earth history: stasis and revolution. Philos Trans R Soc B Biol Sci 361(1470):969–1006. https://doi.org/10.1098/rstb.2006.1842

Chi H et al (2019) Engineering and modification of microbial chassis for systems and synthetic biology. Synth Syst Biotechnol 4(1):25–33. https://doi.org/10.1016/j.synbio.2018.12.001

Cho S, Yang S, Rhie H (2012) The gene encoding the alternative thymidylate synthase ThyX is regulated by sigma factor SigB in Corynebacterium glutamicum ATCC 13032. FEMS Microbiol Lett 328(2):157–165. https://doi.org/10.1111/j.1574-6968.2011.02494.x

Choe D et al (2019) Adaptive laboratory evolution of a genome-reduced *Escherichia coli*. Nat Commun 10(1):935. https://doi.org/10.1038/s41467-019-08888-6

Choi J-M, Han S-S, Kim H-S (2015) Industrial applications of enzyme biocatalysis: current status and future aspects. Biotechnol Adv 33(7):1443–1454. https://doi.org/10.1016/j.biotechadv.2015.02.014

Choi SY et al (2019) Metabolic engineering for the synthesis of polyesters: a 100-year journey from polyhydroxyalkanoates to non-natural microbial polyesters. Metab Eng. https://doi.org/10.1016/j.ymben.2019.05.009

Costanzo G et al (2007) Formamide as the main building block in the origin of nucleic acids. BMC Evol Biol 7(Suppl 2):S1. https://doi.org/10.1186/1471-2148-7-S2-S1

Couto JM et al (2018) The effect of metabolic stress on genome stability of a synthetic biology chassis *Escherichia coli* K12 strain. Microb Cell Fact 17(1):1–10. https://doi.org/10.1186/s12934-018-0858-2

Coyle M, Hu J, Gartner Z (2016) Mysteries in a minimal genome. ACS Cent Sci 2(5):274–277. https://doi.org/10.1021/acscentsci.6b00110

Cravens A, Payne J, Smolke CD (2019) Synthetic biology strategies for microbial biosynthesis of plant natural products. Nat Commun 10(1):2142. https://doi.org/10.1038/s41467-019-09848-w

Cress BF et al (2017) CRISPRi-mediated metabolic engineering of *E. coli* for O-methylated anthocyanin production. Microb Cell Fact 16(1):1–14. https://doi.org/10.1186/s12934-016-0623-3

Csörgo B et al (2012) Low-mutation-rate, reduced-genome *Escherichia coli*: an improved host for faithful maintenance of engineered genetic constructs. Microb Cell Factories 11:1–13. https://doi.org/10.1186/1475-2859-11-11

Curran KA et al (2013) Metabolic engineering of muconic acid production in Saccharomyces cerevisiae. Metab Eng 15(1):55–66. https://doi.org/10.1016/j.ymben.2012.10.003

Dai Y et al (2014) Interplay between oxygen and Fe–S cluster biogenesis: insights from the suf pathway. Biochemistry 53(37):5834–5847. https://doi.org/10.1021/bi500488r

Dailey HA et al (2017) Prokaryotic heme biosynthesis: multiple pathways to a common essential product. Microbiol Mol Biol Rev 81(1):1–62. https://doi.org/10.1128/MMBR.00048-16

Danchin A (1983) L'Oeuf et la poule: histoires du code génétique. Fayard

Danchin A (1989) Homeotopic transformation and the origin of translation. Prog Biophys Mol Biol 54(1):81–86. Accessed 1 June 2019

Danchin A (2007) Archives or palimpsests? Bacterial genomes unveil a scenario for the origin of life. Biol Theory 2(1):52–61. https://doi.org/10.1162/biot.2007.2.1.52

Danchin A (2012) Scaling up synthetic biology: do not forget the chassis. FEBS Lett 586(15):2129–2137. https://doi.org/10.1016/j.febslet.2011.12.024

Danchin A (2017) From chemical metabolism to life: the origin of the genetic coding process. Beilstein J Org Chem 13:1119–1135. https://doi.org/10.3762/bjoc.13.111

Danchin A, Fang G (2016) Unknown unknowns: essential genes in quest for function. Microb Biotechnol 9(5):530–540. 101111/1751-791512384

Danchin A, Sekowska A (2015) The logic of metabolism. Perspect Sci 6:15–26. https://doi.org/10.1016/j.pisc.2015.05.003

Danchin A, Fang G, Noria S (2007) The extant core bacterial proteome is an archive of the origin of life. Proteomics 7(6):875–889. https://doi.org/10.1002/pmic.200600442

de Duve C (1990) The thioester world. In: Frontiers of life. http://www.as.utexas.edu/astronomy/education/spring11/evans/secure/UGS303.Lec10.pdf

Dellomonaco C, Clomburg J, Miller E et al (2011) Engineered reversal of the β-oxidation cycle for the synthesis of fuels and chemicals. Nature 476:355–359. https://doi.org/10.1038/nature10333

Dogovski C, Atkinson SC, Dommaraju SR, Downton M, Hor L, Moore S, Paxman JJ, Peverelli MG, Qiu TW, Reumann M, Siddiqui T, Taylor NL, Wagner J, Wubben JM, Perugini MA (2012) Enzymology of bacterial lysine biosynthesis. In: Ekinci D (ed) Biochemistry. InTech, Croatia. isbn:978-953-51-0076-8. http://www.intechopen.com/books/biochemistry/enzymology-of-bacterial-lysine-biosynthesi

Donati S, Sander T, Link H (2018) Crosstalk between transcription and metabolism: how much enzyme is enough for a cell? Wiley Interdiscip Rev Syst Biol Med 10(1):1–11. https://doi.org/10.1002/wsbm.1396
Draskovits G et al (2017) Genome-wide abolishment of mobile genetic elements using genome shuffling and CRISPR/Cas-assisted MAGE allows the efficient stabilization of a bacterial chassis. ACS Synth Biol 6(8):1471–1483. https://doi.org/10.1021/acssynbio.6b00378
Dyson FJ (1985) Origins of life. Cambridge University Press, Cambridge
Fact C et al (2019) Metabolic engineering of microorganisms for production of aromatic compounds. Microb Cell Fact:1–29. https://doi.org/10.1186/s12934-019-1090-4
Fang G, Rocha E, Danchin A (2005) How essential are nonessential genes? Mol Biol Evol 22 (11):2147–2156. https://doi.org/10.1093/molbev/msi211
Fang G, Rocha EPC, Danchin A (2008) Persistence drives gene clustering in bacterial genomes. BMC Genomics 9. https://doi.org/10.1186/1471-2164-9-4
Fang H, Kang J, Zhang D (2017) Microbial production of vitamin B12: a review and future perspectives. Microb Cell Fact 16(1):1–14. https://doi.org/10.1186/s12934-017-0631-y
Fang H et al (2018) Metabolic engineering of *Escherichia coli* for de novo biosynthesis of vitamin B 12. Nat Commun 9(1):4917. https://doi.org/10.1038/s41467-018-07412-6
Fehér T et al (2007) Systematic genome reductions: theoretical and experimental approaches. Chem Rev 107(8):3498–3513. https://doi.org/10.1021/cr0683111
Feng L et al (2004) Aminoacyl-tRNA synthesis by pre-translational amino acid modification. RNA Biol 1(1):16–20. Accessed 1 June 2019
Foley PL, Shuler ML (2010) Considerations for the design and construction of a synthetic platform cell for biotechnological applications. Biotechnol Bioeng 105(1):26–36. https://doi.org/10.1002/bit.22575
Forterre P, Filée J, Myllykallio H (2007) Origin and evolution of DNA and DNA replication machineries. In: de Pouplana L R (ed) The genetic code and the origin of life. Springer, Boston, MA, pp 145–168. https://doi.org/10.1007/0-387-26887-1_10
Frandsen RJN et al (2018) Heterologous production of the widely used natural food colorant carminic acid in Aspergillus nidulans. Sci Rep 8(1):12853. https://doi.org/10.1038/s41598-018-30816-9
Frank A, Groll M (2017) The methylerythritol phosphate pathway to isoprenoids. Chem Rev 117 (8):5675–5703. https://doi.org/10.1021/acs.chemrev.6b00537
Fredens J et al (2019) Total synthesis of *Escherichia coli* with a recoded genome. Nature 569 (7757):514–518. https://doi.org/10.1038/s41586-019-1192-5
Fuchs G (2011) Alternative pathways of carbon dioxide fixation: insights into the early evolution of life? Annu Rev Microbiol 65(1):631–658. https://doi.org/10.1146/annurev-micro-090110-102801
Gibard C et al (2018) Phosphorylation, oligomerization and self-assembly in water under potential prebiotic conditions. Nat Chem 10(2):212–217. https://doi.org/10.1038/nchem.2878
Gibson DG et al (2010) Supporting online material for creation of a bacterial cell controlled by a chemically synthesized genome. Science 329(May):52–57. https://doi.org/10.1126/science.1190719
Giel JL et al (2006) IscR-dependent gene expression links iron-sulphur cluster assembly to the control of O2-regulated genes in *Escherichia coli*. Mol Microbiol 60(4):1058–1075. https://doi.org/10.1111/j.1365-2958.2006.05160.x
Granick S (1957) Speculations on the origins and evolution of photosynthesis. Ann N Y Acad Sci 69(2):292–308. https://doi.org/10.1111/j.1749-6632.1957.tb49665.x
Hall DO, Cammack R, Rao KK (1971) Role for ferredoxins in the origin of life and biological evolution. Nature 233(5315):136–138. https://doi.org/10.1038/233136a0
Hartman H (1984) The origin of the eukaryotic cell. Specul Sci Technol 7(2):77–81. Accessed 30 June 2019

Hayashi S, Satoh Y, Ujihara T, Takata Y, Dairi T (2016) Enhanced production of polyunsaturated fatty acids by enzyme engineering of tandem acyl carrier proteins. Sci Rep 6:35441. https://doi.org/10.1038/srep35441

Heux S et al (2015) White biotechnology: state of the art strategies for the development of biocatalysts for biorefining. Biotechnol Adv 33(8):1653–1670. https://doi.org/10.1016/j.biotechadv.2015.08.004

Hopfield JJ (1978) Origin of the genetic code: a testable hypothesis based on tRNA structure, sequence, and kinetic proofreading. Proc Natl Acad Sci U S A 75(9):4334–4338. https://doi.org/10.1073/pnas.75.9.4334

Hunter JH et al (2008) Kinetics and ligand-binding preferences of Mycobacterium tuberculosis thymidylate synthases, ThyA and ThyX. PLoS One 3(5):1–10. https://doi.org/10.1371/journal.pone.0002237

Hutchison CA et al (2016) Design and synthesis of a minimal bacterial genome. Science 351(6280). https://doi.org/10.1126/science.aad6253

Jin D, Musier-Forsyth K (2019) Role of host tRNAs and aminoacyl-tRNA synthetases in retroviral replication. J Biol Chem 294(14):5352–5364. https://doi.org/10.1074/jbc.REV118.002957

Juhas M, Eberl L, Glass JI (2011) Essence of life: essential genes of minimal genomes. Trends Cell Biol 21(10):562–568. https://doi.org/10.1016/j.tcb.2011.07.005

Juhas M, Eberl L, Church GM (2012) Essential genes as antimicrobial targets and cornerstones of synthetic biology. Trends Biotechnol 30(11):601–607. https://doi.org/10.1016/j.tibtech.2012.08.002

Jullesson D et al (2015) Impact of synthetic biology and metabolic engineering on industrial production of fine chemicals. Biotechnol Adv 33(7):1395–1402. https://doi.org/10.1016/j.biotechadv.2015.02.011

Kamiryo T, Matsuhashi M (1969) Sequential addition of glycine from glycyl-tRNA to the lipid-linked precursors of cell wall peptidoglycan in Staphylococcus aureus. Biochem Biophys Res Commun 36(2):215–222. https://doi.org/10.1016/0006-291X(69)90317-9

Kanehisa M, Goto S (2000) KEGG: kyoto encyclopedia of genes and genomes. Nucl Acids Res 28 (1):27–30. https://doi.org/10.1093/nar/28.1.27

Kang Z et al (2011) Metabolic engineering to improve 5-aminolevulinic acid production. Bioeng Bugs 2(6):342–345. https://doi.org/10.4161/bbug.2.6.17237

Keller MA, Turchyn AV, Ralser M (2014) Non-enzymatic glycolysis and pentose phosphate pathway-like reactions in a plausible Archean ocean. Mol Syst Biol:1–12. http://onlinelibrary.wiley.com.ep.fjernadgang.kb.dk/doi/10.1002/msb.20145228/full

Kim H-J, Benner SA (2015) Prebiotic glycosylation of uracil with electron-donating substituents. Astrobiology 15(4):301–306. https://doi.org/10.1089/ast.2014.1264

Kim J et al (2016) Properties of alternative microbial hosts used in synthetic biology: towards the design of a modular chassis. Essays Biochem 60(4):303–313. https://doi.org/10.1042/ebc20160015

Koc I, Caetano-Anolles G (2017) The natural history of molecular functions inferred from an extensive phylogenomic analysis of gene ontology data. PLoS One 12(5):e0176129. https://doi.org/10.1371/journal.pone.0176129

Koga Y et al (1998) Did archaeal and bacterial cells arise independently from noncellular precursors? A hypothesis stating that the advent of membrane phospholipid with enantiomeric glycerophosphate backbones caused the separation of the two lines of descent. J Mol Evol 46 (1):54–63. Accessed 1 June 2019

Koonin EV, Wolf YI (2008) Genomics of bacteria and archaea: the emerging dynamic view of the prokaryotic world. Nucleic Acids Res 36(21):6688–6719. https://doi.org/10.1093/nar/gkn668

Kunjapur AM, Hyun JC, Prather KLJ (2016) Deregulation of S-adenosylmethionine biosynthesis and regeneration improves methylation in the *E. coli* de novo vanillin biosynthesis pathway. Microb Cell Fact 15(1):1–17. https://doi.org/10.1186/s12934-016-0459-x

Lachance J-C, Rodrigue S, Palsson BO (2019) Minimal cells, maximal knowledge. eLife 8:8–11. https://doi.org/10.7554/elife.36842

Landauer R (n.d.) Irreversibility and heat generation in the computing process. http://worrydream.com/refs/Landauer – Irreversibility and Heat Generation in the Computing Process.pdf. Accessed 1 June 2019

Lau M, Ferré-D'Amaré A (2016) Many activities, one structure: functional plasticity of ribozyme folds. Molecules 21(11):1570. https://doi.org/10.3390/molecules21111570

Laurino P, Tawfik DS (2017) Spontaneous emergence of S-adenosylmethionine and the evolution of methylation. Angew Chem Int Ed Engl 56(1):343–345. https://doi.org/10.1002/anie.201609615

Layer G et al (2010) Structure and function of enzymes in heme biosynthesis. Protein Sci 19 (6):1137–1161. https://doi.org/10.1002/pro.405

Ledesma-Amaro R, Nicaud JM (2016) Yarrowia lipolytica as a biotechnological chassis to produce usual and unusual fatty acids. Prog Lipid Res 61:40–50. https://doi.org/10.1016/j.plipres.2015.12.001

Lee HH et al (2019) Functional genomics of the rapidly replicating bacterium Vibrio natriegens by CRISPRi. Nat Microbiol 4(7):1. https://doi.org/10.1038/s41564-019-0423-8

Leung EKY et al (2011) The mechanism of peptidyl transfer catalysis by the ribosome. Annu Rev Biochem 80(1):527–555. https://doi.org/10.1146/annurev-biochem-082108-165150

Ling C et al (2019) Engineering self-flocculating Halomonas campaniensis for wastewaterless open and continuous fermentation. Biotechnol Bioeng 116(4):805–815. https://doi.org/10.1002/bit.26897

Liu X, Ding W, Jiang H (2017) Engineering microbial cell factories for the production of plant natural products: from design principles to industrial-scale production. Microb Cell Fact 16 (1):125. https://doi.org/10.1186/s12934-017-0732-7

Locey KJ, Lennon JT (2016) Scaling laws predict global microbial diversity. Proc Natl Acad Sci U S A 113(21):5970–5975. https://doi.org/10.1073/pnas.1521291113

Lohrmann R, Orgel LE (1968) Prebiotic synthesis: phosphorylation in aqueous solution. Science 161(3836):64–66. https://doi.org/10.1126/SCIENCE.161.3836.64

Lorenz P, Zinke H (2005) White biotechnology: differences in US and EU approaches? Trends Biotechnol 23(12):570–574. https://doi.org/10.1016/j.tibtech.2005.10.003

Luo ZW, Cho JS, Lee SY (2019) Microbial production of methyl anthranilate, a grape flavor compound. Proc Natl Acad Sci U S A 116(22):10749–10756. https://doi.org/10.1073/pnas.1903875116

Machnicka MA et al (2012) MODOMICS: a database of RNA modification pathways—2013 update. Nucleic Acids Res 41(D1):D262–D267. https://doi.org/10.1093/nar/gks1007

Mazumder R, Vasudevan S (2008) Structure-guided comparative analysis of proteins: principles, tools, and applications for predicting function. PLoS Comput Biol 4(9):1–12. https://doi.org/10.1371/journal.pcbi.1000151

Miller SL, Urey HC (1959) Origin of life. Science 130(3389):1622–1624. https://doi.org/10.1126/science.130.3389.1622-a

Mori M et al (2017) Quantifying the benefit of a proteome reserve in fluctuating environments. Nat Commun 8(1):1–8. https://doi.org/10.1038/s41467-017-01242-8

Muchowska KB, Chevallot-Beroux E, Moran J (2019a) Recreating ancient metabolic pathways before enzymes. Bioorg Med Chem 27:2292. https://doi.org/10.1016/j.bmc.2019.03.012

Muchowska KB, Varma SJ, Moran J (2019b) Synthesis and breakdown of universal metabolic precursors promoted by iron. Nature 569(7754):104–107. https://doi.org/10.1038/s41586-019-1151-1

Müller A et al (2017) Large-scale reduction of the Bacillus subtilis genome: consequences for the transcriptional network, resource allocation, and metabolism. Genome Res 27:289–299. https://doi.org/10.1101/gr.215293.116.9

Murphy CD (2012) The microbial cell factory. Org Biomol Chem 10(10):1949. https://doi.org/10.1039/c2ob06903b

Nielsen J (2019) Cell factory engineering for improved production of natural products. Nat Prod Rep. https://doi.org/10.1039/C9NP00005D

Nielsen J, Keasling JD (2016) Engineering cellular metabolism. Cell 164(6):1185–1197. https://doi.org/10.1016/j.cell.2016.02.004
Nielsen DR, Moon TS (2013) From promise to practice. The role of synthetic biology in green chemistry. EMBO Rep 14(12):1034–1038. https://doi.org/10.1038/embor.2013.178
Nikel PI, de Lorenzo V (2018) Pseudomonas putida as a functional chassis for industrial biocatalysis: from native biochemistry to trans-metabolism. Metab Eng 50:142–155. https://doi.org/10.1016/j.ymben.2018.05.005
Niu FX, Lu Q, Bu YF, Liu JZ (2017) Metabolic engineering for the microbial production of isoprenoids: carotenoids and isoprenoid-based biofuels. Synth Syst Biotechnol 2(3):167–175. https://doi.org/10.1016/j.synbio.2017.08.001
Nishida H et al (1999) A prokaryotic gene cluster involved in synthesis of lysine through the amino adipate pathway: a key to the evolution of amino acid biosynthesis. Genome Res 9 (12):1175–1183. https://doi.org/10.1101/gr.9.12.1175
Noda S et al (2016) Metabolic design of a platform *Escherichia coli* strain producing various chorismate derivatives. Metab Eng 33:119–129. https://doi.org/10.1016/j.ymben.2015.11.007
O'Toole PW, Paoli M (2017) The contribution of microbial biotechnology to sustainable development goals: microbiome therapies. Microb Biotechnol 10(5):1066–1069. https://doi.org/10.1111/1751-7915.12752
Pascal R, Pross A, Sutherland JD (2013) Towards an evolutionary theory of the origin of life based on kinetics and thermodynamics. Open Biol 3:1–9. https://doi.org/10.1098/rsob.130156
Pauling L, Delbrück M (1940) The nature of the intermolecular forces operative in biological processes. Science 92(2378):77–79. https://doi.org/10.1126/SCIENCE.92.2378.77
Pereira IAC et al (2013) Early bioenergetic evolution. Philos Trans R Soc B Biol Sci 368 (1622):20130088. https://doi.org/10.1098/rstb.2013.0088
Pomiankowski A, Lane N (2014) A bioenergetic basis for membrane divergence in archaea and bacteria. PLoS Boil 12(8):e1001926. https://doi.org/10.1371/journal.pbio.1001926
Porcar M, Peretó J (2018) Creating life and the media: translations and echoes. Life Sci Soc Policy 14(1):19
Pósfai G et al (2006) Emergent properties of reduced-genome *Escherichia coli*. Science 312 (5776):1044–1046. https://doi.org/10.1126/science.1126439
Pretorius IS, Boeke JD (2018) Yeast 2.0—connecting the dots in the construction of the world's first functional synthetic eukaryotic genome. FEMS Yeast Res 18(4). https://doi.org/10.1093/femsyr/foy032
Py B, Barras F (2010) Building Feg-S proteins: bacterial strategies. Nat Rev Microbiol 8 (6):436–446. https://doi.org/10.1038/nrmicro2356
Ramos JL, Duque E (2019) Twenty-first-century chemical odyssey: fuels versus commodities and cell factories versus chemical plants. Microb Biotechnol 12(2):200–209. https://doi.org/10.1111/1751-7915.13379
Reetz MT (2013) Biocatalysis in organic chemistry and biotechnology: past, present, and future. J Am Chem Soc 135(34):12480–12496. https://doi.org/10.1021/ja405051f
Reitzer L (2004) Biosynthesis of glutamate, aspartate, asparagine, L-alanine, and D-alanine. EcoSal Plus 1:1–18. https://doi.org/10.1128/ecosalplus.3.6.1.3
Reuß DR et al (2016) The blueprint of a minimal cell: MiniBacillus. Microbiol Mol Biol Rev 80 (4):955–987. https://doi.org/10.1128/MMBR.00029-16
Ross DS, Deamer D (2016) Dry/wet cycling and the thermodynamics and kinetics of prebiotic polymer synthesis. Life 6(3). https://doi.org/10.3390/life6030028
Rugbjerg P, Sarup-Lytzen K et al (2018a) Synthetic addiction extends the productive life time of engineered *Escherichia coli* populations. Proc Natl Acad Sci U S A 115(10):201718622. https://doi.org/10.1073/pnas.1718622115
Rugbjerg P, Myling-Petersen N et al (2018b) Diverse genetic error modes constrain large-scale bio-based production. Nat Commun 9(1). https://doi.org/10.1038/s41467-018-03232-w

Sander T et al (2019) Allosteric feedback inhibition enables robust amino acid biosynthesis in *E. coli* by enforcing enzyme overabundance. Cell Syst 8(1):66–75.e8. https://doi.org/10.1016/j.cels.2018.12.005

Schindler D, Dai J, Cai Y (2018) Synthetic genomics: a new venture to dissect genome fundamentals and engineer new functions. Curr Opin Chem Biol 46:56–62. https://doi.org/10.1016/j.cbpa.2018.04.002

Schobert M, Jahn D (2002) Regulation of heme biosynthesis in non-phototrophic bacteria. J Mol Microbiol Biotechnol 4(3):287–294

Schroeder S et al (2009) Demonstration that CobG, the monooxygenase associated with the ring contraction process of the aerobic cobalamin (vitamin B12) biosynthetic pathway, contains an Fe-S center and a mononuclear non-heme iron center. J Biol Chem 284(8):4796–4805. https://doi.org/10.1074/jbc.M807184200

Schrumpf B et al (1991) A functionally split pathway for lysine synthesis in Corynebacterium glutamicum. J Bacteriol 173(14):4510–4516

Schwander T, Burgener S, Erb TJ (2016) A synthetic pathway for the fixation of carbon dioxide in vitro. Science 354(6314):900–904

Segré D et al (1998) Graded autocatalysis replication domain (GARD): kinetic analysis of self-replication in mutually catalytic sets. Orig Life Evol Biosph 28(4/6):501–514. https://doi.org/10.1023/A:1006583712886

Sekowska A, Ashida H, Danchin A (2019) Revisiting the methionine salvage pathway and its paralogues. Microb Biotechnol 12(1):77–97. https://doi.org/10.1111/1751-7915.13324

Shapiro R (2000) A replicator was not involved in the origin of life. IUBMB Life 49(3):173–176. https://doi.org/10.1080/713803621

Sheldon RA, Woodley JM (2018) Role of biocatalysis in sustainable chemistry. Chem Rev 118 (2):801–838. https://doi.org/10.1021/acs.chemrev.7b00203

Stairs CW, Leger MM, Roger AJ (2015) Diversity and origins of anaerobic metabolism in mitochondria and related organelles. Philos Trans R Soc B Biol Sci 370(1678):20140326. https://doi.org/10.1098/rstb.2014.0326

Straathof AJJ et al (2019) Grand research challenges for sustainable industrial biotechnology. Trends Biotechnol:1–9. https://doi.org/10.1016/j.tibtech.2019.04.002

Struck AW et al (2012) S-Adenosyl-methionine-dependent methyltransferases: highly versatile enzymes in biocatalysis, biosynthesis and other biotechnological applications. Chembiochem 13(18):2642–2655. https://doi.org/10.1002/cbic.201200556

Suárez RA, Stülke J, Van Dijl JM (2019) Less is more: toward a genome-reduced bacillus cell factory for "difficult Proteins". ACS Synth Biol 8(1):99–108. https://doi.org/10.1021/acssynbio.8b00342

Timmis K et al (2017) The contribution of microbial biotechnology to economic growth and employment creation. Microb Biotechnol 10(5):1137–1144. https://doi.org/10.1111/1751-7915.12845

Timmis K et al (2019) The urgent need for microbiology literacy in society. Environ Microbiol:1–21. https://doi.org/10.1111/1462-2920.14611

Tolborg G et al (2017) Establishing novel cell factories producing natural pigments in Europe. In: Singh OV (ed) Bio-pigmentation and biotechnological implementations. Wiley, Hoboken, NJ, pp 23–60. https://doi.org/10.1002/9781119166191.ch2

Tran Nguyen Hoang P, Ko JK, Gong G, Um Y, Lee SM (2018) Genomic and phenotypic characterization of a refactored xylose-utilizing Saccharomyces cerevisiae strain for lignocellulosic biofuel production. Biotechnol Biofuel 11:268. https://doi.org/10.1186/s13068-018-1269-7

Umenhoffer K et al (2010) Reduced evolvability of *Escherichia coli* MDS42, an IS-less cellular chassis for molecular and synthetic biology applications. Microb Cell Factories 9(1):38. https://doi.org/10.1186/1475-2859-9-38

Van Der Helm E, Genee HJ, Sommer MOA (2018) The evolving interface between synthetic biology and functional metagenomics. Nat Chem Biol 14(8):752–759. https://doi.org/10.1038/s41589-018-0100-x

van Tilburg AY et al (2019) Metabolic engineering and synthetic biology employing Lactococcus lactis and Bacillus subtilis cell factories. Curr Opin Biotechnol 59:1–7. https://doi.org/10.1016/j.copbio.2019.01.007

Venetz JE et al (2019) Chemical synthesis rewriting of a bacterial genome to achieve design flexibility and biological functionality. Proc Natl Acad Sci U S A 116(16):8070–8079. https://doi.org/10.1073/pnas.1818259116

Veronese FM, Boccu E, Conventi L (1975) Glutamate dehydrogenase from Escherichia coli: induction, purification and properties of the enzyme. Biochim Biophys Acta 377(2):217–228

Vévodová J et al (2004) Structure/function studies on a S-adenosyl-L-methionine-dependent uroporphyrinogen III C methyltransferase (SUMT), a key regulatory enzyme of tetrapyrrole biosynthesis. J Mol Biol 344(2):419–433. https://doi.org/10.1016/j.jmb.2004.09.020

Vinokur JM et al (2016) An adaptation to life in acid through a novel mevalonate pathway. Sci Rep 6(Dec):1–11. https://doi.org/10.1038/srep39737

Wächtershäuser G (1988) Before enzymes and templates: theory of surface metabolism. Microbiol Rev 52(4):452–484. Accessed 1 June 2019

Wächtershäuser G (2007) On the chemistry and evolution of the pioneer organism. Chem Biodivers 4(4):584–602. https://doi.org/10.1002/cbdv.200790052

Weber T et al (2015) Metabolic engineering of antibiotic factories: new tools for antibiotic production in actinomycetes. Trends Biotechnol 33(1):15–26. https://doi.org/10.1016/j.tibtech.2014.10.009

Weiss MC et al (2016) The physiology and habitat of the last universal common ancestor. Nat Microbiol 1(9):1–8. https://doi.org/10.1038/nmicrobiol.2016.116

Wieczorek R et al (2013) Formation of RNA phosphodiester bond by histidine-containing dipeptides. ChemBioChem 14(2):217–223. https://doi.org/10.1002/cbic.201200643

Wu RA et al (2017) Telomerase mechanism of telomere synthesis. Annu Rev Biochem 86 (1):439–460. https://doi.org/10.1146/annurev-biochem-061516-045019

Xu H et al (2006) The α-aminoadipate pathway for lysine biosynthesis in fungi. Cell Biochem Biophys 46:43–64

Xu J et al (2019) Prebiotic phosphorylation of 2-thiouridine provides either nucleotides or DNA building blocks via photoreduction. Nat Chem 11(5):457–462. https://doi.org/10.1038/s41557-019-0225-x

Yokoyama K, Lilla EA (2018) C-C bond forming radical SAM enzymes involved in the construction of carbon skeletons of cofactors and natural products. Nat Prod Rep 35(7):660–694. https://doi.org/10.1039/c8np00006a

Yu L, Wu F, Chen G (2019) Next generation industrial biotechnology-transforming the current industrial biotechnology into competitive processes. Biotechnol J:1800437. https://doi.org/10.1002/biot.201800437

Zargar A et al (2017) Leveraging microbial biosynthetic pathways for the generation of "drop-in" biofuels. Curr Opin Biotechnol 45:156–163. https://doi.org/10.1016/j.copbio.2017.03.004

Zhang LY, Chang SH, Wang J (2010) How to make a minimal genome for synthetic minimal cell. Protein Cell 1(5):427–434. https://doi.org/10.1007/s13238-010-0064-4

Zhang J et al (2015) Optimization of the heme biosynthesis pathway for the production of 5-aminolevulinic acid in *Escherichia coli*. Sci Rep 5:1–7. https://doi.org/10.1038/srep08584

Zhang Y, Nielsen J, Liu Z (2017) Engineering yeast metabolism for production of terpenoids for use as perfume ingredients, pharmaceuticals and biofuels. FEMS Yeast Res 17(8). https://doi.org/10.1093/femsyr/fox080

Zhao XR, Choi KR, Lee SY (2018) Metabolic engineering of *Escherichia coli* for secretory production of free haem. Nat Catal 1(9):720–728. https://doi.org/10.1038/s41929-018-0126-1

Zhou J, Du G, Chen J (2014) Novel fermentation processes for manufacturing plant natural products. Curr Opin Biotechnol 25:17–23. https://doi.org/10.1016/j.copbio.2013.08.009

Resource Allocation Principles and Minimal Cell Design

David Hidalgo and José Utrilla

Abstract Most natural organisms are generalists, as they deploy cellular resources for growth and survival under changing environments. Minimal cells are thought to be specialists; therefore, they should display specialized behaviors for very specific functions. Depending on the required function to display, the cellular resources should be differentially allocated, generating an optimal resource use that maximizes its designed function. Recently, many studies have focused on the economy of cellular resource allocation in different environments. With several tools and approaches, resource allocation has been extensively studied in natural and engineered cellular systems. These approaches have generated genome-scale models, coarse-grained models, and growth laws that may be used in minimal cell design. In this chapter, we will review the recent advances in econometric approaches to study and engineer resource allocation. We will propose design principles for cell minimization focusing on the cellular resource allocation framework to maximize the functions that they are designed to display.

Keywords Resource allocation · Proteome · Efficiency · Trade-off · Minimal cell · Bacteria · Design

1 Cellular Resources

Quantitative studies on the macromolecular composition of cells started as early as the 1950s. These studies were made for exponentially growing cultures. After measuring the cell's total quantities of DNA, RNA, and protein, it was seen that at a given temperature, the RNA/protein ratio of a culture is a direct function of the growth rate (Fig. 1a). It was determined that these variables present a linear relation (Kjelgaard and Gausing 1974). Importantly, it was also observed that during fast growth, 86% of

D. Hidalgo · J. Utrilla (✉)
Programa de Biología de Sistemas y Biología Sintetica, Cengro de Ciencias Genómicas, Universidad Nacional Autónoma de México, Cuernavaca, Morelos, México
e-mail: utrilla@ccg.unam.mx

A. R. Lara, G. Gosset (eds.), *Minimal Cells: Design, Construction, Biotechnological Applications*, https://doi.org/10.1007/978-3-030-31897-0_8

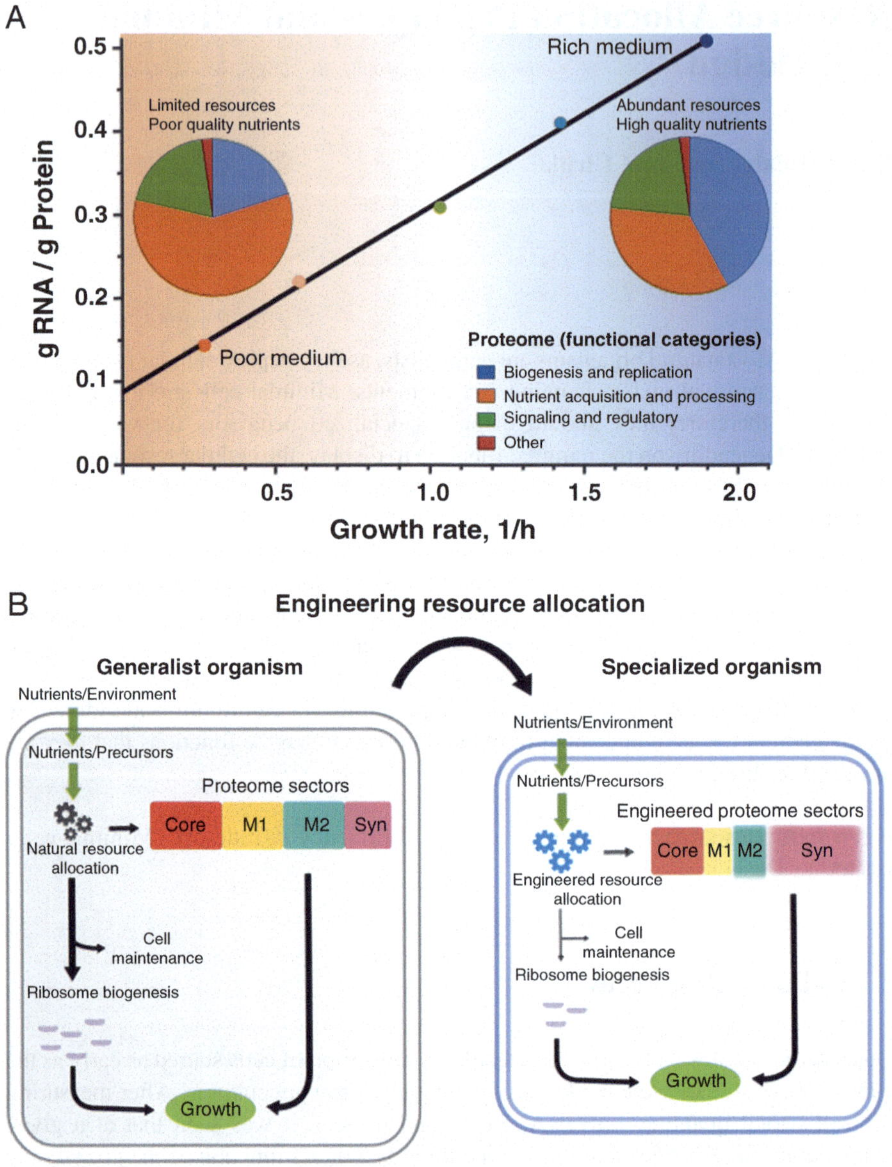

Fig. 1 (**a**) Increasing RNA/protein ratio with growth rate and differential proteome allocation to each coarse-grained sector at two growth conditions in *E. coli*, *a* cartoon based on experimental measurements. (**b**) Engineering resource allocation for a minimal specialized organism. Engineered proteome sectors should allow a larger synthetic (Syn) sector for a designed function. M1 and M2 are general and specialized metabolic sectors respectively

RNA is ribosomal (Neidhardt and Magasanik 1960; Shepherd et al. 1980) and that ribosomes constitute around 30% of the cell's dry mass (particularly for *E. coli*). More recently (Wilson and Nierhaus 2007), it was seen that about 40% of the cell's energy turnover is due to protein synthesis, which is one of the main processes that ensure a cell's growth and adaptation (fitness) to changing environments.

Ribosomes consist of three different RNAs (16S, 23S, and 5S) and 52 proteins. After a single transcript (35S) is processed from one of the seven *rrn* transcription units (rrn operons, independent from each other), the three species of rRNA are derived (Dennis and Bremer 2008). The genes for the 52 different ribosomal proteins (r-proteins) are located in different transcription units located at 14 different positions on the *E. coli* chromosome (Bachmann 1990). According to quantitative proteomics, up to 40% of the proteome is allocated to the Clusters of Orthologous Groups (COGs) translation, ribosomal structure, and biogenesis under fast growth in lysogeny broth (LB) medium (Schmidt et al. 2015). Five ATP molecules are consumed per peptide bond formation, and the rate of translation is approximately 20 amino acids per second (Kudva et al. 2013).

1.1 Energy

The detailed advances in molecular biology and biochemistry have allowed us to quantify the energetic needs of many cellular processes. However, even whole cell models show differences between calculated energy production and energy consumption; a whole cell model of *Mycoplasma genitalium* shows 44% of discrepancy (Karr et al. 2012). A classic study is the Pirt maintenance energy definition (Pirt 1965). The maintenance energy is a concept still used these days to balance energy production and consumption in models aiming to account for all cellular energy. In his studies, Pirt made two important observations: one is that substrate consumption per mass unit of an organism increased with the growth rate. The second refers to the substrate consumption even when the culture is not growing. He, then, called these processes growth-associated maintenance and non-growth-associated maintenance, respectively (GAM and *n*-GAM). For example, for *E. coli*, the *n*-GAM for ATP consumption is 7.6 mmol per gram of dry cell weight per hour (Selvarasu et al. 2009). These two simple parameters are of critical importance due to the fact that many computational modeling approaches need to perform a mass and energy balance in order to correctly output a certain phenotype prediction or cell composition.

1.2 Regulation of Cellular Resource Allocation

Bacteria have regulatory mechanisms that allow them to adjust their growth rate according to the nutritional environment and the availability of cellular resources.

When the cell senses an optimal nutritional environment, the expression of ribosomal RNA operons is favored, increasing growth rate. Several intracellular signals regulate the distribution of RNA polymerase (RNAP), the use of different sigma factors, and other elements of the regulatory network to promote or silence the transcription of biosynthetic, ribosomal, or stress operons (de Jong et al. 2017).

This extracellular nutrient availability is generally translated into intracellular signals such as charged tRNAs, 2-oxoglutarate/glutamine ratio, concentration of energetic molecules (ATP, GTP), and amino acid concentration, among others, which regulate the formation of (p)ppGpp through the following described mechanisms. During periods of nutrient starvation, particularly amino acids, (p)ppGpp is synthesized via the *relA*-dependent system, which induces the well-characterized stringent response. Normally, (p)ppGpp levels are low. When cells grow exponentially, the basal levels of ppGpp primarily come from a *relA*-independent system, where the *spoT* gene is involved, such that (p)ppGpp levels decrease approximately from 55 to 6 pmol per OD 460 unit when the exponential growth rate increases from 0.6 to 3.0 doublings/h (Dennis and Bremer 2008). It is important to consider this regulation because it severely impacts one of the most costly processes of the cell: ribosome biosynthesis. The seven rRNA (*rrn*) operons have the same two promoters, P1 and P2, in tandem. The single 35S transcript is expressed from them. The upstream region of the P1 promoter is heavily regulated. There are three binding sites for the protein Fis which is known to stimulate expression from the P1 promoter. On the other hand, ppGpp and DksA are known to negatively impact transcription from P1 due to its property to bind to the β' subunit of the RNAP, close to the active site (Artsimovitch et al. 2004). For the case of the P2 promoter, it has been observed that when isolated from P1, there seems to be a growth-dependent regulation in which ppGpp is not necessary (Murray et al. 2003). Additionally, in the presence of other rRNA operons, stringent regulation occurs, but when rRNA transcription is mostly repressed (through P1), a minimal (basal) or "background" rRNA transcription is achieved by the activity of P2 (Baracchini and Bremer 1988). For the case of the RNAP and ribosomal proteins (r-proteins), in bacteria such as *E. coli*, the genes for the RNAP subunits (*rpoA*, *rpoB*, *rpoC*, *rpoD*) are in three different operons along with some of the r-protein genes. Since mRNAs from these operons are subject to degradation (r-proteins bind to their mRNA when not bound to rRNA) in the absence of rRNA, cells modulate ribosome production, at least in part, through modulation of RNA polymerase and r-protein synthesis. These regulatory systems allow to balance the cellular composition and the proteome to the external nutritional environment; therefore, the growth rate is determined mainly by the resource availability through the regulatory mechanisms described above (Scott et al. 2014).

1.3 Insights into Natural Resource Allocation

The genes for rRNA typically exist in various copies in the form of operons along bacterial chromosomes (Klappenbach et al. 2000). The survival of a certain species in changing environments or in a microbial community depends on its ability to grow fast (faster than the other members of the consortium), as well as its ability to quickly adapt to changing conditions. Both features rely on ribosome availability and activity in order to synthesize the necessary proteins to carry out these processes. It can then be said that more copies of rRNA genes ensure species robustness, in other words, what makes it a generalist organism. Although ribosome biosynthesis is essential, it is also one of the most energetically expensive processes of the cell (as mentioned above). The essentiality of rRNA genes does not necessarily imply that all copies present in the chromosome are needed in all environments. In this context, studies have been made to test the effects of ribosomal (*rrn*) operon deletions. A study (Condon et al. 1993) reported the effects of such deletions in *E. coli*. In that work, it was seen that on rich medium (LB + glucose) both the growth rate and the ribosome concentration per cell did not decrease severely until inactivation of at least three rRNA operons. However, the mutants showed a diminished ability to transition to new nutritional sources, thus reflecting specialization rather than generalism (Condon et al. 1995) (discussed further in Sect. 4).

In this section, we have reviewed the knowledge on cellular resource allocation. The protein synthesis is the cellular processes that drain more cellular resources (estimated 40% of energy turnover as mentioned above); thus great attention is provided to it. A reference minimal organism to compare how many resources are channeled to each cellular process is the synthetic organism JCVI-Syn 3.0 with the smaller genome of any known organism assembled by Hutchinson and co-workers (Hutchison et al. 2016). In JCVI-Syn 3.0 genome analysis, the largest functional category is translation with 89 of 373 genes (24%), the next category is RNA (not mRNA) with 35 genes, and together they sum 124 genes and account for one third of all the genes in JCVI-Syn 3.0 (33%) (Glass et al. 2017).

2 Coarse-Grained Approaches to Study Resource Allocation

In this section, we review recent advances on how bacteria allocate resources according to different modes of growth. For well-known organisms such as *E. coli*, of which many kinetic and molecular parameters have been determined, one might be inclined to use a bottom-up computational model to perform phenotypical predictions, simulations, and strain design. However, even with this model bacteria, there are still many molecular mechanisms (in particular, those regarding gene expression regulation) to be elucidated to full detail. Coarse-grained models that describe biological systems contrast with genome-scale approaches in the sense that they require few parameters to work (e.g., growth rate, RNA/protein ratios, and knowledge of growth dynamics in response to nutrient or environmental perturbations) and that they are, fundamentally, phenomenological. These top-down approaches, particularly those that consider robust correlations in growth rate-dependent macromolecular parameters and proteome partition, have been successfully implemented to understand and explain resource allocation strategies used by cells (You et al. 2013; Scott et al. 2014; Basan et al. 2015; Mori et al. 2016; Dai et al. 2016). Some of which have been long-known biological conundrums, such as overflow metabolism. For well-known organisms such as *E. coli*, of which many kinetic and molecular parameters have been determined, one might be inclined to use a bottom-up computational model to perform phenotypical predictions, simulations, and strain design. However, even with this model bacteria, there are still many molecular mechanisms (in particular, those regarding gene expression regulation) to be elucidated to full detail. Coarse-grain models that describe biological systems contrast with genome-scale approaches in the sense that they require few parameters to work (e.g. growth rate, RNA/Protein ratios and knowledge of growth dynamics in response to nutrient or environmental perturbations) and that they are, fundamentally, phenomenological. These top-down approaches, particularly those that consider robust correlations in growth rate-dependent macromolecular parameters and proteome partition, have been successfully implemented to understand and explain resource allocation strategies used by cells (You et al. 2013; Scott et al. 2014; Basan et al. 2015; Mori et al. 2016; Dai et al. 2016). Some of which have been long-known biological conundrums, such as overflow metabolism.

The growth rate-dependent parameters considered in top-down approaches typically are the cellular amounts of macromolecules: DNA, RNA, and protein content. Over 50 years ago, there were already studies regarding the medium-dependent variation in the growth rate of exponentially growing cells (Neidhardt and Magasanik 1960), as mentioned before in this chapter. When the medium becomes richer, more doublings per hour can be achieved by the culture. Also, when the total amounts of macromolecules were measured, it was seen that the RNA to protein ratio displayed a positive and linear correlation with the growth rate (Kjelgaard and Gausing 1974). In an important work (Scott et al. 2010), this correlation was used as a fundamental basis to introduce the concept of proteome partition as a constraint that shapes resource allocation. The core of this concept is the consideration of a finite proteome partitioned into fractions (functional categories), whose sizes can change at the expense of the others in response to the environment. In the simplest example, three fractions were considered: a relatively fixed fraction (its size is not growth-dependent), a ribosomal fraction (ribosomal and ribosome-related proteins), and a metabolic fraction (catabolism and anabolism). In a rich medium, having non-limiting amounts of nutrients and added amino acids, cells can reach higher growth rates. This is because, in this scenario, the proteome fraction related to metabolism (particularly anabolism) can be diminished due to the amino acid synthesis enzymes being unnecessary. As a consequence, the ribosomal fraction is increased. On the contrary, at low growth such as those achieved in minimal medium with poor carbon sources, the ribosomal fraction should decrease to allow the metabolic fraction to expand (here, nutrient acquisition and processing become a priority).

With this simple consideration of how cells allocate resources, various works, particularly those from the Terrence Hwa research group (You et al. 2013; Scott et al.

2014; Basan et al. 2015; Dai et al. 2016; Mori et al. 2017), have been able to establish "growth laws" that describe these strategies. Also, they can better explain and reconcile experimental data from some of the most long-known and studied biological processes. The two important growth laws emanating from this approach are established from the cellular response upon the presence of free amino acids (Scott et al. 2014). Both of them refer to regulatory loops; one ensures a steady-state growth by supplying amino acids and is regulated by end-product inhibition. The other is denominated by the authors as a "supply-driven" activation that acts upon an imbalance between the use and the consumption of amino acids as molecular building blocks. As a result, the protein synthesis-dependent amino acid flux is balanced.

Among the biological phenomena explained by coarse-grained models is catabolic repression mediated by cyclic AMP (cAMP). This regulatory process inhibits the expression of many catabolic genes in the presence of a rapidly metabolizable sugar, such as glucose. After analyzing the correlation of growth rate and the expression of catabolic and biosynthetic genes in various media with carbon and nitrogen limitation, a study revealed linear correlations between them (You et al. 2013). For catabolic genes, there was a negative effect on the expression with the increasing growth rate. For biosynthetic genes, the correlation was positive. Under the consideration of proteome partition as a resource allocation constraint, it was possible to determine that the way cAMP acts on gene expression guarantees that the proteome demands for each sector are supplied in accordance to the particular growth medium. Interestingly, this work also pointed out α-ketoacids as one of the key catabolites mediating this regulatory response.

Another prominent example of the usefulness of coarse-grained models is the case of overflow metabolism. This phenomenon widely studied in many microbes and cell lines occurs when cells perform fermentation instead of respiration even in the presence of oxygen, therefore seen traditionally as an inefficient strategy to generate energy. In the study by Basan (Basan et al. 2015), gene expression, acetate excretion, and growth rates were analyzed. The authors found that the proteome cost to generate energy by oxidative phosphorylation exceeds that of fermentation. The growth-dependent demand for energy, protein synthesis, and biomass, according to the authors, is dealt with through a global strategy reflected in the amounts of acetate excretion. Since overflow metabolism has been observed in organisms from bacteria to yeast and even cancer cells, it is likely that this type of analysis will be extended to gain critical information for the chassis design beyond prokaryotes.

Extending the applicability of these models, a recent study (Dai et al. 2016; Mori et al. 2017) has pointed out that the translational efficiency and therefore the capacity for fast growth during the shift from famine to feast (nutrient upshift) are closely related to the use of inactive ribosomes. These inactive ribosomes appear to be latent in order to cope with fast growth upon a nutrient upshift. However, there is a high cost for ribosome synthesis, so this extra ribosomal capacity creates a trade-off in terms of fast growth in static versus changing environments.

In conclusion, coarse-grained approaches to study biological systems show, with relative ease, broad regulatory mechanisms that govern resource allocation. On a general view, the synthesis of ribosomes (rRNA transcription and r-protein translation), the growth rate, and the levels and availability of metabolites used as biomass building blocks create a control circuit signaled by ppGpp levels. The proteome partitions are controlled in a supply and demand manner. Such partitioning is a constraint that increases the mass of the currently limited proteome fraction at the expense of decreasing the mass of the others. Such scheme suggests that biological systems can be viewed as divided into coarse-grained modules (Fig. 1a).

3 Resource Allocation at the Genome Scale

Resource allocation can be studied using high-throughput technologies that provide detailed information per each gene in a genome. Many recent studies have applied proteomics to study resource allocation (Valgepea et al. 2013; Hui et al. 2015; Peebo et al. 2015). A recent compelling data set used state-of-the-art quantitative mass spectrometry techniques to measure protein abundance for *E. coli* in 22 growing conditions (Schmidt et al. 2016). In addition to protein abundance, they also identify methylation and acetylation not previously studied in bacteria. This data set is a valuable source of information to correlate condition-dependent protein abundances, elucidating the allocation of protein sectors (e.g., metabolic, ribosomal, and housekeeping) over a wide range of environments. Together with the classically used parameter of RNA/protein ratio, several cellular strategies for the specific use of nutrients can be elucidated.

A powerful tool to study resource allocation is the recently developed ribosome profiling method (Riboseq) (Ingolia et al. 2009). Essentially, Riboseq implies the sequencing of mRNA but only those molecules that are being translated (protected by the ribosome) at a certain point in time. This fundamental difference is what gives this technique a huge advantage over other RNAseq since the correlation of mRNA to protein quantities has shown to be low (Greenbaum et al. 2003; Ebrahim et al. 2016). By knowing the exact mRNAs being translated, one can quantify the rates or protein synthesis with high coverage and with reproducibility similar to that of mRNA abundances (Li et al. 2014). These important kinetic parameters can be beneficial to computational models in the sense that they can be used to elucidate cellular strategies for gene expression control as well as to help make better predictions.

Fine-grained omics studies have documented the excess of ribosomal capacity beyond bacteria. In the case of *Saccharomyces cerevisiae*, Metzl-Raz and colleagues studied proteomics, RNAseq, and Riboseq in a wide variety of growth conditions. In accordance with bacterial growth laws, they found that the higher the growth rate, the more ribosomes per cell were present. Interestingly, when they measured the fraction of inactive ribosomes, they found that a constant fraction of the proteome encodes for ribosomal proteins that are not actively translating (~8%), and even in fast growth, cells still maintain a ~25% of their ribosomal proteins inactive. Such a population of ribosomes seems to point out that cells prepare for possible changes or fluctuations in the environment. This consistently explains why there is a higher delay in the recovery from starvation in cells with a diminished excess ribosome pool.

To understand cellular processes that contribute to observed phenotypes and to guide resource minimization projects, it is necessary to make genome-wide evaluations of genetic function (functional genomics). To perform such a task, transposon mutagenesis and sequencing approaches, such as barcoded Tn-seq, are very useful. It allows the high-throughput generation of mutants which help to measure and track fitness for a mutant library of an organism. In this technique, transposons are used to insert randomly in the genome. The generated library of mutants is then sequenced, and depending on the position where the transposons are inserted, assessment of gene function and essentiality can be done on a large scale (Deutschbauer et al. 2014). In this study, they reported that using DNA barcoded Tn libraries for quantitative parallel analysis, 89% of the genes in *Zymomonas mobilis* ZM4 have an associated phenotype. Thus, a comprehensive and genome-wide relation between genotype and phenotype can now be assessed for this important organism. This approach should prove useful for exploring genotype-phenotype relations for other industrially important organisms, and it is helpful to assess gene essentiality experimentally. Price and co-workers combined Tn-seq with ribosome profiling data, and they found which genes are highly translated in *E. coli* and do not contribute to fitness in glucose minimal media; these genes are ideal candidates to eliminate since they consume many resources and do not have an apparent fitness contribution, at least under laboratory conditions (Price et al. 2016).

Genome-scale data sets require modeling tools and frameworks to be thoroughly analyzed and understood. The representation of biological systems with a mathematical approach has been mainly done by bottom-up computational approaches. These approaches account for all known enzyme quantities, reaction rates, and mass fluxes and, more recently, also include processes associated with the gene expression machinery at a genome scale. Their usage enables one to predict phenotypes, based on the input of certain parameters such as gene deletions, the maximization of growth rate, or biomass production, by defining a mass flux or by constraining the model via the fractionation of protein modules. A new generation of genome-scale models includes the gene expression process (transcription and translation) through the reconstruction of a stoichiometric matrix of these processes (Thiele et al. 2009). Coupled to classic genome-scale metabolic models, these models of Metabolism and Expression (ME-models) calculate the macromolecule need of an organism; they capture several phenomena mentioned here such as growth-dependent biomass composition, the demand for higher ribosomes, and the specific needs for gene expression (O'Brien et al. 2014). These genome-scale models allow the calculation of the needed proteome in a per gene basis for a particular growth condition. These detailed molecular predictions make ME-model an excellent tool to study resource allocation and to predict the minimal resource needs of a self-replicating organism.

4 Suboptimal Resource Allocation Revealed by ALE

Adaptive laboratory evolution (ALE) is a widely used tool to evolve specialist genotypes. Using this approach, evolved populations accumulate mutations that confer them a fitness benefit to the specific environment to which they have adapted, generally causing a fitness trade-off for other environments that were not adapted to. Using whole-genome resequencing of recurring mutations in an evolutionary experiment, the causal mutations can be easily identified (LaCroix et al. 2015). While there are many studies on ALE for a wide array of applications (Dragosits and Mattanovich 2013), few of them have focused on the resource allocation matter. It is pertinent to expect that specialist phenotypes will allocate their resources better and that generalist capacity will come with a fitness cost for the organisms that express them. The fitness benefit of expressing a non-immediately used proteome (a reserve or standby proteome) for generalist organisms has been documented by several authors (Condon et al. 1995; Utrilla et al. 2016; Price et al. 2016). Constraint-based metabolic models are optimality models; it has been observed that the ALE endpoint approach to the model predicted optimal phenotype (Edwards et al. 2001; Lewis et al. 2010). Using compelling data sets such as the Schmidt et al.'s (2016) proteomic resource (Schmidt et al. 2016), one can compare measured protein expression to a genome-scale model of metabolism and gene expression (ME model) predicted optimal proteomes. O'Brien et al. (2016) showed that the overexpression of the core proteome (compared to theoretical computed needs) enables a fitness benefit upon encountering an environment that supports a faster growth rate. The upregulation of stress resistance and nutrient readiness functions maximizes survival under changing and harsh environments; however, these generalist proteomes may result in a large proteome burden. In such study, they found a clear correlation between the growth rate and the unused protein fraction. This can explain why there are similar environments (i.e., growth on galactose minimal media vs. glucose minimal media) that display large growth rate variation.

Many commonly occurring mutations in ALE experiments are pleiotropic regulatory mutations. These kinds of mutations are expected to shift the global state of the cell to a new one, reprogramming several functions through changes in the regulatory network. Mutations in the RNAP genes of *E. coli* are very often found. A recent study (Utrilla et al. 2016) focused on the resource allocation of the reprogrammed phenotype. Utrilla et al. (2016) observed that *E. coli* strains with a single amino acid change in the beta subunit of the RNAP shifted the cellular state to a specialized proteome for growth in glucose minimal media increasing growth rate by 25% and biomass yield by 11%. The regulatory response is a consistent shift of gene expression from stress preparing functions such as acid resistance, motility, DNA repair, and nutrient scavenging, among others (also called "hedging functions") to growth functions. ME-model analysis shows 2–5% reduction of gene expression allocated to non-ME genes and about one-third of reduction of maintenance energy (Utrilla et al. 2016). In this case, a single amino acid change in an

RNAP subunit rewires the regulatory network and creates a global resource reallocation for a specialist reduced proteome.

In order to study bacterial resource allocation at the protein level, Price and co-workers combined ribosome profiling data to a library of transposon mutants. They found that many proteins are being expressed but not used. Up to 13% of the *E. coli*-expressed proteome is in "standby," in case conditions change, and 22% of the expressed proteome did not show any benefit for glucose minimal media growth (Price et al. 2016). In a related approach, to improve ME-model predictions, Yang and co-workers (2016) used the Schmidt et al. (2016) proteomics data to generate a generalist ME model that includes proteome sectors that constrain growth. Using this approach, they create constraints that force the expression of specific functional protein groups to account for the costs of non-ME-model protein production. First, they identified sectors, based on Clusters of Orthologous Groups (COGs), whose measured mass fractions were greater (over-allocated) or smaller (under-allocated) compared to the optimal proteomes across different growth conditions. The addition of proteome sector constraints greatly improved growth rate predictions. The ME-model analysis showed that at low growth rates (0.1 h^{-1}), up to 95% of the proteome is not used for growth. They discuss that much of the unused proteome is for stress-related and hedging functions. For the scope of this chapter, this means that *E. coli* cells use much of their resources in stress and change preparation. These resource allocation principles need to be addressed for minimal cell designed to grow and thrive in specific environments.

5 Design of a Minimal Cell (From a Resource Allocation Standpoint)

We foresee the design of a minimal cell as a two-major-step process: (a) Define the minimal set of genes needed to grow/replicate/thrive in the desired conditions (the core functions or proteome) and (b) design the minimal amount of resources to allocate to carry out cell growth and the desired function, in other words the amount of each gene to be expressed to carry out the desired functions in a proper way (Fig. 1b). The first part of this process will be thoroughly reviewed elsewhere in this book.

5.1 *The Core Proteome Definition*

First, we need to review the core functions since the minimal proteome needs to be synthesized at least once per every cell division. This will set a constraint on the minimal requirement on the protein synthesis capacity of a dividing cell. So this raises the question, what is the minimal proteome needed? Several studies have focused on the quest of defining the minimal genome or minimal proteome in this case. Recently, Yang and co-workers identified the core proteome of *E. coli* (Yang et al. 2015). They used genome-scale models to define a core proteome with the aim of computationally supporting basic cellular functions. Comparing computational simulations in 333 conditions to expression data sets, they defined a functional core proteome that consists of 356 proteins consistently expressed in all conditions. Those proteins account for 44% of the proteome of *E. coli* by mass, based on proteomics data. According to the Yang et al. core proteome definition, it is not the smallest set of genes to allow growth, but it is the set of genes that are consistently used across a large number of conditions. This means that the core proteome defined by Yang et al. is the central core that needs to be expressed, but this set alone will not support growth.

5.2 The Non-core Proteome

If the core proteome is comprised of those functions being expressed consistently, there is a proteome that is necessary in a condition-specific manner. There have been a few approaches to define the condition-specific proteome. The most straightforward manner to define the non-core proteome is using a genome-scale model, since they capture the specific needs of gene expression requirements to grow on a certain nutritional environment (Monk et al. 2017; Lloyd et al. 2018). Here we focus on the resource allocation studies of the non-core proteome in different environments. Yang et al. (2015) clustered growth conditions into 18 niches depending on the carbon, nitrogen, phosphorus, and sulfur sources (C/N/P/S). They identified condition-specific genes, adding 160 genes to the core data set. This data set shows a close approximation to the genomes of minimal organisms, such as *Buchnera aphidicola* and *Mycoplasma genitalium*, when comparing orthologous genes among them. O'Brien and collaborators (2016) classify the non-core proteome into element-dependent proteome, C for carbon, N for nitrogen, P for phosphorus, and S for sulfur. These are proteins used for growth under alternative C, N, P, and S sources and are largely of catabolic function. They found that approximately 6% of the proteome is non-core across the conditions profiled with proteomics by Schmidt et al. (2016). Comparing the ME-model prediction of proteome needs versus the proteomics data, they show that most of the non-core proteome is un-utilized. Also, in those profiled environments, the largest non-core proteome segment is the C-proteome. The C-proteome is enriched in targets of the transcription factor CRP, and under some non-preferred carbon sources, the C-proteome represents a larger fraction of the proteome, presumably as a regulatory response for the cellular preparation to change in carbon sources. These responses show a proteomic cost of specific carbon source catabolism and a nutrient generalist cost and partially explain why similar carbon sources may result in large growth rate differences (glucose vs. galactose growth in *E. coli*). In the O'Brien et al. study (2016), it was shown that the C-proteome abundance decreased linearly with the growth rate. This has been also demonstrated in other studies (You et al. 2013). These observations raise a question, are fast-growing cells expressing a minimal set of functions? Again comparing theoretical model prediction and actual gene expression has showed that fast-growing cells meet the theoretical prediction, while slow-growing cells have a very poor prediction of growth rate (O'Brien et al. 2016) (Fig. 2a). The correlation between proteome use and proteome requirement has shown that in certain conditions, nearly half of the proteome mass is unused. A minimal cell should thus express a minimal specialized proteome; regulatory responses such as nutrient generalism reduce resource availability for other proteome sector.

5.3 The Proteome Catalytic Efficiency

We can discuss proteome efficiency in two ways: (a) as efficiency in proteome allocation (Basan et al. 2015) and (b) as efficiency in catalytic rates of the enzymes comprising the proteome (Heckmann et al. 2018). So far we have discussed proteome allocation; however, the proteome catalytic efficiency is a critical parameter since the amount of enzyme needed to carry a flux will be reduced if that enzyme is being used at its highest rate (Fig. 2b). Two recent studies have compared the proteome allocation dynamics to changing growth rate. Valgepea et al. (2013) studied transcriptomics and proteomics in steady-state growth in an accelerostat going from 0.11 h^{-1} to 0.49 h^{-1}. They found a fivefold growth range in *E. coli*, which was due to a 3.7-fold increase of apparent catalytic rates of enzymes and a 2.5-fold increase of translation rates, demonstrating the importance of the posttranslational regulation of the proteome efficiencies as a mechanism of growth rate increase (Valgepea et al. 2013). In a similar study, Peebo et al. (2015), performed proteomics in an accelerostat going from 0.2 h^{-1} up to 0.9 h^{-1}. They showed that *E. coli* achieves faster growth by increasing the catalytic efficiency and the allocation efficiency (to overflow metabolism) of the proteome (Peebo et al. 2015).

Many studies of the Terence Hwa group have determined the relations between growth, proteome composition (allocation), and proteome efficiency. As we have reviewed in this chapter, the growth laws show a linear correlation in the ribosome content of cells and growth rate (Scott et al. 2010); however, the crowding theory suggests a limitation in ternary complexes (TCs, comprised by aminoacyl-tRNA, elongation factor Tu, and GTP), which are the substrates of ribosomes. The crowding theory postulates that TCs diffuse slowly in a crowded cytoplasm, and this limits translation efficiencies (Klumpp et al. 2013). In its recent study, they show a growth rate dependence of protein elongation rate in translation. They show that the translation efficiency and the protein elongation rate show a Michaelis–Menten-like dependency on growth rate (Dai et al. 2016). The latter theory reconciles the difference between ribosomal content and elongation rate. The recent evidence is clear in that the elongation rate is not constant and the proteome increases its efficiency as growth rate increases. As mentioned previously, studies have focused on ribosome use and translation efficiency showing, for example, that at fast growth

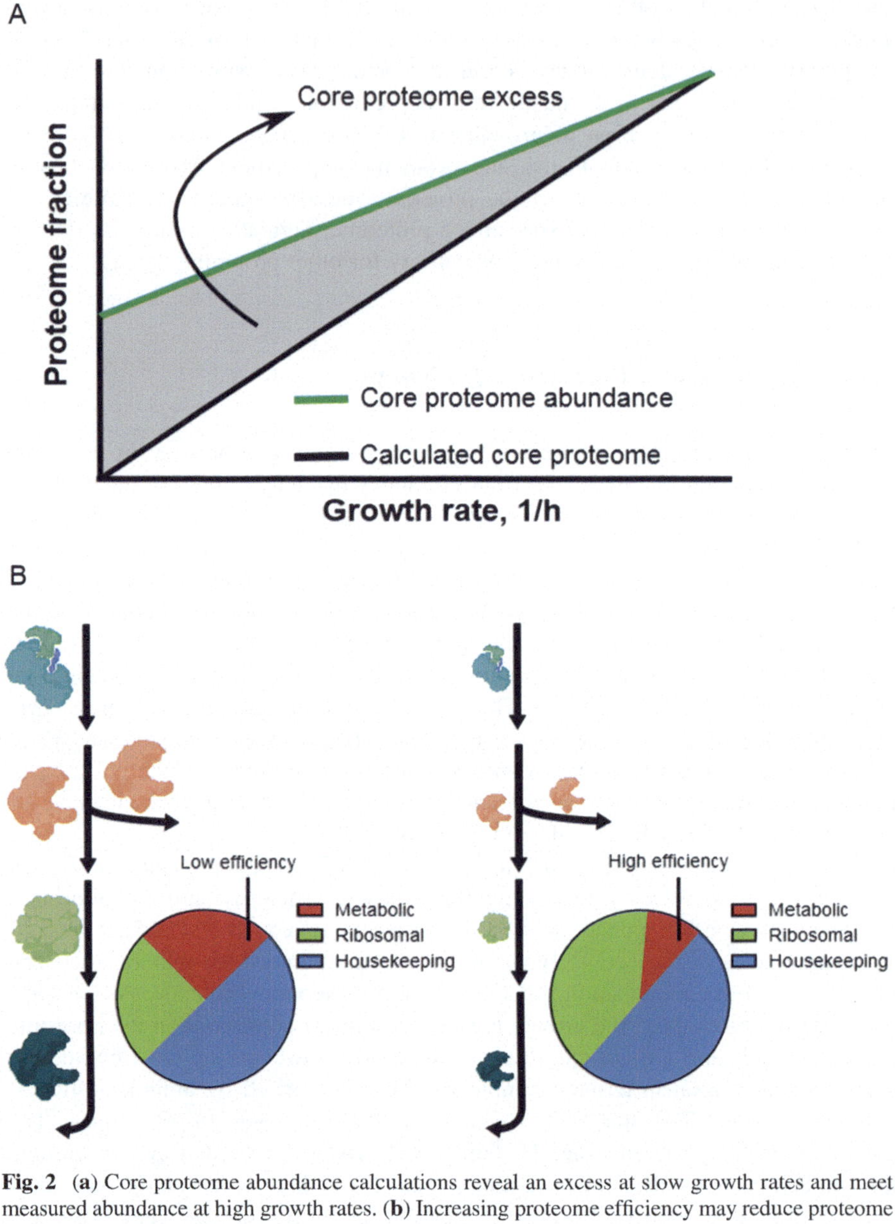

Fig. 2 (**a**) Core proteome abundance calculations reveal an excess at slow growth rates and meet measured abundance at high growth rates. (**b**) Increasing proteome efficiency may reduce proteome needs by using more efficient enzymes in a proteome sector

rate, 90% of the ribosomes are active, whereas at slow growth, only less than 20% of ribosomes are active. When comparing measured proteomes to theoretical ME-model predictions, O'Brien et al. (2016) documented a higher abundance of the core proteome than calculated needs (Fig. 2a). These observations point to an inactive ribosomal fraction or to low efficiency of the translation process. As the growth rate increases, ME-model predictions get closer to the actual core proteome fraction, meaning that ME-model shows a better accuracy at high growth rates and its theoretical calculations match the measured proteome at fast growth. The relationship between proteome composition and growth rate is affected by the effective rates in which biological processes occur (Barenholz et al. 2016). Nutrient availability produces the most efficient proteome; therefore, a cell growing on rich media achieves higher growth rate and higher catalytic efficiency; thus it may be expressing an efficient proteome that is naturally reduced since all its capacity is used in full.

To design a cell with minimal proteome we therefore should achieve the best proteome allocation and the highest catalytic efficiency of the enzymes in that proteome. Fast-growing cells display the highest proteome efficiency; this observation has been done under broad nutrient availability, and thus it is based on the external supply of cellular building blocks such as amino acids, vitamins, cofactors, etc. However, as we have shown in this chapter, we sustain the hypothesis that a minimal proteome should be the most efficient one, meaning that it should be constituted by the most catalytically efficient set of enzymes; this will ensure the minimal amount of protein needed to perform the necessary functions.

A very interesting bacterium to review for the scope of this chapter is *Vibrio natriegens*. It has been recently considered as the new molecular biology host for its fast duplication time (<10 min) (Weinstock et al. 2016). Some groups are developing molecular tools and protocols claiming it will speed molecular biology workflows. Recently, a quantitative study of its metabolic flux distributions was published. The study found that *V. natriegens* is able to grow at a specific rate of 1.7 h^{-1} in glucose minimal media, whereas even laboratory-adapted strains of *E. coli* seem to reach a limit around 1.0 h^{-1} (LaCroix et al. 2015). It also has a much higher glucose uptake rate (21 vs 8.5 mmol/gDW h), oxygen uptake rate (28 vs. 12 mmol/gDW h), and cell yield (0.52 vs. 0.48 g/g) on glucose minimal media (Long et al. 2017). *V. natriegens* do not have a reduced genome; it is even larger than *E. coli* (5.1 vs. 4.6 Mbps), and it has more ribosomal operon copies than *E. coli* (12 vs. 7). Its genome is divided into two chromosomes (Lee et al. 2019; Aiyar et al. 2002; Wang et al. 2013; Maida et al. 2013). Using a functional genomics assay with CRISRPi, it was found that *V. natriegens*' non-essential core genes were enriched for respiration, and this may explain its high growth and respiration capacity (Lee et al. 2019). As we have argued here, we hypothesize that the proteome of *V. natriegens* may be catalytically more efficient than that of *E. coli*. Unfortunately, at this time there are no resource allocation studies for this organism to compare with the vast information available on *E. coli*.

6 Biotechnological Applications

Minimal specialized cells hold a promise to be a leap of technology for biotechnological applications; however, they have not been sufficiently developed nor widely adopted. In terms of resource allocation, there are many considerations to focus on. Several studies have focused on characterizing the cost of expressing genes and to optimize gene circuit design to minimize burden (Ceroni et al. 2015, 2018). Fewer studies have focused on the cell resources that need to be allocated to correctly display a synthetic function (Liao et al. 2017), and no engineering efforts have been done to remodel the proteome allocation for synthetic biology applications. Expressing an unused protein, metabolic pathway, or synthetic gene circuit drains cellular resources for growth (Tan et al. 2009). Weiße et al. (2015) used a trade-off model to explain how a synthetic circuit, the repressilator (Elowitz and Leibler 2000), competes with host resources. They found that the expression of the circuit proteins imposes a burden on the cell because they do not contribute to growth or survival. Their model predicts a sigmoidal decrease in growth for stronger induction of circuit genes. Comparing the results of simulation of the circuit in isolation and within the cell (taking in account the competition for cell resources), they predicted induction regions where cells are overloaded (Weiße et al. 2015). Cells reduce their growth rate when forced to express high amounts of unused proteins. A recent modeling approach showed that at weak heterologous gene expression, the amount of protein output can be increased at the expense of growth reduction. In their work, Bienick and coworkers (2014) showed a theoretical calculation for a "critical capacity" above which heterologous protein production and host growth decrease sharply. With the aid of their modeling approach, they showed that those regions of sharp decrease are a result of ribosomal scarcity (Nikolados et al. 2019). The cost of protein production in yeast (*Saccharomyces cerevisiae*) depends on the growth conditions, while in nitrogen limitation the cost is likely on the shortage of amino acids. During slow growth on a non-fermentable carbon source, protein production was limited by ribosome content; however, under conditions of rapid growth, ribosome content was not limiting (Kafri et al. 2016). In the case of *E. coli*, Frumkin et al. (2017) showed that there are gene architectures that minimize the cost of gene expression. Such mechanisms are the reduced macromolecular machinery use (RNAP and ribosomes), slow translation speed, and the use of cheap amino acids. They showed that these mechanisms are selected to reduce the expression costs of natural genes and highly expressed genes evolved toward lower expression costs. In other production models such as Chinese hamster ovary (CHO) cells, a recent study showed the ribosome profiling-guided silencing of the unnecessary resistance marker NeoR improved production of a therapeutic antibody (IgG) and growth of a CHO cell line (Kallehauge et al. 2017).

Detailed studies of protein cost show that regardless of the initial conditions, when the unused heterologous protein reaches 15% of the total cell mass, growth rate is reduced at around half. Of course, this is dependent on the other fractions of the proteome (Bienick et al. 2014), and it is explained by a simple equation derived from the Scott et al. (2010) growth laws. They showed a theoretical calculation for a metabolic engineering application in which a six-enzyme heterologous pathway, assuming typical parameters of protein size and efficiency (K_{cat} 2.6 s^{-1}, molecular weight 40 kDa), may reduce growth rate to 50% of the non-producing strain; this reduction is calculated just taking into account the cost of enzyme production, not considering carbon loss to the product of interest. If we can minimize the resource allocation to unused sectors of the proteome or use the most efficient set of enzymes, then we may increase the fraction of the proteome that is feasible to be used for engineered synthetic functions and biotechnological applications. Many bioprocesses are carried out in well-controlled fermenters where the environment is somehow homogeneous, and depending on the bioprocess scale and conditions, some of the environmental hedging functions and regulatory responses may be eliminated from the host cell. However, many bioprocesses are carried out in large-scale vessels (from thousands to million liters). We have to consider that in such large-scale bioreactors, heterogeneities may be formed. For example, heterogeneous regions of dissolved oxygen, low or high pH regions, or substrate concentration gradients, all these kinds of stress source need to be taken in account for minimal cell design to be used in such large-scale vessels (Wehrs et al. 2019). If we know beforehand the nature and scale of our process, we can design specific stress response mechanisms for specific stressors. These factors need to be addressed to design the allocation of resources in minimal cells aimed at specific biotechnological applications.

7 Concluding Remarks

A minimal cell design should take into account the cellular resource need of each process. Protein synthesis and translation machinery synthesis are the most resource-extensive cellular processes. Therefore, special attention has to be paid to the proteome allocation for the desired condition. Cells living in natural changing environments need to deploy a large number of resources for preparation to non-ideal conditions, be prepared to change to a wide variety of environments, and protect themselves from harsh conditions. However, a minimal cell designed for a specific function can express a minimal specialized proteome. By reviewing recent literature on resource allocation, we hypothesize on how to reduce cellular resource consumption and use liberated resources to express an engineered cellular function.

Acknowledgment Support from grants UNAM-PAPIIT-IA201518 and Newton Advanced Fellowship Project NA 160328 is acknowledged.

References

Aiyar SE, Gaal T, Gourse RL (2002) rRNA promoter activity in the fast-growing bacterium *Vibrio natriegens*. J Bacteriol 184:1349–1358

Artsimovitch I, Patlan V, Sekine S et al (2004) Structural basis for transcription regulation by alarmone ppGpp. Cell 117:299–310. https://doi.org/10.1016/S0092-8674(04)00401-5

Bachmann BJ (1990) Linkage map of *Escherichia coli* K-12, edition 8. Microbiol Rev 54:130–197

Baracchini E, Bremer H (1988) Stringent and growth control of rRNA synthesis in *Escherichia coli* are both mediated by ppGpp. J Biol Chem 263:2597–2602

Barenholz U, Keren L, Segal E, Milo R (2016) A minimalistic resource allocation model to explain ubiquitous increase in protein expression with growth rate. PLoS One 11:e0153344. https://doi.org/10.1371/journal.pone.0153344

Basan M, Hui S, Okano H et al (2015) Overflow metabolism in *Escherichia coli* results from efficient proteome allocation. Nature 528:99–104. https://doi.org/10.1038/nature15765

Bienick MS, Young KW, Klesmith JR et al (2014) The interrelationship between promoter strength, gene expression, and growth rate. PLoS One 9:e109105. https://doi.org/10.1371/journal.pone.0109105

Ceroni F, Algar R, Stan G-B, Ellis T (2015) Quantifying cellular capacity identifies gene expression designs with reduced burden. Nat Methods 12:415–418. https://doi.org/10.1038/nmeth.3339

Ceroni F, Boo A, Furini S et al (2018) Burden-driven feedback control of gene expression. Nat Methods 15:387–393. https://doi.org/10.1038/nmeth.4635

Condon C, French S, Squires C, Squires CL (1993) Depletion of functional ribosomal RNA operons in *Escherichia coli* causes increased expression of the remaining intact copies. EMBO J 12:4305–4315

Condon C, Liveris D, Squires C et al (1995) rRNA operon multiplicity in *Escherichia coli* and the physiological implications of rrn inactivation. J Bacteriol 177:4152–4156. https://doi.org/10.1128/JB.177.14.4152-4156.1995

Dai X, Zhu M, Warren M et al (2016) Reduction of translating ribosomes enables *Escherichia coli* to maintain elongation rates during slow growth. Nat Microbiol 2:16231. https://doi.org/10.1038/nmicrobiol.2016.231

de Jong H, Geiselmann J, Ropers D (2017) Resource reallocation in bacteria by reengineering the gene expression machinery. Trends Microbiol 25:480–493

Dennis PP, Bremer H (2008) Modulation of chemical composition and other parameters of the cell at different exponential growth rates. EcoSal Plus. https://doi.org/10.1128/ecosal.5.2.3

Deutschbauer A, Price MN, Wetmore KM et al (2014) Towards an informative mutant phenotype for every bacterial gene. J Bacteriol 196:3643–3655. https://doi.org/10.1128/JB.01836-14

Dragosits M, Mattanovich D (2013) Adaptive laboratory evolution – principles and applications for biotechnology. Microb Cell Factories 12:64. https://doi.org/10.1186/1475-2859-12-64

Ebrahim A, Brunk E, Tan J et al (2016) Multi-omic data integration enables discovery of hidden biological regularities. Nat Commun 7:13091. https://doi.org/10.1038/ncomms13091

Edwards J, Ibarra R, Palsson B (2001) In silico predictions of *Escherichia coli* metabolic capabilities are consistent with experimental data. Nat Biotechnol:125–130

Elowitz M, Leibler S (2000) A synthetic oscillatory netwrok of transcriptional regulators. Nature 403:335–338

Frumkin I, Schirman D, Rotman A et al (2017) Gene architectures that minimize cost of gene expression. Mol Cell 65:142–153. https://doi.org/10.1016/j.molcel.2016.11.007

Glass JI, Merryman C, Wise KS et al (2017) Minimal cells-real and imagined. Cold Spring Harb Perspect Biol 9:a023861. https://doi.org/10.1101/cshperspect.a023861

Greenbaum D, Colangelo C, Williams K, Gerstein M (2003) Comparing protein abundance and mRNA expression levels on a genomic scale. Genome Biol 4:117. https://doi.org/10.1186/gb-2003-4-9-117

Heckmann D, Lloyd CJ, Mih N et al (2018) Machine learning applied to enzyme turnover numbers reveals protein structural correlates and improves metabolic models. Nat Commun 9:5252. https://doi.org/10.1038/s41467-018-07652-6

Hui S, Silverman JM, Chen SS et al (2015) Quantitative proteomic analysis reveals a simple strategy of global resource allocation in bacteria. Mol Syst Biol 11:784. https://doi.org/10.15252/msb.20145697

Hutchison CA, Chuang R-YR-Y, Noskov VN et al (2016) Design and synthesis of a minimal bacterial genome. Science 351:aad6253. https://doi.org/10.1126/science.aad6253

Ingolia NT, Ghaemmaghami S, Newman JRS, Weissman JS (2009) Genome-wide analysis in vivo of translation with nucleotide resolution using ribosome profiling. Science 324:218–223. https://doi.org/10.1126/science.1168978

Kafri M, Metzl-Raz E, Jona G, Barkai N (2016) The cost of protein production. Cell Rep 14:22–31. https://doi.org/10.1016/J.CELREP.2015.12.015

Kallehauge TB, Li S, Pedersen LE et al (2017) Ribosome profiling-guided depletion of an mRNA increases cell growth rate and protein secretion. Sci Rep 7:40388. https://doi.org/10.1038/srep40388

Karr JR, Sanghvi JC, MacKlin DN et al (2012) A whole-cell computational model predicts phenotype from genotype. Cell 150:389–401. https://doi.org/10.1016/j.cell.2012.05.044

Kjelgaard N, Gausing K (1974) Regulation of biosynthesis of ribosomes. Cold Spring Harb Monogr Arch 4:369–392

Klappenbach JA, Dunbar JM, Schmidt TM (2000) rRNA operon copy number reflects ecological strategies of bacteria. Appl Environ Microbiol 66:1328–1333

Klumpp S, Scott M, Pedersen S, Hwa T (2013) Molecular crowding limits translation and cell growth. Proc Natl Acad Sci U S A: 110(42):16754–16759. https://doi.org/10.1073/pnas.1310377110

Kudva R et al (2013) Protein translocation across the inner membrane of Gram-negative bacteria: the Sec and Tat dependent protein transport pathways. Res Microbiol 164:505–534

LaCroix RA, Sandberg TE, O'Brien EJ et al (2015) Use of adaptive laboratory evolution to discover key mutations enabling rapid growth of *Escherichia coli* K-12 MG1655 on glucose minimal medium. Appl Environ Microbiol 81:17–30. https://doi.org/10.1128/AEM.02246-14

Lee HH, Ostrov N, Wong BG et al (2019) Functional genomics of the rapidly replicating bacterium *Vibrio natriegens* by CRISPRi. Nat Microbiol 4(7):1105–1113. https://doi.org/10.1038/s41564-019-0423-8

Lewis NE, Hixson KK, Conrad TM et al (2010) Omic data from evolved *E. coli* are consistent with computed optimal growth from genome-scale models. Mol Syst Biol 6:390. https://doi.org/10.1038/msb.2010.47

Li G-W, Burkhardt D, Gross C, Weissman JS (2014) Quantifying absolute protein synthesis rates reveals principles underlying allocation of cellular resources. Cell 157:624–635. https://doi.org/10.1016/j.cell.2014.02.033

Liao C, Blanchard AE, Lu T (2017) An integrative circuit–host modelling framework for predicting synthetic gene network behaviours. Nat Microbiol 2:1658–1666. https://doi.org/10.1038/s41564-017-0022-5

Lloyd CJ, Ebrahim A, Yang L et al (2018) COBRAme: a computational framework for genome-scale models of metabolism and gene expression. PLoS Comput Biol 14(7):e1006302. https://doi.org/10.1371/journal.pcbi.1006302

Long CP, Gonzalez JE, Cipolla RM, Antoniewicz MR (2017) Metabolism of the fast-growing bacterium *Vibrio natriegens* elucidated by 13C metabolic flux analysis. Metab Eng 44:191–197. https://doi.org/10.1016/J.YMBEN.2017.10.008

Maida I, Bosi E, Perrin E et al (2013) Draft genome sequence of the fast-growing bacterium *Vibrio natriegens* strain DSMZ 759. Genome Announc 1:e00648–e00613. https://doi.org/10.1128/genomeA.00648-13

Monk JM, Lloyd CJ, Brunk E et al (2017) iML1515, a knowledge base that computes *Escherichia coli* traits. Nat Biotechnol 35:904–908. https://doi.org/10.1038/nbt.3956

Mori M, Hwa T, Martin OC, De Martino A, Marinari E (2016) Constrained allocation flux balance analysis. PLoS Comput Biol 12:e1004913

Mori M, Schink S, Erickson DW et al (2017) Quantifying the benefit of a proteome reserve in fluctuating environments. Nat Commun 8:1225. https://doi.org/10.1038/s41467-017-01242-8

Murray HD, Appleman JA, Gourse RL (2003) Regulation of the *Escherichia coli* rrnB P2 promoter. J Bacteriol 185:28. https://doi.org/10.1128/JB.185.1.28-34.2003

Neidhardt FC, Magasanik B (1960) Studies on the role of ribonucleic acid in the growth of bacteria. Biochim Biophys Acta 42:99–116. https://doi.org/10.1016/0006-3002(60)90757-5

Nikolados E-M, Weisse AY, Ceroni F, Oyarzun DA (2019) Growth defects and loss-of-function in synthetic gene circuits. bioRxiv:623421. https://doi.org/10.1101/623421

O'Brien EJ, Lerman JA, Chang RL et al (2014) Genome-scale models of metabolism and gene expression extend and refine growth phenotype prediction. Mol Syst Biol 9:693–693. https://doi.org/10.1038/msb.2013.52

O'Brien EJ, Utrilla J, Palsson BO (2016) Quantification and classification of *E. coli* proteome utilization and unused protein costs across environments. PLoS Comput Biol 12:e1004998. https://doi.org/10.1371/journal.pcbi.1004998

Peebo K, Valgepea K, Maser A et al (2015) Proteome reallocation in *Escherichia coli* with increasing specific growth rate. Mol BioSyst 11:1184–1193. https://doi.org/10.1039/C4MB00721B
Pirt SJ (1965) The maintenance energy of bacteria in growing cultures. Proc R Soc Lond B Biol Sci 163(991):224–231
Price MN, Wetmore KM, Deutschbauer AM, Arkin AP (2016) A comparison of the costs and benefits of bacterial gene expression. PLoS One 11:e0164314. https://doi.org/10.1371/journal.pone.0164314
Schmidt A, Kochanowski K, Vedelaar S et al (2015) The quantitative and condition-dependent *Escherichia coli* proteome. Nat Biotechnol 34:104–110
Schmidt A, Kochanowski K, Vedelaar S et al (2016) The quantitative and condition-dependent *Escherichia coli* proteome. Nat Biotechnol 34:104–110. https://doi.org/10.1038/nbt.3418
Scott M, Gunderson CW, Mateescu EM et al (2010) Interdependence of cell growth and gene expression: origins and consequences. Science 330:1099–1102. https://doi.org/10.1126/science.1192588
Scott M, Klumpp S, Mateescu EM, Hwa T (2014) Emergence of robust growth laws from optimal regulation of ribosome synthesis. Mol Syst Biol 10:747–747. https://doi.org/10.15252/msb.20145379
Selvarasu S, Ow DS-W, Lee SY et al (2009) Characterizing *Escherichia coli* DH5α growth and metabolism in a complex medium using genome-scale flux analysis. Biotechnol Bioeng 102:923–934. https://doi.org/10.1002/bit.22119
Shepherd N, Churchward G, Bremer H (1980) Synthesis and function of ribonucleic acid polymerase and ribosomes in *Escherichia coli* B/r after a nutritional shift-up. J Bacteriol 143:1332–1344
Tan C, Marguet P, You L (2009) Emergent bistability by a growth-modulating positive feedback circuit. Nat Chem Biol 5:842–848. https://doi.org/10.1038/nchembio.218
Thiele I, Jamshidi N, Fleming RMT, Palsson BØ (2009) Genome-scale reconstruction of *Escherichia coli*'s transcriptional and translational machinery: a knowledge base, its mathematical formulation, and its functional characterization. PLoS Comput Biol 5:e1000312. https://doi.org/10.1371/journal.pcbi.1000312
Utrilla J, O'Brien EJ, Chen K et al (2016) Global rebalancing of cellular resources by pleiotropic point mutations illustrates a multi-scale mechanism of adaptive evolution. Cell Syst 2:260–271. https://doi.org/10.1016/j.cels.2016.04.003
Valgepea K, Adamberg K, Seiman A, Vilu R (2013) *Escherichia coli* achieves faster growth by increasing catalytic and translation rates of proteins. Mol BioSyst 9:2344–2358. https://doi.org/10.1039/c3mb70119k
Wang Z, Lin B, Hervey WJ et al (2013) Draft genome sequence of the fast-growing marine bacterium *Vibrio natriegens* strain ATCC 14048. Genome Announc 1(4):e00589–13. https://doi.org/10.1128/genomeA.00589-13
Wehrs M, Tanjore D, Eng T et al (2019) Engineering robust production microbes for large-scale cultivation. Trends Microbiol:1–14. https://doi.org/10.1016/j.tim.2019.01.006
Weinstock MT, Hesek ED, Wilson CM, Gibson DG (2016) *Vibrio natriegens* as a fast-growing host for molecular biology. Nat Methods 13:849–851. https://doi.org/10.1038/nmeth.3970
Weiße AY, Oyarzún DA, Danos V, Swain PS (2015) Mechanistic links between cellular trade-offs, gene expression, and growth. Proc Natl Acad Sci U S A 112:E1038–E1047. https://doi.org/10.1073/pnas.1416533112
Wilson DN, Nierhaus KH (2007) The weird and wonderful world of bacterial ribosome regulation. Crit Rev Biochem Mol Biol 42:187–219. https://doi.org/10.1080/10409230701360843
Yang L, Tan J, O'Brien EJ et al (2015) Systems biology definition of the core proteome of metabolism and expression is consistent with high-throughput data. Proc Natl Acad Sci U S A 112(34):10810–10815. https://doi.org/10.1073/pnas.1501384112
Yang L et al (2016) Principles of proteome allocation are revealed using proteomic data and genome-scale models. Sci Rep 6:36734
You C, Okano H, Hui S et al (2013) Coordination of bacterial proteome with metabolism by cyclic AMP signalling. Nature:1–6. https://doi.org/10.1038/nature12446

The manufacturer's authorised representative in the EU is Springer Nature Customer Service Centre GmbH, Europaplatz 3, 69115 Heidelberg, Germany. If you have any concerns regarding our products, please contact ProductSafety@springernature.com

Printed and bound by CPI Group (UK) Ltd, Croydon, CR0 4YY
05/10/2026
02228123-0008